SCIENTIFIC MATERIALISM
IN NINETEENTH CENTURY GERMANY

STUDIES IN THE HISTORY
OF MODERN SCIENCE

Editors:

VOLUME 1

FREDERICK GREGORY

SCIENTIFIC MATERIALISM IN NINETEENTH CENTURY GERMANY

D. REIDEL PUBLISHING COMPANY
DORDRECHT–HOLLAND / BOSTON–U.S.A.

Library of Congress Cataloging in Publication Data

Gregory, Frederick, 1942–
Scientific materialism in nineteenth century Germany.

(Studies in the history of modern science; 1)
Bibliography: p.
Includes index.
1. Materialism History. 2. Philosophy, German 19th century. 3. Science History Germany. I. Title. II. Series.
B3188.G73 146'.3'0943 76 56077
ISBN 90 277 0760 X
ISBN 90 277 0763 4 pbk.

Published by D. Reidel Publishing Company,
P.O. Box 17, Dordrecht, Holland

Sold and distributed in the U.S.A., Canada, and Mexico
by D. Reidel Publishing Company, Inc.
Lincoln Building, 160 Old Derby Street, Hingham,
Mass. 02043, U.S.A.

Printed in The Netherlands

to my mother, Eleanor R. Gregory,
and the memory of
Richard P. Gregory, my father

TABLE OF CONTENTS

PART 1. BACKGROUND

PART 2. THE SCIENTIFIC MATERIALISTS AND THEIR WORKS

PART 3. ISSUES

PREFACE

A comprehensive study of German materialism in the second half of the nineteenth century is long overdue. Among contemporary historians the mere passing references to Karl Vogt, Jacob Moleschott, and Ludwig Büchner as materialists and popularizers of science are hardly sufficient, for few individuals influenced public opinion in nineteenth-century Germany more than these men. Büchner, for example, revealed his awareness of the historical significance of his *Kraft und Stoff* in comments made in 1872, just seventeen years after its original appearance.

> A philosophical book which has undergone twelve big German editions in the short span of seventeen years, which further has been issued in non-German countries and languages about fifteen to sixteen times in the same period, and whose appearance (although its author was entirely unknown up to then) has called forth an almost unprecedented storm in the press, ... such a book can be nothing ordinary; the world-calling it enjoys at present must be justified through its wholly special characteristics or by the merits of its form and content.[1]

Vogt, Moleschott and Büchner explicitly held that their materialism was founded on natural science. But other materialists of the nineteenth century also laid claim to the scientific character of their own thought. It is likely that Marx and Engels would have permitted their brand of materialism to have been called scientific, provided, of course, that 'scientific' was understood in their dialectical meaning of the term. Socialism, Engels maintained, had become a science with Marx. Given Engels concern with demonstrating the dialectics of nature through natural science, it is not difficult to imagine that he also might have labeled Marx's materialism scientific. In that case what Plekhanov later dubbed dialectical materialism would now be known as scientific materialism.[2] In fact, the materialism of Marx and Engels did not become known under this or any other label in their lifetimes, in spite of the fact that both men noted the significant place natural science occupied in their philosophical outlook.

In addition to the Marxist version there were other variants of materialism in Germany during the nineteenth century. Feuerbach's critique of Hegelian idealism, which dominated the theological scene in the 1840's, was received as a materialistic correction of the Master. Although

Feuerbach's thought was an important influence on the works of Vogt, Moleschott and Büchner, his materialism was not generally regarded as a doctrine which owed its inspiration to the natural sciences. The same cannot be said, however, about the reductionistic program of the students of Johannes Müller, which also was being formulated in the 1840's, nor about the later materialism of Ernst Haeckel.

My reason for excluding Haeckel from this study of German scientific materialism is that his work was not part of the flurry of materialistic activity of the 1850's. Unlike the materialists of the fifties, who incorporated and shaped Darwin's contribution to fit their already formulated message, Haeckel, a younger man than Büchner by more than a decade, took his cue from Darwin. Haeckel's reputation did not become firmly established until the Stettin Congress of 1863. In 1859, the year of Darwin's *Origin*, Vogt said that he had tired of the debate between science and religion, and hoped that he would not have to write about it anymore. What was drawing to a close in the mind of one had not yet been broached in the life of the other.

While it is true that Büchner later expressed agreement with Haeckel's emphasis on monism, none of the earlier materialists could stomach his anti-Semitism or his notion of religion. They were more sympathetic to the goals of the German liberal tradition, and had no use for the revival of romanticism in the late nineteenth century. In spite of conceptual similarities in message, Haeckel's fame belongs to a slightly later period of German history than the one which Vogt, Moleschott and Büchner flourished.

It remains to distinguish the thought of Vogt, Moleschott and Büchner from that of biological mechanists such as Brücke and DuBois-Reymond. This is best accomplished by noting the purpose each group of scientists possessed. Müller's students were concerned with a specific problem of biology; viz., the explanation of organic phenomena. They chose to address this problem by means of the then radical suggestion that life could be explained mechanistically, thereby eliminating the need for a special vital force. Since this approach was not intended to serve as a metaphysical explanation of life, they cannot properly be called materialists at all. As a noted scholar has pointed out, materialism is a metaphysical position, and in the nineteenth century it often entailed the following tenets: (1) that there is an independently existing world; (2) that human beings, like all other subjects, are material entities; (3) that the human mind does not exist as an entity distinct from the human body; (4) that there is no God (nor any other nonhuman being) whose mode of existence is not that of material entities.[3] These are metaphysical postulates which are not necessarily implied by mechanism or reductionism.

Vogt and Moleschott were physiologists, Büchner a medical doctor. The subjects they wrote about clearly reflected the interests of men of science, and the methodology they claimed was modelled on their approach to scientific research. On the other hand, the message they proclaimed was metaphysical materialism. The term 'scientific materialism,' then, best expresses what their works represent. Vogt, Moleschott and Büchner were not the only scientists to preach materialism to their age, but they were among the first, and in Germany they were among the most popular.

A word about my working assumptions in the task at hand. I take it for granted that the period of disenchantment with idealism in the German intellectual theater of the middle years of the nineteenth century reflects changes on many fronts, including the social, political and economic. My concern here is to deal with the rise of scientific materialism as an aspect of intellectual history; i.e., I want to examine scientific materialism as a second wave of the criticism of idealism that surfaced in the late thirties and early forties. Among the critical authors most read in Germany in the forties were David Strauss and Ludwig Feuerbach. Feuerbach quickly became known as a materialist. His works, largely carried out in the context of theological criticism, helped to establish the naturalistic attitude proclaimed a decade later in Germany by the scientific materialists.

Further, I should like to remind the reader that it will be difficult to work our way back into the minds of men whose thought patterns are foreign to us today. Their denials to the contrary, the scientific materialists were certainly metaphysicians by twentieth-century standards. Grand issues were true or false for them. With this in mind, I have chosen an approach which immerses the reader in the issues rather than one which becomes preoccupied with ready-made analyses. In this way I hope to communicate the assumptions, logic, values, and dispositions of the major figures, thereby providing the best possible understanding of why they behaved as they did. To accomplish this a conscientious attempt has been made to explore the wealth of primary material available, and to examine the accounts and evaluations of scientific materialism of writers who experienced the debates first hand. The dependence on older historians is also necessary because of the lack of interest in the subject on the part of modern scholars.

I should like to express my special thanks to several individuals without whose assistance this book would not have been completed. To Professor

Erwin Hiebert of Harvard University I owe a great deal. Not only has he been a ready and willing source of advice, but he has taught me by example the value of careful and thorough scholarship. It was in his seminars that he passed on his enthusiasm for the history of science, and it was there I first became familiar with the nature of science in the nineteenth century. Many of the questions I have pursued in this study occurred to me originally in Professor Hiebert's seminars. Naturally he bears no responsibility for any errors I may have committed.

To Mr. Nelious Brown I am also indebted. From the beginning he has been a real source of encouragement, and his continuing interest and support over the years has not gone unnoticed. Professor Mario Bunge of McGill University examined the manuscript and made several helpful suggestions which have been adopted here. Some oversimplifications in the original draft concerning Feuerbach were pointed out to me by Professor Marx Wartofsky of Boston University, whose own work on Feuerbach was not yet published at the time of writing. I am grateful to him for his kind assistance. Without the help of the Reference Staff of Widener Library, Harvard University, and that of my proofreader and typist, Ms. Judy Meyers from the Office of Student Affairs at Eisenhower College, my job would have been hopeless. Nor can I fail to register my appreciation to Mr. David Traumann, who checked the manuscript.

Finally, I should like to break tradition where my wife Tricia is concerned. Her plight no doubt has been similar to that of countless other spouses, but, being the woman she is, a generous compliment about her supportive role would neither suffice nor be tolerated. It is enough simply to say that "that damn book," which has demanded so much of the family, is finished.

May, 1975 F.G.G.

NOTES

[1] *NW*, I, p. 3.
[2] For a discussion of the background of the phrase 'dialectical materialism' see Loren Graham, *Science and Philosophy in the Soviet Union* (New York, 1972), p. 475, n. 1.
[3] Maurice Mandelbaum, *History, Man, and Reason* (Baltimore, 1971), p. 22.

FOREWORD

Two related theses emerge more and more clearly from contemporary studies in the history and philosophy of science: first, that scientific understanding is an historical understanding; that is, scientific modes of thought, scientific discovery, the work of science itself is marked by its social and historical context. Second, that the understanding of the scientific enterprise is an historical understanding; that is, the ways in which we comprehend science change with our historical reconstructions and evaluations of major periods of scientific achievement. So, for example, we understand Newton's science differently now from the way in which it was understood, say, in the eighteenth century. In effect, our Newton is different from theirs (or, more accurately, our Newtons are different from theirs: for among contemporary views there are alternative and sometimes conflicting interpretations of Newton's achievement, of his method, of the very nature of the scientific revolution which his work exemplified.) Moreover, we are forced to a reconsideration of the very status of Newton's science – i.e. of his mechanics, his optics, his color theory, etc. – with respect to its scientific truth. Shall we take Newton's mechanics as an approximation, at the limit, of relativistic mechanics, for sufficiently small velocities, and for sufficiently short distances? Or shall we take it as false, and superseded by relativistic mechanics?

In all these ways, the scientific enterprise comes to be understood more deeply – both internally, in its truth claims and its methods, and externally, in its relation to its time and place, and to the personalities of its practitioners – as a complex human enterprise. The norm of scientific objectivity is not thereby given up for some mess of relativist pottage, but is rather seen in its fuller problematical form, and no longer as simple-minded animal faith, or as methodological dogma.

Frederick Gregory's study of scientific materialism in nineteenth century Germany is a model of this deeper and more complex understanding of science. He reveals for us the intimate connections among science, philosophy, religion (or anti-religion), and politics. And he does so with detailed attention to the scientific issues and movements which were central at the time. Unlike more literary historians of scientific ideas, Gregory does not merely sum up, or give us popular thumbnail sketches of the scientific issues, but places them in the center of his study, so that we may see how these other contexts impinged upon them, and how the scientific ideas interacted with the social, political, religious and philosophical movements of the time.

Why is Gregory's work important? What is original about it? He has chosen a relatively coherent group of scientists who played a unique role in nineteenth century thought, and whose *Problematik* turns out to be relevant to philosophical and scientific issues of our own time. Although the group is relatively little known and little studied, yet it figures in at least three major traditions in the history and philosophy of science: most clearly, in the tradition of reductive materialism – that view which sees reduction to a materialist basis, whether physicalist or physiological – as the most fruitful program for scientific research; secondly, in the Marxist philosophy of science which has traditionally adopted Marx's own critical stance toward this school as "vulgar materialists"; and finally, to the neo-Kantian movement of the late nineteenth century – the "back to Kant" movement from which so much of the dominant philosophy of science of our own times has derived. What is noteworthy is that the group of "scientific materialists", as Gregory characterizes them, are understood not simply in their relation to these other traditions, but in their distinction from these alternative views. Thus Gregory makes clear that, for all the reductionist elements in their thought, the scientific materialists are to be distinguished in several important respects from other reductionist materialists. So too, he reveals, with great perspicuity and in rich detail, the ideological, philosophical and political differences between these scientific materialists, and Marx and Engels. Similarly, Gregory shows how neo-Kantian philosophies of science were viewed and attacked by the scientific materialists. In short, Gregory has rescued a significant scientific movement from the footnotes and peripheral treatments to which they had been assigned, and uses his reconstruction of this movement to teach us some lessons on how to do the history of science in a more integrated way.

Who are the "scientific materialists"? What is their connection with the major scientific movements of the nineteenth century? Gregory focuses on the four major figures: Karl Vogt, Jacob Moleschott, Ludwig Büchner and Heinrich Czolbe. They stand in a double connection: first, to German science – in direct personal and intellectual relation to Liebig, Mayer, Helmholtz, Clausius, Virchow, Mohr – and to the Swiss Agassiz, as well to DuBois-Reymond and others; second to German philosophy and politics, and to the radical, anarchist, and socialist movements of their time and to such leading figures as Feuerbach, Marx and Engels, Bakunin, Herzen, and Herwegh.

Gregory's thesis concerning the interaction between scientific theory and the social philosophical views of these scientific materialists, is an interesting and provocative one. In the final chapter, 'Materialism and Society', Gregory writes:

> Even if we are successful in locating points at which the materialists appeared to reason directly from science to society, we cannot on this account conclude simply that their ideas about science were the basis of their social convictions and actions. It is not immediately obvious which influenced which. Might it not be equally possible that the social and political turmoil of the years following 1830 was more influential in shaping the materialists' conception of science than science was in shaping their conception of society?

What is at issue in Gregory's formulation are two related questions: first, what is the relation of science, as an institution, to society? Second, what is the relation of the beliefs, methods and theoretical conceptions of working scientists, concerning their science itself, to their philosophical and socio-political views? The virtue of Gregory's historical study is that these questions are dealt with concretely, and not as matters of abstract reflection. We are led into the grain and the details of the scientific (and popular-scientific) writings of this group, as well as into their considerable personal involvement in the heated politics and polemics of the *Vormärz* period, the abortive 1848 Revolution, the Frankfurt parliament and the reaction that followed. The importance of Gregory's method of reconstructing the interplay of philosophical, political, and scientific views may be measured by the clarity with which the scientific work itself is revealed in this light.

The scientific issues at stake turn out to be extraordinarily timely and philosophically important. They derive from the nineteenth century developments in organic chemistry, in biology, in physiology and neurophysiology, and in evolutionary theory. The materialists probed such questions as that concerning the relation of physico-chemical processes to the origins of life, and to the nature of consciousness; the evolution of biological species, and genetic theory; animal behavior and social organization in animal life; the principle of conservation of energy in thermodynamics. True, not each member of this group addressed all of these questions equally. But as a group, with all the individual differences in emphasis, competence, and in the conclusions reached, they provide a vivid image of the major questions in nineteenth century science, as these were filtered and refracted through the medium of a more or less common philosophical outlook.

It is in the reconstruction of that outlook that Gregory provides us with the framework for this scientific materialism. The towering and abiding philosophical figure, whose views dominated and influenced these scientists, was Ludwig Feuerbach. In 1841, Feuerbach's major work, *The Essence of Christianity*, provided the radical philosophical critique of religion and theology which was to become the rallying cry of the left-Hegelian movement. Feuerbach's message was simple, and powerful: human beings create the gods in their own image, and as an esoteric understanding of their own, human nature. Therefore, the "secret of religion" was revealed as, in effect, a projection and a reification of human needs, capacities, wishes. Religion is "nothing but" (Feuerbach's favorite reductionist locution) the alienated form of human self-consciousness, of human knowledge and belief about human species-nature itself. "The secret of theology is anthropology", wrote Feuerbach. In subsequent works, notably in *The Essence of Religion*, Feuerbach went on to interpret or translate religious beliefs as expressions of the recognition of the human dependence on nature, as the source of biological, material existence, and on other human beings, as the requisites for social existence. Feuerbach's naturalism had two philosophical consequences: it directly challenged the reigning authority of the church, and of idealist philosophical views, such as Hegel's, which saw the material world as dependent on divine creation, or on the activity of consciousness (as the "othersideness" or objectification of Spirit); and further, it proposed that an account of human belief, of human thought and action, had to be grounded on an account of the material conditions of this-wordly existence, i.e. on the biological, social and psychological prerequisites of everyday life. As Gregory shows, the anti-clerical, anti-authoritarian and humanist thrust of Feuerbach's work suited both the political and the scientific temper of this group of materialists. Feuerbach, however, was neither a natural scientist nor a reductive materialist in the physicalist sense. His interest in the natural sciences was strong, but remained at the level of an intense amateur interest in geology, and fairly wide reading in contemporary scientific works in physiology and chemistry. He had studied physiology and zoology, as a student, but was hardly a trained scientist. His "materialism" is problematic, but its emphasis is upon the importance of the material – i.e. physico-chemical and biological – conditions of human existence, and the forms in which human consciousness in general, and religious belief in particular, express an awareness of these conditions as life-needs. His materialism may therefore be characterized as anthropological or humanist rather than physicalist. Yet, Feuerbach's achievement lay in relating the material life of human beings to their consciousness, their belief, their theoretical thought, (whether in science or in theology); and further, in making of this humanist or anthropological materialism a critical instrument in the political struggle against the reactionary clerical-monarchist political forces of the time. He provided, therefore, that link between the natural sciences and politics which the scientific materialists required.

Marx and Engels were to transcend Feuerbach's anthropological materialism in their development of an historical theory of human social evolution. Marx in particular, criticized Feuerbach's emphasis on the mere *reflection* of human needs in consciousness, and insisted on a more dynamic and central role for the concrete practical activity of satisfying these needs by labor. Vogt, Moleschott, Büchner, Czolbe, by contrast, emphasized the notion of the material, bodily processes that characterize human life and thought, and they tended towards a physiological, if not fully physicalistic reductionist account of human function and of consciousness. But such a general characterization does violence to the subtlety of the views which the scientific materialists developed. It is true that Moleschott's work on nutrition, *Die Physiologie der Nahrungsmittel* (and its popular version, *Die Lehre der Nahrungsmittel: für das Volk*, 1850), was, in effect, an elaboration of the German pun, *Der Mensch ist was er isst* ("Man is what he eats"), and was popularized in this vein, (e.g. in Feuerbach's enthusiastic 1850 review, provocatively entitled 'The Natural Sciences and the Revolution.') Yet, the intent of Moleschott's work was to provide a scientific argument for the material conditions of human well-being, as well as a self-help guide (in its popular version) for the oppressed and deprived, as to how to make the best use of available foodstuffs. Moreover, Moleschott was advancing a philosophical materialist view of monist sort, (closely related to the 18th century "transformism" of Diderot,) which held that life and consciousness were forms which arose by transformations of matter. This is especially clear in Moleschott's vastly popular work, *Das Kreislauf des Lebens* (1852), subtitled *Physiologische Antworten auf Liebig's Chemische Briefe.* Whether such transformism is reductionist or emergentist depends, of course, on the specific formulation. Vogt, as Gregory shows, spoke sometimes in a radically physicalist reductionist way, e.g., in his statement that "since belief is only a characteristic of the body's atoms, a change in belief depends only on the manner and kind of replacement of the atoms of the body." Similarly, Gregory points to Czolbe's *Neue Darstellung des Sensualismus* and his *Grundzüge einer Extensionalen Erkenntnisstheorie* as attempts at a reductive account of consciousness in terms of neurophysiology, or to a fully mechanical reduction (under the influence of Müller's work on specific nerve energies).

The issue of reductionism and emergence is most fully joined in the debate which the four scientific materialists waged against vitalism in biology, and in their (differing) acceptances and criticisms of Darwin's *Origin of Species*. In a richly detailed account, Gregory traces the philosophical and scientific issues in the debate, on the questions of spontaneous generation, preformationisms, speciation, and the transformation of species, natural selection, etc. Gregory's account is a significant addition to the literature on the reception of Darwinism. He writes, "Darwin did not create the sensation in Germany he had in England . . . Once Germans had been told that man's mind could be compared to urine, it came as no shock that man was now supposedly related to apes."

In these and many other contexts, Gregory's study introduces us to a complex intellectual and social movement in science, one whose ramifications for our own time become more evident as one reflects on the details. There is a pseudo-scientific and pseudo-philosophical tradition, in which scientific theories are projected onto society, or are extrapolated as metaphysical truths. So, for example, late nineteenth century Social Darwinism, so-called, read back into society the struggle for survival of the fittest as a biological model of superior and inferior races and social classes. (By "read back into society", I am suggesting that the social theory and ideology of competition, and of the justification of domination by the "winners" in the social and economic struggle may have had its influence on the very categories and formulations of the biological theory.) Similarly, the second law of thermodynamics has been projected into fantastic political and aesthetic interpretations; and quantum "indeterminacy" has been offered as an argument for the existence of free will (much as Epicurus had used this argument from the postulated *clinamen* or random "swerve" of his atoms).

The question is: Do scientific theories provide valid heuristics or analogies for social and metaphysical theories? And conversely, do social beliefs and metaphysical commitments have a useful, or rather a deleterious effect on scientific theory formation? I believe these are serious questions and deserve serious examination. But they cannot even be raised as questions, in any serious way, unless we first recognize the fact that, historically, as well as at present, there *is* such an interaction of science, politics, and philosophy. I do not believe there is a general answer to these questions. It all depends on how the social projection of a scientific theory is worked out concretely in a specific case, and on how political or philosophical ideas are seen to be related to science. The failure to recognize that *in fact* they are thus related in the working life of the scientist; or the pretence that science can be immunized or isolated from its broader social and intellectual contexts –

both lead to falsely one-sided and abstract conceptions of science, which are condemned to remain uncritical of the very distortions which social, political or philosophical views may impose on science, or which result from unexamined projections of scientific models onto society.

Gregory's historical study, beyond its intrinsic value as a reconstruction of a hitherto neglected chapter in the history of nineteenth century science, is also an antidote to such uncritical philosophy and history of science. By virtue of his artful and thorough analysis of the interplay of science with philosophy and politics, he permits us to make concrete and critical judgments about the important questions raised above. The scientific materialists whom Gregory examines were concerned with issues which come before us again, in genetics, in theory of evolution, in sociobiology, in neurophysiology – and certainly, in sociology and political theory. That the philosophical issues of materialism, reductionism, mind-body relation, emergence, remain live questions in contemporary philosophy is evident from any cursory review of the literature. Just as clearly, the issues of the growth of scientific knowledge, and of the relation between theory-change in science and social change, between science "proper" and science "improper" (metaphysics) are once again central issues for the philosophy and history of science. If Gregory concludes that "the overwhelming trademark of the scientific materialists, as far as the historian is concerned, is not their materialism, but their atheism, more properly, their humanistic religion", he is not yet giving us a normative judgment of how this affects the value, the quality, or the originality of their scientific thought. But he is providing for us the context within which we can begin to approach such questions critically, and begin to understand what a complex human enterprise science is.

MARX WARTOFSKY

ABBREVIATIONS

Feuerbach:

Werke, I–XII	*Sämtliche Werke*, 2d ed. Reprint, 1959–1964

Marx and Engels:

Werke, I–XXXIX	*Werke*, 1957–1968

Vogt:

AA	*Zur Anatomie der Amphibien*, 1839
AL	*Aus meinem Leben. Erinnerungen und Rückblicke*, 1896
AN	*Altes und Neues aus Thier- und Menschenleben*, 1859
AS	*Der achtzehnte September in Frankfurt a.M.*, 1848
BT	*Bilder aus dem Thierleben*, 1852
ED	*Eduard Desor: Lebensbild eines Naturforschers*, 1883
ES	*Embryologie des Salmones*, 1842
FA	*Ein frommer Angriff auf die heutige Wissenschaft*, 1882
GG	*Im Gebirge und auf den Gletschern*, 1843
GK	*Untersuchungen über die Entwicklungsgeschichte der Geburtshelf Kröte*, 1842
HS	*Über den heutigen Stand der beschreibenden Naturwissenschaften*, 1847
KW	*Köhlerglaube und Wissenschaft*, 1855
LG I, II	*Lehrbuch der Geologie und Petrefactenkunde*, 1846
MP	*Mein Prozess gegen die Allgemeine Zeitung*, 1859
NF	*Nord-Fahrt entlang der Norwegischen Küste*, 1863
NG	*Natürliche Geschichte der Schöpfung des Weltalls*, 1851
NR	*Beiträge zur Neurologie der Reptilien*, 1840
OM I, II	*Ocean und Mittelmeer*, 1848
PB	*Physiologische Briefe für Gebildete aller Stände*, 1845–47
PBK	*Politische Briefe an Friedrich Kolb*, 1870
PM	'Die Physiologie des Menschen,' in *Die Gegenwart*, **IV**, 1850
SGL	*Studien zur gegenwärtigen Lage Europas*, 1859
UT	*Untersuchungen über die Thierstaaten*, 1851
VM, I, II	*Vorlesungen über den Menschen*, 1863
VT	*Vorlesungen über nützliche und schädliche . . . Thiere*, 1864
W	'Weltanschauungen,' in *März*, I, 1910
ZB, I, II	*Zoologische Briefe*, 1851

Moleschott:

EL	*Die Einheit des Lebens*, 1864
FMF	*Für meine Freunde*, 1894
FW	*Zur Feier der Wissenschaft*, 1888
GF	*Georg Forster, der Naturforscher des Volks*, 1854
GM	*Die Grenzen des Menschen*, 1863

HM	*Hermann Hettner's Morgenroth*, 1883
KL	*Der Kreislauf des Lebens*, 1852
KRD	*Karl Robert Darwin*, 1883
LL	*Licht und Leben*, 1856
LN	*Die Lehre der Nahrungsmittel*, 1850
NH	*Natur und Heilkunde*, 1864
PN	*Physiologie der Nahrungsmittel*, 1850
PP	*Pathologie und Physiologie*, 1865
PS	*Physiologie der Stoffwechsels in Pflanzen und Thieren*, 1851
PSe	*Eine physiologische Sendung*, 1864
PSk	*Physiologisches Skizzenbuch*, 1861
SLM	*Von der Selbststeuerung im Leben des Menschen*, 1871
US	'The Unity of Science,' in *Popular Science Monthly*, 1888
UW	*Ursache und Wirkung in der Lehre vom Leben*, 1867

Büchner:

BFL	*Meine Begegnung mit Ferdinand Lassalle*, 1894
DS	*Darwinismus und Sozialismus*, 1894
FE	*Fremdes und Eigenes aus dem geistigen Leben der Gegenwart*, 1890
FNG	*Der Fortschritt in Natur und Geschichte*, 1894
GB	*Der Gottesbegriff und dessen Bedeutung in der Gegenwart*, 1874
GT	*Aus dem Geistesleben der Thiere*, 1876
GZ	*Das goldene Zeitalter*, 1891
HL	*Beiträge zur Hall'schen Lehre*, 1848
ID	*Im Dienst der Wahrheit*, 1900
KL	*Das künstige Leben und die moderne Wissenschaft*, 1889
KS	*Kraft und Stoff*, 1855
LA	*Herr Lassalle und die Arbeiter*, 1863
LL	*Licht und Leben*, 1882
LT	*Liebe und Liebesleben in der Thierwelt*, 1879
MA	*Die Macht der Vererbung*, 1882
MSN	*Der Mensch und seine Stellung in der Natur*, 1870
NG	*Natur und Geist*, 1857
NH	*Der neue Hamlet*, 1885
NW, I, II	*Aus Natur und Wissenschaft*, 1862, 1884
PB, I, II	*Physiologische Bilder*, 1861, 1875
RWW	*Über religiöse und wissenschaftliche Weltanschauung*, 1887
SJ	*Am Sterbelager des Jahrhunderts*, 1898
SV	*Sechs Vorlesungen über die Darwin'sche Theorie*, 1868
TT	*Thatsachen und Theorien aus dem naturwissenschaftlichen Leben der Gegenwart*, 1887
ZF	*Zwei gekrönte Freidenker*, 1890

Czolbe:

ES	*Entstehung des Selbstbewusstseins*, 1856
GE	*Grundzüge einer extensionalen Erkenntnistheorie*, 1875
GU	*Die Grenzen und der Ursprung der menschlichen Erkenntnis im Gegensatz zu Kant und Hegel*, 1865
NS	*Neue Darstellung des Sensualismus*, 1855

LIST OF FIGURES

INTRODUCTION

For those who had been involved in the uprisings of 1848–1849, the situation in Germany at mid-century was anything but heartening. Friedrich Wilhelm IV's refusal to "pick up the crown from the gutter" was the final blow to the helpless Frankfurt Parliament. By April of 1849 the Prussian army was in firm control, as it in fact had been for some time, and the German flirtation with revolution was history. It was a time to admit failure, to drop utopian illusions and to face the world soberly and realistically for what it was.

Not that the goals and aspirations of Germans changed radically after 1848–1849. Indeed, they did not. Unification was still a dream, and for many some form of representative government was not ruled out of the hopes for the future.

What had changed was the perspective from which the goals were envisioned. No longer did Germans of the 1850's view their ends idealistically; rather, they concentrated on a realistic means for achieving them. And, as Breunig notes, "they very often reached the conclusion that abstract ideals and principles were less important in securing their goals than were power and force."[1] Now was not the time for sermonizing.

In this context the German scientific materialists were exceptions, not the rule. Karl Vogt, Jacob Moleschott, and Ludwig Büchner refused to forsake idealistic rhetoric and capitulate to a praxis based on pragmatism. To them it was necessary first to educate the people by raising the level of popular consciousness. Only then would change come about. They held firmly to the notion that ideas change people and people change history, and their goal was to carry out this sorely needed education of the masses.

The primary task, as they saw it, was to correct the one great error of the past – unquestioned acceptance of and belief in authority. Although they sometimes approached the matter theoretically, denouncing the very principle of authority, they were more concerned to oppose the specific political and religious institutions of their time. Their task was a negative one: to expose the arbitrary and illusory foundations of the existing authority structure. But their spirit was markedly optimistic, in sharp contrast to the general tenor of the immediate post-revolutionary years.

They were not deterred by persecution from the authorities they were fighting; in fact, persecution was proof to them that they were succeeding. As edition after edition of their works came forth, their fame spread until they became known, to use Büchner's words, as "a kind of underground trinity,"[2] this in spite of the fact that they had never joined forces formally.

Although the rise of scientific materialism with its message of hope and optimism seems out of place amidst the pessimism of the early 1850's, it is not impossible to locate the factors which gave rise to its appearance. Briefly stated, the scientific materialists picked up where the Young Hegelians, in particular Ludwig Feuerbach, had left off, but their critique of Hegel, their atheism, their criticism of authority, and their monism were proclaimed as the results of science, not as the musings of philosophers or radical theologians.[3]

Feuerbach's rise to fame peaked in the mid-1840's, after which time he quickly was forgotten. Natural science in Germany, on the other hand, was undergoing a rapid expansion in the 1840's, an expansion clearly evident in the proliferation of popular science. The respect enjoyed by philosophers in the time of Hegel was being replaced by the prestige associated with scientific achievement. When Vogt, Moleschott, and Büchner, who were associated with the scientific community, adopted Feuerbach's radical conclusions, modified them to fit their new context, and proclaimed them as if they were scientific results, they were blending together Feuerbach's sensationalized message with the authority of science. Given the confluence of Feuerbach's message with the rise of popular scientific literature in Germany, the popularity and optimism of the scientific materialists is not surprising. A brief look at Feuerbach's conclusions themselves, and at the relatively sudden appearance of German popular science in the 1840's and 1850's will make this clear.

The Young Hegelians reformulated the philosophy of Hegel into a far more radical position than anything Hegel himself had said. Prior to Ruge, Strauss, Feuerbach and the Bauers there had been few who dared to criticize Hegel's speculative thought for its *a priori* character, and those who did were largely unheard.[4]

By far, however, the most devastating critique of Hegel came from Ludwig Feuerbach in three works, *Zur Kritik der Hegelschen Philosophie* of 1839, *Das Wesen des Christentums* of 1841, and *Grundsätze der Philosophie der Zukunft* of 1843. Although Feuerbach was no materialist, he could sound like one in his denunciations of idealism. In fact, he was called the crassest of materialists by his contemporaries.

Feuerbach came to believe that man's conception of nature was dependent upon an act of human experience equally as primary as self-consciousness. This was the act of sensation, *Sinnlichkeit* in his terminology. With this realization Feuerbach pointed out that the failure of speculative philosophy to acknowledge the integrity of sensation meant that the senses were being subordinated to the intellect, or the Idea, and that in so doing sense experience was not being understood for what it was in reality, but that it was being treated as a purely intellectual phenomenon dependent solely on the mind.

To an idealist, sensations were in part the product of the intellect. For Feuerbach, however, nature was as much given to man as his own mental activity, his senses were as primary and as *a priori* as his thinking.[5] Given this, any attempt to account for sensations in terms of the intellect amounted to rejecting sensation for what it was. This tendency of idealists had the effect of discounting the actual experience of sensation in favor of reflections *about* sense experience. According to Feuerbach, reflections about sense experience could not do justice to the actual experiencing of sensations themselves. Reflections were not, as speculative philosophers seemed to purport, equivalent to actual sensations.

Once he had demonstrated to his own satisfaction that a wedge could be driven between the experience of sensation and reflection about the experience, Feuerbach began to exploit the distinction for what it was worth in other, more general contexts. He extended and generalized the notion of experience beyond sensation itself to 'experience of the world' and 'experience of human needs'; i.e., experience in the sense of man's existential confrontation with that which is outside himself.

Feuerbach's reason for extending the referent of experience beyond sensation was to expose the disastrous result that followed from Hegel's refusal to acknowledge the primary role of the experience of sensation in his totally intellectual system. What happened, explained Feuerbach, was inevitable. By intellectualizing the experience of sensation Hegel and other idealists had severed sensation's roots in the real world and made it possible to bestow upon sensation an illusory, false, and merely imagined foundation, one that existed only in the mind. The abstractions of speculative philosophy were no more than real experiences transferred to the realm of thought and there made into a separate, ideal reality.[6]

This philosophical objection to Hegel was the position of Feuerbach's *Kritik der Hegelschen Philosophie* of 1839. In his next book, *Das Wesen des Christentums*, the disaster became more apparent, for Feuerbach here exposed the intellectualization not of the experience of sensation, but of the experience of human needs. When these were intellectualized, said Feuerbach, the result was theology.

In *Das Wesen des Christentums* Feuerbach explained that historically Christian thinkers had operated as idealists; i.e., they had dealt with the genuine experience of human needs in terms of intellectualized symbols of those needs. Throughout the first half of the book, Feuerbach labored to demonstrate that the various theological doctrines of Christianity were intellectualized forms of authentic human experiences, and that as such these intellectualizations had been enjoying a separate existence of their own. The appearance and evolution of theology as a system of doctrine represented a theorization of man's consciousness of his species-being (*Gattungswesen*). Man became aware of his nature as a social creature, he realized the sensual needs he possessed. But in theological doctrine man was expressing his awareness theoretically. In summary, Feuerbach declared that theology and speculative philosophy, being guilty of the same error, should be viewed as qualitatively similar endeavors.

Among the unfortunate results of the intellectualization of human needs (theology) was the development of a false sense of transcendence. To Feuerbach, intellectualized experience, in part summarized and symbolized by the idea of God, could not possess the characteristics of complete transcendence. To conceive of God as the totally Other was misleading, for to do so neglected that the reality of intellectualized experience consisted in its being an image. It was, namely, an image of human experience.

It should be noted that Feuerbach was not an ardent opponent of religion in general. His criticism was aimed specifically at speculative philosophy and theology. Throughout his life his preoccupation with religion was for the most part positive and constructive. He was concerned first and foremost to find out the real human needs religion expressed. In this he was greatly misunderstood, for his work was perceived, as was that of Strauss, to be wholly negative.

Needless to say, he resented it. In the preface to volume one of the collected works, first published in the 1840's, Feuerbach must have confused many when he declared: "He who says no more of me than that I am an atheist, says and knows nothing of me."[7] He was the enemy not of religion, but of theology, of religion reflecting upon itself. "The element of alienation, of setting something over and above man that should be man's, increases as religion comes to reflect on itself, . . . as it becomes *theology*."[8] For its own part, religion was based on man's recognition of his dependency. Man tried to negate his dependency with the help of his imagination; hence traditional religion was a stop-gap measure, but one which prevented man from ever really overcoming his alienation. It was, as Marx would put it, an opium.

Feuerbach resented enormously, to the point of anger, that theologians and philosophers had, over the course of history, successfully exploited the hypostatization upon which theology and speculative philosophy thrived. Man had come to believe that things which he himself created in his mind (such as God) had an independent existence, and mankind had as a result been kept in an unnecessary state of indignity and humility.

Feuerbach's message was designed to expose and correct this situation. *'Homo homini Deus est,'* Man is the God of man, Feuerbach declared to his age. God was the creation of man, He was the projection of human needs into the heavens, and one could read *Das Wesen des Christentums* to find out the how and why. Feuerbach thought, as did the materialists of science later, that he was basing his views on the reality of facts, and that he was bringing together into one humanistic message the truth about religion. *German* materialism, Feuerbach said later, had its roots in the Reformation, in a tradition that took its stand against false, unreal, 'theological' authority.[9]

Christianity no longer corresponds either to the theoretical or the practical man; it no longer satisfies the mind, nor does [it satisfy] the heart, for we have other interests for our heart than eternal heavenly bliss.[10]

What were these other interests?

Unbelief has taken the place of belief, reason the place of the Bible, politics the place of religion and the church, earth the place of heaven, work the place of prayer, material want the place of hell, man the place of the Christian.[11]

There was to be a new religion – politics, with its own official principle. "This principle is no other – negatively expressed – than atheism; i.e., the surrender of all gods different from man."[12]

If there was to be new religion, there was also to be a new philosophy, a 'philosophy of the future'. Four essays on philosophy appeared within two years of *Das Wesen des Christentums*· (1) 'Uber den Anfang der Philosophie' (1841): (2) 'Die Notwendigkeit einer Reform der Philosophie' (1842); (3) 'Vorläufige Thesen zur Reform der Philosophie' (1842); (4) 'Grundsätze der Philosophie der Zukunft' (1843). Although they contained much that had already been said in the critique of Hegel, they did attempt to spell out some of the implications of Feuerbach's new appreciation of the empirical world for philosophy.

Feuerbach, then, reacted against Hegel because he understood the result of Hegelianism to be an illusory and unnecessary transcendence. This is not to say that Feuerbach interpreted Hegel as a conscious proponent of dualism.[13] We need but recall that the concern with alienation among Marx and the Young Hegelians derived from Hegel. Rather, Hegel's monism had left something out, a whole realm of human experience, which, when alloted its place of primacy alongside the traditional primacy of self-consciousness, revealed that speculative philosophy was one-sided and without roots in reality.

The elimination of transcendence had repercussions beyond philosophy and theology. Feuerbach's sensualism was one expression of the general disposition towards realism which appeared in the wake of, and in reaction to, the successive idealistic systems of philosophy of the first thirty years of the nineteenth century. His notoriety in the forties and the evaluation of his work by Marx are proofs that here was the first critique of Hegel that gave voice successfully to this disposition. It was he upon whom the scientific materialists ultimately depended when they railed against the existence of a vital force, or a human soul. Heinrich Czolbe, for example, began his *Neue Darstellung des Sensualismus* with the declaration that the basic principle of his system required "the elimination of the assumption of suprasensual things in all thinking."[14] Each of the other scientific materialists insisted over and over again that what he was doing was replacing supernatural explanations in science with natural ones. In this endeavor each was representative of the age of realism and naturalism that Ludwig Feuerbach symbolized.

Within natural science, the shift of the focus in scientific activity to Germany was becoming evident in the forties. Most well known of the German scientists in this decade was Alexander von Humboldt, whose *Kosmos* came out in 1845, and who at one time had been, according to Agnus Clerke, "the most famous man in Europe with the exception of Napoleon Bonaparte."[15] Humboldt's *Kosmos* soon became more widely read than any book except the Bible.[16]

Second in international reputation probably was Justus von Liebig at Giessen. His chemical laboratory attracted students from all over Europe and from America. The 'Chemische Briefe' that Liebig commenced publishing anonymously in the *Allgemeine Zeitung* in 1842 were gathered and printed in book form first in English in 1843. Within approximately a decade this often revised and extended book had been published in three English editions, and in six Italian, three German, four French, three Russian, three Spanish, three Danish, two Dutch, two Swedish, one American, and two Polish editions.[17]

Characteristic of the new German scientific community on the philosophical front was a replacement of the older *Naturphilosophie* with no philosophy at all. The successes of experimental science seemed to be proving that it at least could do quite well without traditional philosophy. By 1840 the older school of Schelling and Oken had been largely displaced by younger experimentalists like Matthias Schleiden, whose critical work, *Über Schellings und Hegels Verhältnis zur Naturwissenschaft* came out in 1844.[18] In place of philosophy there was left a gap; scientists simply began to shy away from philosophy and retreat into experimentation.[19]

If anything took the place of philosophizing about science it was popular scientific literature. Besides Humboldt and Liebig, Rossmässler, Burmeister, Cotta, and Vogt[20] all had written material which had reached the public before 1848. Although there were some pronouncements in these works that associated natural science with materialism, namely in the writings of Vogt, it was not until after the Revolution that this association became the basis of an entire movement.[21] What these works did communicate was that science dealt with the real and true in contrast to the unproven claims of tradition, be they religious or philosophical.

Beginning in 1847, Vogt had been the first to write material in which radical implications were drawn from the study of natural science, and Moleschott's first such book came out in 1850. These were not meant to be systematic defenses of materialism in the sense of the later *Kreislauf des Lebens* or *Köhlerglaube und Wissenschaft*. These early works belonged more to the popularization of science brought about by and indicative of the increased interest in the natural sciences in the Germany of the 1840's.

Important for the materialism in the fifties was the birth of several popular journals of science, at least two of which, *Die Natur* and *Das Jahrhundert*, had scientific materialism as a primary reason for existence. Popular journals of natural science were not unknown even in eighteenth-century Germany,[22] but there, as in the first half of the nineteenth century, natural science was generally treated, if treated at all, along with the other *Wissenschaften*. This practice also continued throughout the nineteenth century; Vogt, Moleschott and Büchner, for example, were frequent contributors to such journals as *Westermann's*

Monatshefte, Gartenlaube, Nord und Süd, Stimmen der Zeit, Die Gegenwart, and others.

The last periodical mentioned, for which Vogt and Moleschott wrote long articles on science, was the fourth supplemental series of an old popular *Conversationslexikon* whose purpose was to increase the general educational level of the populace, and thereby provide a solution to the social problem. *Die Gegenwart*, begun in 1848, represented the best in the German liberalism of the day, and while tolerating the materialistic views of Vogt and Moleschott, the editors always printed a disclaimer of responsibility for ideas expressed.[23]

There were, however, other popular journals which began to appear in the early 1850's and which were particularly devoted to natural science. The first of these commenced in 1850 with the same name as Oken's journal, *Isis*, and carried the subtitle, *Encyclopädische Zeitschrift vorzüglich für Naturgeschichte, Physiologie, etc.*[24] In 1852 three more popular journals were launched, the Dutch *Album der natuur, een werk ter verspreiding van natuurkennis onder beschaafde lezers van allerlei stand*, and two German journals, *Aus der Natur. Die neuesten Entdeckungen auf dem Gebiete der Naturwissenschaften,* and *Die Natur. Zeitung zur Verbreitung naturwissenschaftlicher Kenntniss und Naturanschauung für Leser aller Stände.*[25] This last one was a veritable organ of scientific materialism, and was edited by Rossmässler, Karl Müller, and Moleschott's brother-in-law Otto Ule. Later Rossmässler left the journal's staff and founded his own popular scientific periodical in 1859, entitled *Aus der Heimat. Ein naturwissenschaftliches Volksblatt.*

Die Natur was primarily the brainchild of Ule and Müller. Volume one, number one was dated January 3, 1852, and for fifty years Germany was provided weekly with hymns to the glories of science and the scientific world view. On each title page there appeared the trademark of the new journal, a sketch of an exploding volcano, which was explained by the editors in terms that could only be understood politically: "The fire inside is not extinguished; its passion still breaks forth in catastrophic flaming streams from the pores of the earth."[26]

Another journal dedicated to materialism began in 1856. Although its title, *Das Jahrhundert, Zeitschrift für Politik und Literatur*, indicated that it was not a journal of natural science, many of its articles were about the influence of science on politics, especially those by the Jewish socialist Moses Hess. Martin May, businessman, publicist, and one of the editors, explained in a letter to Hess the way in which the periodical had come about.

A union of men was formed in Hamburg for the purpose of founding a journal, which should take as its task the advancement and diffusion of proper views in the area of politics and in that of natural science.[27]

Within a year the publisher could tell Hess that Vogt and Moleschott had promised articles and that circulation was good enough to support the venture by itself.[28] The strong materialistic position of the editors and the publisher became evident when Arnold Ruge submitted a series of articles under the title 'Der Geist unserer Zeit,' which contained forceful antimaterialistic tenets. Somehow the first number was printed before the "error," as the publisher put it, was spotted. A letter to Ruge explaining that the journal's stance was strictly materialistic and humanistic brought a reply denouncing the work of Büchner and Hess. The outcome was Ruge's break with the periodical because of its clear identification with materialism.[29]

Feuerbach's critique of idealism on the intellectual front, and the proliferation and popularization of natural science on the practical side made an impressive tandem. Their combined effect was to cast suspicion on philosophy as it had been known in Germany, i.e., on idealistic philosophy. Before the Revolution Feuerbach had tried to make it clear that there was a natural relationship between the *a priori*, authoritative, illusory and artificial categories of idealism, and the qualitatively similar justifications of religious, political, and social authority. After the Revolution this message was taken up by two separate schools of thought, dialectical and scientific materialism, each of which was rooted in Feuerbach's work.

For the scientific materialists, and to some extent for Marx as well, opposition to groundless authority was the task and natural science was its justification. Science could not be restricted by political force no matter how threatening science might become. Vogt condemned authority in the name of anarchism, and Feuerbach, in his review of Moleschott's *Lehre der Nahrungsmittel* of 1850, proclaimed that the truths of natural science were the real revolutionary forces of history. When Büchner finally emerged from his period of personal crisis and depression it was with the cry, "No longer do I acknowledge any human authority over me."[30]

All in all the scientific materialists had their own scepter hovering over the land. Reviewing the onslaught of materialism in Germany at mid-century, Karl Fortlage ominously summarized the situation as follows:

> A new world view is settling itself into the minds of men. It goes about like a virus. Every young mind of the generation now living is affected by it. . . . Only this much is certain: the old has become obsolete and the new presses powerfully forward, as if mankind were pregnant with a new humanity.[31]

Small wonder that there was such an outcry against the materialism of natural science on the part of the established powers.

PART 1

BACKGROUND

CHAPTER I

LUDWIG FEUERBACH: FATHER OF GERMAN MATERIALISM

The decade of seniority Ludwig Feuerbach held over the scientific materialists is significant. His published works were already known to literary circles in the 1830's, to be sure as ardently Hegelian pieces, but even his philosophy of sensualism was in the hands of the public by 1845. Further, the ten years separating Feuerbach's university days from those of Karl Vogt correspond to noteworthy changes in German academe, changes symbolized by the death of Hegel in 1831. Feuerbach's university career began in 1823 and ended the year Hegel died, while Vogt did not begin his studies at Giessen until 1833. In order to become a proponent of materialism, Feuerbach had to undergo a conversion that was unnecessary for Vogt or Moleschott. Unlike Feuerbach, they never experienced the intellectual commitment to the idealistic, speculative philosophy of the 1820's.

Ludwig Andreas Feuerbach was born in 1804 in Catholic Bavaria, fourth son of Anselm Feuerbach, the noted jurist.[1] Anselm was a liberal Protestant who accepted the invitation of Crown Prince Max Joseph to assume the chair of law at Landshut in 1803. Barely a year after Ludwig's birth, Anselm left Landshut for Munich because of a dispute with colleagues. The family remained in Munich, where Anselm served in the Ministry of Justice until 1813, Ludwig attending primary school there. When the time came in 1817 to enter the *Gymnasium*, Ludwig left his mother's house to go to Ansbach, where his father, now formally separated from his mother, was living with another woman.

Feuerbach's letters to his mother reveal a fiery temperament. The sixteen year old youth wrote of his trip to see his older brother in Baden, and of the return by way of Mannheim where he visited the grave of the theological student Karl Sand. Sand had been executed for assassinating Kotzebue, a dramatist who had publicly scoffed at those Germans demanding freedom. Included in his letter were blades of grass from Sand's grave, with the revealing comment: "Because you too loved this German youth."[2] These early letters also displayed a decided interest in religion.[3]

Before his *Gymnasium* years terminated in 1822, the parents were reunited, and Ludwig spent a year with them before beginning university studies in theology at Heidelberg. At Heidelberg Feuerbach heard lectures of the Kantian H. E. G. Paulus, called "the patriach of theological rationalism" by Schnabel.[4] He also attended the course in speculative theology taught by Karl Daub, who was at the time strongly influenced by Hegel.[5] By the end of one year Feuerbach knew that he wanted to attend Hegel's lectures. After he secured the permission of his father, the twenty year old youth arrived in Berlin in the spring of 1824.

No sooner had Ludwig arrived in Berlin than the Prussian police exhibited for him the general paranoia of the post-Napoleonic German officials. A secret investigative commission had received a report that several Feuerbach brothers were members of a secret society. Karl Feuerbach, known in mathematics for the theorem which bears his name,[6] was teaching mathematics at Erlangen when he was arrested and imprisoned. Ludwig was barred from matriculation in Berlin because he was under police surveillance. The matter was not cleared up until July 28, when Feuerbach finally was admitted to theological studies. Karl, who was moved from prison to prison, attempted to commit suicide in December of 1824, and eventually was released without evidence of criminal behavior.[7]

It was not long before Ludwig realized what Daub had suspected all along; namely, that his heart was not in theology, but in philosophy. He had attended Hegel's lectures even as a theology student. To switch faculties he would have had to overcome a major roadblock – his father. He obtained the consent he needed, though not without difficulty,[8] and became in 1825 a full-time student of philosophy.

Just as things began to settle down, Feuerbach again was forced to change his plans. King Maximilian I of Bavaria died on October 13, 1825, and with him went the stipend he had granted to the Feuerbach sons for their university education. In April of the following year Ludwig left Berlin for the less expensive Erlangen, where he received his doctorate in 1828. His study from 1826 to 1828 was interspersed with several visits to his home in Ansbach, where he occupied himself with the history of philosophy and the writings of Aristotle.

Feuerbach had spent two years with Hegel, one as a student of theology. Four weeks after beginning courses in Berlin he wrote enthusiastically about Hegel to his father:

> What was dark and incomprehensible, or at least what seemed unfounded with Daub I have already seen through clearly and recognized in its necessity as a result of the few of Hegel's lectures I have heard; what only glowed in me as kindling I see already bursting into bright flames.[9]

Hegel was clearer in his lectures than he was in his writing, said Feuerbach. There was no substitute for being in his presence. But although Feuerbach felt Hegel's impact upon him for the rest of his life, he was not one of those who uncritically worshipped the Master. Two years under anyone was enough. The young philosopher of twenty-two had gone through the whole of Hegel with the exception of the Aesthetics. He had heard all the lectures, those on logic twice,[10] and he had been very impressed. But he had not been enticed into thinking that the last word had been said. As he left Berlin for Erlangen he took his leave of Hegel, a proud and almost insolent youth: "I have heard you for two years, I have been faithfully dedicated to your philosophy for two years; now I need to plunge into the direct opposite. Now I shall study anatomy."[11]

Unfortunately we know little of Feuerbach's work while at Erlangen. Bolin reports, as already noted, that he divided his time between Erlangen and his home at Ansbach. While at home he studied the history of philosophy. At Erlangen he pursued the natural sciences,[12] presumably in fulfillment of his declaration to Hegel. At any event we do know that the years at Erlangen are given the title 'Zweifel' in the 'Fragmente zur Charakteristik meines philosophischen Entwicklungsganges,' which Feuerbach put together for the second volume of his collected works in 1846.[13] Apparently his involvement in natural science, "the direct opposite," caused Feuerbach to think seriously about the place science had in Hegel's system. Doubts expressed at this point in his life demonstrate clearly that nature, the object of natural science, had not been treated fairly by Hegel. The doubts came out as questions:

> How does thinking relate to being, and logic to nature? Is the transition from the former to the latter justified? ... Can therefore that which is other than logic be deduced not from logic, not logically, but unlogically? .... Were there no nature the unspotted maiden of logic would never produce one out of itself.[14]

Nature, in other words, possessed an integrity all its own, one that Hegelian logic could not construct by itself.

De ratione una, universali, infinita was the title of the doctoral dissertation Feuerbach submitted to the philosophical faculty at Erlangen in 1828. Not only did its title betray Feuerbach's Hegelianism, but the covering letter that accompanied Hegel's copy of the work was yet more explicit. Feuerbach wrote Hegel concerning his dissertation: "It is the product ... of a study which was sustained by a free, ... living, and essential appropriation and absorption of those ideas and concepts which form the content of both your works and oral discourse."[15] Still, Feuerbach had been influenced by his contact with the experimental sciences, as was evident from his qualification of the significance of *Geist*. *Geist* should not remain suspended above the sensuous and the phenomenal, but should descend down to the particular.[16]

With the doctorate successfully completed, Ludwig received the right to teach as a *Privatdozent* at Erlangen in the winter of 1829. It began to look as if Ludwig would follow the lead of his older brother Eduard at Munich, who had been promoted from *Privatdozent* to *ausserordentlicher Professor* of law in 1828. On the basis of the lectures Feuerbach gave at Erlangen from 1829 to 1832 there would be every reason to suspect that a promotion was in the offing. There was a gradual disenchantment with Hegel evident in his lectures (a fact that one scholar associates with the fast growing prestige of natural science in Germany[17]), but there was no reason to believe that this would forestall his promotion even at conservative Erlangen.[18] But Feuerbach sealed his fate with the anonymous publication in 1830 of *Gedanken über Tod und Unsterblichkeit*. It did not take the censors long to identify the author who depicted the Christian heaven as a kind of insurance company,[19] and whose purpose in writing the work was summarized in the proposition: "To cancel above all the old cleavage between this side and the beyond in order that humanity might concentrate on itself, its world and its present with all its heart and soul."[20] The work itself was strongly Hegelian in form and language. Although Hegel's philosophy was being given a conservative interpretation by most thinkers of the day, here, clearly, was a new and radical Hegelianism.

When it became obvious to Feuerbach that he was not going to be promoted, he left the University in 1832, and went to stay with his father's younger sister in Frankfurt. Back in Erlangen in the same year at brother Eduard's urging, Feuerbach did not claim his right to teach, but worked on putting his lectures on the history of philosophy into book form. The first fruits of these efforts, his *Geschichte der neueren Philosophie von Bacon bis Spinoza*, was published in the early summer of 1833 and was warmly received among Hegelians.[21] The leading thought of the work, according to Jodl, was a rationalistic idealism. Yet Feuerbach had high praise for Bacon and Gassendi, noting that the former had laid the foundation for natural science.[22]

In 1834 Feuerbach met Bertha Löw, whom he eventually married in 1837. One result of the union was an improvement in his financial situation, for his wife had inherited part ownership in a porcelain factory fifteen years earlier. The couple retired to a castle at Bruckberg, living on Bertha's income and whatever Ludwig could earn from his literary productions, which were on the increase. The major works on Leibniz and Bayle of 1837 and 1838 met with general approval,[23] but as far as a post was concerned, they were not sufficient to overcome the stigma that the *Gedanken* of 1830 had placed upon him. Feuerbach understandably grew bitter towards the university system and university philosophy, but he remained unwilling to break with Hegel. His writings of the early and middle 1830's did not significantly challenge Hegelian doctrine.

Although the 1830's was still too early for a systematic critique of Hegelianism from the side of the natural sciences, there were some voices of opposition from among the various philosophy faculties, although they were voices crying in the wilderness.[24] The most significant of these for our purpose were those dealt with by Feuerbach, namely those of Karl F. Bachmann and Franz Dorguth.

Bachmann had, like Feuerbach, begun in theology and switched to philosophy. He studied at Jena, where he attended Hegel's lectures, and rose through the ranks of *Privatdozent, ausserordentlicher Professor* to *ordentlicher Professor*. In 1813 he assumed the directorship of the Jena Mineralogische Societät, succeeding the University's noted mineralogist, Johann Lenz. Already in 1828 Bachmann had criticized Hegel in his *System der Logik*, but in 1833, two years after Hegel's death, he made him the target of specific criticism in his *Über Hegels System und die Nothwendigkeit einer nochmaligen Umgestaltung der Philosophie.*

The first part of this work was an exposition of Hegel's system, the second part a critique. It was in the latter section that Bachmann exposed what he considered to be a contradiction. The Hegelian postulate, "Whatever is reasonable is real, and whatever is real is reasonable," would be perfectly acceptable, Bachmann said, were it not that Hegel distinguished real existence from phenomenal existence. Phenomena for Hegel represented fleeting existence, entities without significance, subjective, and purely accidental. Phenomena did not deserve the name reality.[25] Bachmann next cited a place in Hegel's works where the Berlin philosopher had celebrated phenomena as a category which expressed the full extent of real existence. He showed that Hegel had alleged that there was nothing in the essence of an object that was not manifested in the phenomena. Why then, Bachmann asked, did Hegel carry out the polemic against phenomena?[26]

Bachmann conceded that the Hegelian critique of phenomena might well be used against the crass empiricism that was seeping into Germany from England and France.[27] Empiricism was but the beginning, the lowest level of reasoning, at the top of which philosophy reigned. But how could Hegel argue that experience was the point of departure for philosophy and then proceed to begin his system with the Logic?[28]

Bachmann's work came under the fledgling eye of the budding Hegelian Karl Rosenkranz, and the latter's critique of Bachmann provoked a reply entitled *Anti-Hegel: Antwort an Herrn Professor K. Rosenkranz in Königsberg*. In this work Bachmann revealed his dislike of the Hegelians. He often departed from philosophical analysis to argue *ad hominem* against Rosenkranz, at the same time defending himself against the charge that he had slandered Hegel's personality. But when he finally did get down to business, it became obvious why this work caught Feuerbach's eye. Bachmann's bone of contention with Hegel was cast in a similar vein to doubts Feuerbach had harbored as early as 1827: "The question is simply: how is one to think of the relation of the higher level to the lower one from which it comes?"[29] Bachmann's answer was to argue that the lower level, phenomena, was eliminated by the higher level of the rational in Hegel's system.

Feuerbach entered the fray with a review of Bachmann's *Anti-Hegel* in 1835. After a brief tirade against those, including Bachmann, who criticized from a vantage point of misunderstanding, Feuerbach betrayed what really upset him; namely, Bachmann's mockery of Hegel's idea of the identity of logic and metaphysics. The law of identity had offended better minds than Bachmann's, Feuerbach commented, and one need only point out in its defense that

> If the laws of the world are not also the laws of our thinking, and vice versa, if the general and essential forms in which we think are not themselves also the general and essential forms of things themselves: then no real knowledge, no metaphysics at all is possible, then there is in the world . . . an absolute vacuum, an absolute absurdity, and this absolute absurdity, this existent nothing, this rotten spot is our mind itself. Further, our reason would be nothing but a mental conception of an absolute irrationality.[30]

For Feuerbach at this point in his life, the guarantee that no such hiatus existed rested with his belief that human beings formed a juncture of the external and internal worlds: "Are we not while in ourselves also in the external world, and in our inner world are we not at the same time in the real world?"[31] The identity of logic and metaphysics was no more than an expression of the nature of thought, said Feuerbach, and Hegel had brought this fact to consciousness. This equation of the world of reality and the world of thought represents a permanent fixture in the Feuerbachian perspective.

Throughout his entire review Feuerbach was preoccupied with Bachmann's "platitudes against the identity of thought and being."[32] His disgust at Bachmann's distortion of Hegelian Being to mean not that which resolved the contradictions between spirit and nature, but "being without spirit and soul, being negated by the idea, being that is pure appearance,"[33] resulted from his own frustration vis-à-vis his ignorant opponent. Feuerbach's polemical eloquence was more inspired by his unconscious realization that he could not answer Bachmann, that Bachmann was harping on those very doubts he once had expressed himself, that there was a basic difference in attitude between Bachmann and Hegel which showed itself in different presuppositions. He surely went overboard, for example, when he associated Bachmann's view with materialism, or when he criticized Bachmann as "a loyal empiricist who supports himself solely on facts."[34] Bachmann's own aversions to empiricism and materialism were clear. Feuerbach's rage was a direct measure of his own discomfort; he was protesting too much.[35]

That was 1835. In 1837 Feuerbach published his work on Leibniz. In section 18 of this book, entitled "Einleitung zur Leibniz'schen Pneumatologie: Kritik des Empiricismus," Feuerbach continued to hammer away at nonidealistic philosophy. But each new work contained hints that Feuerbach was changing his mind, even if he himself had not yet admitted it. The Leibnizian critique of Locke's superficiality was balanced here with an appraisal of the historical significance of empiricism. Empiricists pointed out, he explained, that the senses were the initial source of knowledge. Pursuit of the empirical has been of great advantage to humanity.

> Empiricism has facilitated the freedom and independence of thought – it has delivered us from the bonds of belief in authority by referring us to the holy, inalienable natural right of autopsy and self-examination.[36]

The expected criticism of empiricism was here, but in such a guise that the very same words could later be used in *defense* of his new position. It was a change in attitude, in values, and in goals that Feuerbach as yet lacked. He criticized the notion that the senses were the source of ideas. The senses might mediate true perception, but they could not themselves produce it – that required the active participation of the mind. The empiricist, he said, treated seeing and hearing as irreducible fact, never recognizing the need to ask, "How are seeing and hearing possible?" For Feuerbach, the act of perceiving was not the same as that of sensing. It required the participation of the mind. "Perception of the object outside me as object is . . .*already* a purely mental act, already consciousness, already thinking."[37]

Feuerbach brought into the discussion the concept of intuitive understanding (*Anschauung*), a notion which in a revised form proved crucial in the work of Heinrich Czolbe. Intuitive understanding here was distinguished from thinking or the perception of reason. It was associated with sensation, but not with perception.

Everything lies in the intuition (*Anschauung*), correct; but to find it and to see it one must think . . .The senses here (in man as opposed to animal) are already primordially emanations of the theoretical *capacity*. *Man* is born to theory. The senses are the means of his knowledge, but means which are only active, indeed which are only means when one presupposes the existence of their inner purpose – the theoretically active, thinking capacity. The senses illumine the world to us, but their light is not their own, rather, it comes from the central sun of the mind.[38]

It was the idea that perception is *already* thinking that could be appealed to by both the early and the late Feuerbach. Feuerbach the idealist could give it the emphasis he has above in his work on Leibniz; viz., that the senses are the *occasion* for thinking, but that thought itself is a purely intellectual activity. Feuerbach the sensualist could and later did interpret this very same notion to means that "Nature is *as much given* to man as his own mental activity; his senses are as primary, as *a priori* (for him), as his thinking."[39] Feuerbach later explained the rise of idealism as the result of man's need for certainty, in which the self-certainty of consciousness and of the ego was chosen as the point to which all experience must be referred.[40] What if, then, man's experience of the external world by means of his senses was taken as equally certain, and thinking was taken to include and be part of this aspect of man as well as part of his conscious activity?[41]

The final expression of Feuerbach's idealism came in 1838 in his review of Franz Dorguth's materialistic criticism of idealism. The significance of this work has largely been overrated, for Feuerbach here simply stated a position he held throughout his life; i.e., he opposed crass empiricism. Jodl's contention that Dorguth influenced Feuerbach, and that the review may be seen as a precursor of the *Philosophie der Zukunft* is mitigated by the erosion already observed in Feuerbach's idealism, and by the fact that Dorguth's alleged reductionistic materialism[42] was less likely to have influenced Feuerbach than a work like Bachmann's, where the doubts Feuerbach himself experienced were underscored.[43] The work is rather a marker, for it is the last thing Feuerbach published which had a criticism of empiricism as its specific purpose.[44]

The review is an entirely predictable set of idealistic arguments against the materialistic emphasis that mind is dependent on body. Feuerbach called this dependent relationship in Dorguth a euphemism for the ruthlessness of the absolute materialism to which such a position leads.[45] Feuerbach called attention to the difference between the claim, "thinking is *only* an activity of the brain," and the assertion, "thinking is *also* an activity of the brain." The former statement was meaningless, argued Feuerbach, for unless the brain itself were identical to understanding, the activity of the brain would be without understanding (*verstandlos*), and thinking, the act of understanding, would also be devoid of understanding, an absurd conclusion.[46] Further, to call thinking an activity of the brain tells us nothing, for an activity is only known by its product, by what it does, and the product of thinking is thought. Hence only from thought can we know what thinking is.[47]

Reiterating his idealistic assumption that sensation is devoid of thinking, Feuerbach struck out at Dorguth's empiricism. The senses, he said, give us only appearance, unreality, illusion. Thought is the thing as it really is. Had Copernicus been foremost an empiricist, he never would have come to the true, heliocentric state of the solar system. "The Copernican system is the most glorious victory idealism and reason have achieved over empiricism and the senses."[48]

How, queried Feuerbach, did materialists come to the idea of mind at all? Or, for that matter, how did they know matter? Where there was only matter, there was no concept of matter. Matter was that which confirmed the mind as mind, and thinking was the activity by which we distinguish each from the other.[49] Here thinking was placed on a level above both mind and body, and Feuerbach's idealistic persuasion was evident. Thinking was no object of physiology, said Feuerbach, only of philosophy. To make it an object of physiology was an error, similar to the mistake of equating reading with seeing letters on a page, or the act of reading with understanding what was read. We read the book of nature by means of the senses, true, but we do not understand it by them.[50]

From this it is seen that, in so far as Feuerbach was opposed to reductionistic materialism, he need later apologize for nothing in the review. But in so far as he characterized thinking as solely an intellectual activity, independent of the body and nature, there was still an adjustment necessary. In 1838 he published both his last article in the Hegelian *Jahrbücher für wissenschaftliche Kritik* and his first in Ruge's new journal of opposition, the *Hallesche Jahrbücher*.[51]

In his letter to Feuerbach thanking him for his review of *Die Lehre der Nahrungsmittel*, Jacob Moleschott declared that Feuerbach was the first one to cancel the old relationship between empiricism and speculative philosophy, one "in which one holds empiricism up to the dubious light of speculative philosophy."[52] Whether or not Moleschott had in mind Feuerbach's *Zur Kritik der Hegelschen Philosophie* is uncertain, but the remark was surely applicable to it. The *Kritik* has been discussed thoroughly by others,[53] and it is not my intent to reiterate here what has been said elsewhere. All of the discussions noted emphasize that Feuerbach's criticism centred around the fact that Hegel had failed to consider man's undeniable and primary experience of nature and of the sensuous. In other words, Feuerbach's attitude had changed. No longer did he glory to see thought in its difference from sensation; rather, he opposed all who refused to acknowledge that sensation was an equally primary activity to thought.

Much of the critique was negative, being an exposition of why and how Hegel had erred. For example, Feuerbach explained that Hegel might have imagined that he was dealing with the other-than-thought, but he was not. Hegel began "not with being other than thought, but with the thought of being other than thought."[54] The same was true of Hegel's treatment of sense certainty in the *Phenomenology*. Hegel there claimed to have shown, as we have seen Feuerbach himself contend in his critique of empiricism, that the senses give us no knowledge by themselves. The certainty of the senses is only that of immediacy; it is empty, and without quality, a "here" and a "now." In order for there to be knowledge, a *mind* must be present.

Feuerbach criticized Hegel in the *Kritik* in a manner similar to that we have already observed in the 1847 edition of the book on Leibniz; i.e., he demonstrated that Hegel's "here" and "now" were empty and without quality because they were not the "here" and "now" of real sense experience, but of Hegel's abstraction. "Hegel refutes . . . the logical here, the logical now. He refutes the thought of here-and-now being (*Diesseins*)."[55] Anyone who spoke as one caught up in the real immediacy of sense experience would not talk of an abstraction, but of a sensation which already possessed a quality. The traditional distinction between sensation, as experience devoid of thought, and perception was rejected.

Feuerbach cried out in frustration: "Words, words. *Show* me what you are saying."[56]

If, as Feuerbach claimed, Hegel never took seriously the actual experiencing of sensation, but dealt only with the thought of such an experience, then the Hegelian Absolute Idea had failed in its task of resolving the opposition between itself and its Other, and any alleged proof that it had done so was carried out within the realm of thought alone.

The alienation of the Idea is only a *pretence*, so to speak; it [alienates itself from itself], but not seriously; it is merely playing. . . . The Idea begets and testifies to itself not in a real other – which could only be the empirical, concrete perception of the understanding – it begets itself out of a formal, apparent contradiction.[57]

All of this was a far cry from English empiricism, but Feuerbach had pointed up in a convincing way the necessity for philosophy to include the very experiencing of nature, and this in a country that was ripe for such an idea.

The critique of Hegel was in 1839. Feuerbach's next major work, the one for which he is known, was *Das Wesen des Christentums*, published in 1841. Feuerbach saw this work as an application of his changed philosophical attitude, as "an empirical or historical-philosophical analysis," in which there were "no *a priori*, self-invented [propositions], no products of speculation." His general propositions, he said, were "the actual utterances of the human essence, which were apprehended in general terms and thereby brought to the understanding."[58] Clearly Feuerbach was going out of his way to repudiate unconditionally Hegelian speculation.[59] Proudly he identified with the natural scientist: "I am nothing but a natural philosopher (*Naturforscher*) of the mind."[60] Throughout his work he often alluded to the natural sciences, and he repeatedly emphasized the factual nature of his material.[61] "My only purpose was to 'unveil existence,' not to invent – to discover."[62]

What had he discovered? He had found that religion was one grand anthropomorphism. God was not *really* there; He was a projection of man. *Man* was God. By carefully working through the major doctrines of Christianity, and by showing the real and immediate human needs from which they had sprung, Feuerbach exposed Christianity. He argued that the theologian was not concerned with the real stuff of religion, but an ideal expression of it.

The impact of the work was enormous. No one could ignore it, regardless of persuasion. Friedrich Engels, speaking of the young German communist movement of the 1840's, said that after the appearance of Feuerbach's book "the enthusiasm was general: we all immediately became Feuerbachians."[63] The theologians raised such a stir and demonstrated such ignorance that Feuerbach in 1842 published, again in Ruge's *Jahrbücher*, "Zum Beurtheilung vom Wesen des Christentums."[64] The *Wesen* brought him admirers from all ranks of society, though no doubt some of the tales of his fame have taken on a legendary quality.[65]

The peak of Feuerbach's fame came in the mid-forties. His *Grundsätze der Philosophie der Zukunft* and three other short pieces followed the *Das Wesen des Christentums* as attempts to elucidate the nature of "sensualism."[66] In these philosophical writings of the early 1840's Feuerbach was still reacting to Hegelian philosophy as much as he was replacing it. The result was that the language he employed was chosen to redress the one-sidedness of idealism, a practice which in turn produced a one-sidedness of its own: the all pervasive glorification of sensation (*Sinnlichkeit*). Although there was nothing contradictory in this, that is, although it was only a one-sidedness of language, it tended to give the impression that thinking had been forgotten, or, that thinking could indeed be separated from sensation.

One effect this had was to bring to the surface the naive realism Feuerbach had been assuming all along. In the position he had taken against Bachmann regarding the identity of the laws of thought and the laws of being, Feuerbach had maintained that a denial of this axiom was equivalent to denying the possibility of knowledge itself. This position he had not changed. What had changed in his new philosophy was his view of the nature of thought. It was no longer Hegelian thought, or thought which does not take reality seriously, but sensation, whose virtue was precisely that it did meet reality, comprising both external and internal experience, on its own terms. Feuerbach was very explicit: "The laws of reality are also the laws of thought."[67] "Philosophy is the knowledge of what is. The highest law, the highest task of philosophy is to think of and to recognize things and essences as they are."[68]

Finally, in a moment of self-correction, Feuerbach recast his own words from his critique of Dorguth's materialism. There he had said that man reads the book of nature by the senses, but he does not understand it by them. Ten years later, in his 'Kritische Bemerkungen zu den Grundsätzen der Philosophie,' he said of this:

Correct. But the book of nature does not consist of a chaos of wildly strewn letters or characters to which the understanding first brings order and connection, as if their union into an intelligent proposition were subjective and made arbitrarily by the understanding. No! We distinguish and unite things through the understanding on the grounds of distinguishing and connecting marks given by the senses; we separate what nature has separated, unite what it has united, and we subordinate the phenomena and the things of nature to one another in relation to their foundation and sequence, their cause and effect because things factually, sensually, objectively, and really stand to one another in such a relation.[69]

Meanwhile Feuerbach had been establishing contacts for himself by moving beyond Bruckberg. In 1842 he met D. F. Strauss, whose name was constantly linked with his. In 1843 he finally became personally acquainted with Ruge. In a letter dated September 3, 1842, the revolutionary poet Georg Herwegh, friend and companion of Karl Vogt, wrote to Feuerbach from Switzerland about plans to start a new journal of politics and science to replace the suppressed *Hallesche Jahrbücher*, and invited him to contribute.[70] This began a long lasting friendship between Feuerbach and Georg and Emma Herwegh. Ludwig later wrote to his brother Anselm of Herwegh, "I have revered him, now I love him."[71] In the summer of 1845 Feuerbach travelled to Heidelberg where he stayed with his friend Christian Kapp. Here he made contact with J. R. Blum, the mineralogist to whom he owed much of his knowledge of geology; Jakob Henle, the well known anatomist; and Karl von Pfeufer. It was in Kapp's house that the Heidelberg student Jacob Moleschott first met Feuerbach.

Feuerbach maintained a sporadic interest in the natural sciences. In an early meeting in Berlin he was greatly impressed by both Alexander von Humboldt and Leopold von Buch.[72] At Erlangen he had his initial first-hand contact with the *Erfahrungswissenschaften.* Again after his marriage he took up science once more, this time to help redress his feeling of deficiency.[73] Throughout the forties he displayed a considerable enthusiasm for the natural sciences, and especially for geology. In 1840 he wrote to Kapp: "Incidentally, it is incomprehensible to me how one can live without geology, although I know it only through the eyeglasses of books."[74] From the reply it is evident that Feuerbach had asked Kapp to inform him about the major trends of the day in geology, which Kapp promptly did.[75] In a letter to Herwegh in 1845, when the latter was on a scientific excursion with Karl Vogt, Feuerbach lamented that he could not be with them.

How gladly would I stroll with you in Paris, namely in the area of natural science! What I could see and learn there! How beneficent these riches would be to my perception.[76]

In 1850 Feuerbach's review of Moleschott's *Lehre der Nahrungsmittel* helped to solidify his relationship to the scientific materialists.

With the outbreak of the Revolution of 1848 Feuerbach went to Frankfurt, not as a delegate (although he had been proposed as one), but as an observer. No doubt the dramatic open letter to him from several Heidelberg students, among whom Moleschott was most likely included, had struck home.

> Noble sir! You are one of those rare persons in whom the spirit of the new times began to dawn. You must not be absent from the structure which is supposed to be erected for the world, and particularly for our long enslaved people's everlasting well being.[77]

In Frankfurt Feuerbach associated with the delegates he already knew, namely Kapp and Ruge, and they, in turn, introduced him to other members of the Left. In this context Feuerbach established, in Bolin's words, "an extremely gratifying acquaintance" with Karl Vogt.[78] But by the end of the summer he had become so disgusted with the Assembly's impotence, and with the September riots, that he left the city with the complaint, "Politics has driven me from Frankfurt."[79] While in Frankfurt he had taken no active part whatever. In fact, according to Ludwig Bamberger, he and Ferdinand Freiligrath, revolutionary poet and key figure in the later intrigues between Marx and Vogt, belonged "to the race of the silent ones."[80]

Once again the Heidelberg students called for Feuerbach, this time as lecturer. The one hour public lectures which appeared in 1851 as volume eight of the collected works under the title *Vorlesungen über das Wesen der Religion*, were delivered Wednesday, Friday and Saturday evenings in the main lecture room of the City Hall. Those attending ranged from educated workers to *Privatdozenten*, in which latter category came Jacob Moleschott. Two other noteworthy figures who heard Feuerbach here were the art and literary historian Hermann Hettner, and Gottfried Keller the poet. Moleschott, Hettner, and Keller were frequent visitors to the house of Christian Kapp, where they met Feuerbach personally.[81]

The lectures lasted from December 1, 1848 to March 2, 1849. Feuerbach returned to Bruckberg and watched as the forces of reaction wiped out all hope for truly representative government in Germany. No doubt goaded by this state of affairs, and by recent discussions at Kapp's home with Moleschott, he published his enthusiastic review of the latter's materialistic *Lehre der Nahrungsmittel*, entitling it 'Die Naturwissenschaft und die Revolution.' In it he subscribed to a crass materialism totally out of character with his earlier, more balanced position. Later he would regain his perspective,[82] but for the moment he vented the frustration his helplessness had provoked. Without a job, but with the memory of lecturing fresh before him, he had to stand by at Bruckberg with no prospect for the future while society relapsed into authoritarianism.

If the forties represented his day in the sun, the fifties would soon forget him. His quick rise to fame was matched by an even more rapid demise. For one thing, new critics of religion were appearing, and for another Marx had taken over the stage where philosophical and political liberation was concerned.[83] Popular attention, however, was lost to the young materialists from the world of science. Having been replaced in this fashion, Feuerbach's brand of philosophical sensualism was soon forgotten. His reputation as an atheist, of course, lingered on, but the new materialistic popularizers of science who so often quoted him were much easier to understand than Feuerbach himself. His Hegelian language, and his lack of systematic approach to philosophy explain why, according to Kamenka, he 'was not able to retain leadership of the movement he founded.'[84]

The publication of Feuerbach's Heidelberg lectures did not result in renewed clamour. It was as if Feuerbach had said his piece in the forties and now had nothing new to add. This same attitude greeted the *Theogonie* of 1857. Feuerbach considered a move to America, but the porcelain factory was going downhill, and soon he could not afford even the exploratory trip such a venture seemed to require. By 1860 the factory was liquidated, and Feuerbach faced dire financial straits. The family was forced to move, but had no money to do even that. A collection was raised among Feuerbach's followers, and the Feuerbachs relocated at Rechenberg near Munich.

In October of 1862 Feuerbach developed a warm friendship with a farmer named Konrad Deubler, whose admiration for the works of Feuerbach (and the scientific materialists) earned him the title *Bauern-philosoph*. A year after the appearance of the tenth and final volume of the collected works, which went unnoticed by friend and foe alike, Feuerbach suffered a light stroke, resulting in a slight paralysis of his face and organs of speech. Deubler, who had visited Rechenberg in 1865, invited the ailing philosopher to recuperate with him at his home in the Austrian Alps. Vogt visited Feuerbach soon after his stroke, and wrote to Deubler in May that he had urged Feuerbach to accept his invitation, and that in addition he had requested Feuerbach's doctor to instruct him to go.[85]

In July of 1870 Vogt's fears were confirmed; the first stroke had only been a warning. Feuerbach's second stroke took a greater toll than the first, leaving him totally incapacitated. It was not long before the lack of income threatened ominously once more. This time it was the Sozialdemokratische Partei Deutschlands that came to the rescue. An article published in the Würzburg Party paper described Feuerbach's sickness and material misfortune. Money came from Germany, Austria, Belgium, England, and America, thus ending the worries of the Feuerbach family from then on. Appropriately Feuerbach in 1870 became a member of the Nürnberg section of the *SPD*.

His condition improved slightly in the spring of 1871. He was able to write a letter, the final product of his pen, to his friend Deubler. But by year's end he had progressively worsened. With the new year a request went out from Bertha Feuerbach to Deubler to come before he died. Deubler eventually did come, and provides us with a moving account of the final meeting before Feuerbach died September 13, 1872.

The genesis of the German version of scientific materialism in the nineteenth century was often thought by its opponents to have come from Ludwig Feuerbach's literary efforts. Since Feuerbach was educated under Hegel, it took a thorough defection from the ranks to make Feuerbach into a materialist. The accumulation of doubts throughout Feuerbach's apprenticeship finally produced, in Engels' parlance, a transformation from quantity into quality. By 1839 Feuerbach had done a turnabout in his position vis-à-vis Hegel. One must not forget, therefore, that the formulations of scientific and dialectical materialism, both rooted in the impact Feuerbach's work had on German society in the 1840's, grew originally from deep within German idealism. Long before our study of Vogt, Moleschott, Büchner and Czolbe is over, it will be clear that they were never able to rid their materialism of its idealistic heritage.

CHAPTER II

REACTION IN THE FIFTIES

"Take any three doctors, and two will be atheists every time." In mid-nineteenth century Germany this old maxim read: "Take any three natural scientists, and two will be atheists and materialists every time."[1] During the 1850's there was heard in Germany a loud and sudden outcry against materialism in general, and scientific materialism in particular. Of the antimaterialistic books and tracts that were published in the 1850's, the majority appeared in 1856.[2] In 1855 two of the three most well known formulations of German scientific materialism were penned: Karl Vogt's *Köhlerglaube und Wissenschaft*, which went through four editions between January and June, and Ludwig Büchner's *Kraft und Stoff.*[3]

Most of the issues raised by critics of materialism in the 1850's proved to be variations and restatements of a few basic objections. The critical literature falls into four classifications. (1) There were, first of all, professional philosophers and theologians who found it impossible to remain silent in the face of what they considered a crude, long dead philosophical position. This group was made up of men of different stripes, but all shared a sympathy for some kind of idealism. (2) Next, there were a few natural scientists who publicly dissented from the inferences the scientific materialists were drawing from their profession. These men were not, with the exception of Liebig, Rudolph Wagner, and later Schleiden, the better known scientists of the day. (3) A third class comprised orthodox theologians and pastors. Here were included those clerics who did not voice their religious opposition in scholarly terms, but in sermons. (4) Finally, there were laymen who were fearful of materialism. These persons often did not identify themselves beyond being "friends of science and truth." Nor were their works always as much intended to refute materialism as to offer alternatives. Karl Fortlage, whose review of materialism[4] in the 1850's has been the source of information for several of these, included them in his survey no doubt as responses to materialism, if not refutations.

Already in 1853 there were some who realized that a trend was fast developing. Karl Fischer, professor of philosophy at Erlangen and right-wing Hegelian, sounded the alarm in his *Die Unwahrheit des Sensualismus mit besonderer Rücksicht auf die Schriften von Feuerbach, Vogt, und Moleschott bewiesen*. The work mentioned Vogt's *Physiologische Briefe* only in passing.[5] It was mainly concerned with Feuerbach's *Philosophie der Zukunft* and Moleschott's *Kreislauf des Lebens*, both of which, the author explained, were written to accomplish the same end.[6]

Fischer lost no time in identifying Feuerbach as the major representative of this new philosophical trend rooted in natural science.[7] His book was written in the language and categories of Hegelian philosophy. An idealist, he could not acknowledge that the 'I', which for him existed only in knowing and willing, had its basis in the activity of the brain. The 'I' might be conditioned and mediated through the body and brain, but it was not grounded or caused there.[8] Were this not the case, as in Feuerbach's position, then, according to Fischer, man would be no more self-conscious than the animals. In fact, he continued, Feuerbach and Moleschott both concluded that man and animal were qualitatively similar.[9] In this connection Fischer allowed that Feuerbach's position was not as naive as Vogt's atomistic and mechanical view of the world. Feuerbach, Fischer thought, believed in a living nature.[10] Yet in statements such as: "The body in its totality is my ego (*Ich*), my very essence," he was clearly a materialist.[11]

Moleschott fared no better than Feuerbach. But if Feuerbach had taken the brunt of the attack on consciousness, Moleschott was the object of Fischer's charge regarding the basic incompatibility between determinism and morality. Fischer defined materialism as the view that matter is the true cause of all things, that actions and phenomena are explained as modifications or effects of matter. According to Fischer, this meant that the world was determined. Therefore freedom of the will and capacity for all moral choice had been cancelled. Once again man and animal were joined. Moreover, Fischer denied that Moleschott had the right to defend, as he had done in the *Kreislauf*, values such as the love of one's neighbor, for he had no grounds whatsoever to do so.[12] At the end of the *Kreislauf* Moleschott had denounced certain customs as pharisaical. Fischer wanted to know how this could be, given that Moleschott had no moral grounds on which to speak?[13]

Nor was Fischer any happier with the materialists' view of the state as an institution dedicated to the equalization of particular interests. Fischer conceded that poverty was rampant, but not even the best theories of economy, he declared, could alone suffice to alleviate the situation. Bodily life, he said, was but an organ of mental life. Free will determined the true fate of man, and Fischer called for "the spiritual rebirth of our age" to help the people.[14]

Fischer wanted, in his words, to know (*erkennen*), not to probe (*ergründen*) nature. How could Moleschott, he queried, claim that he *understood* the life of nature better by believing in blind necessity? Far more preferable were those scientists who concluded from nature's lawful and purposeful structure that there was an absolute Intelligence in the background. For an idealist nature's purposeful constitution simply could not be explained by means of blind natural necessity.[15] The Erlangen philosopher resented Moleschott's practice in the *Kreislauf* of separating science from religion and making them incommensurate. Far better, in his view, was the approach of Moleschott's arch enemy, Justus Liebig, who sanctioned inferences from natural science to the realm of religion.[16]

Fischer, as a right-wing Hegelian, was not the only idealistic opposition that arose in the face of scientific materialism. As a noted scholar has pointed out, it was the superficial philosophical position of scientific materialism which caused the return to epistemological questions at mid-century in Germany.[17] The clearest indication of the revived interest in epistemology is exhibited in the Neo-Kantian movement in philosophy, led by such figures as O. Liebmann, F. Lange, and J. B. Meyer.

One of the earliest Neo-Kantian critics of materialism, writing considerably *before* the *Geschichte des Materialismus* of Lange in 1865, was Julius Frauenstädt. Frauenstädt studied theology, then philosophy at Berlin, and spent much of his early career in the 1840's as a private tutor. His intellectual development progressed from an early phase of Hegelianism through Kant to Schopenhauer, with whom he became enamored in the 1850's, and whose ideas he helped to revive. Frauenstädt wrote two works dealing with materialism in the 1850's, both of which clearly indicated his dependence on Kant and Schopenhauer.

With the return to the philosophy of Kant came a change of perspective, and therefore change of attitude towards materialism. To a Kantian, the difficulties with materialism were not that it endangered morality and justice, or that it reduced man to animal, or even that it was so often proclaimed in overly simple formulae. These things, explained Frauenstädt, were not part of materialism as such; even Feuerbach and the materialists themselves acknowledged as much.[18] His critique, he promised, would go to the heart of the matter. He would expose the qualitative similarity between theology and materialism. Thinkers of both stripes assumed that they were talking about the world as it really was; hence, both theology and materialism were forms of metaphysical realism. Frauenstädt, based on Kant, claimed that the error here lay in not distinguishing between the world of appearance and things-in-themselves.[19] Since the latter were inaccessible to human thought, realism was an unsatisfactory starting point.

The first forty pages of Frauenstädt's book were devoted to the positive points that materialism contained. These generally revolved around the service materialists had rendered by attacking the dogmatism of theology.[20] In the rest of the work, however, Frauenstädt went on to show why scientific materialism *à la* Vogt, Moleschott, Büchner, and Czolbe was no better. These men, for example, believed that their categories of force and matter were real things-in-themselves, and not reflections of the real world. Materialism did have an edge over theology in rejecting God as a starting point, but by choosing the external world and not consciousness as their point of departure materialists threw even that advantage away. As Kant had shown, pure sensation without understanding (*Verstand*) was an impoverished tool.[21]

Frauenstädt pointed out instances in the works of the materialists in which they had depended on more than immediate sense perception; e.g., their belief in the unalterability and omnipresence of natural law.[22] He exposed that they unwittingly assumed a principle that gave order and form to organic matter, and demonstrated that as materialists they had not escaped teleological development.[23] Finally, he denied the materialists the right to speak of ethical responsibility unless they admitted that man had free will.[24]

In another work, Frauenstädt set himself the task of contrasting materialism not with theology, but with the dualistic position Rudolph Wagner had taken in a Göttingen lecture of 1854. Wagner had said there that science and faith were such distinct aspects of human experience that they never intersected nor mutually affected one another, and that he personally defended the existence of an immortal soul and the origin of man from a single pair not because these were the conclusions of science, but because their denial would be morally reprehensible.[25] While Frauenstädt agreed with Wagner that natural science would never lead to the destruction of moral and social order, he could not, as a good idealist, permit human experience to be compartmentalized into separate realms of truth.

In chapters dealing with the influences of natural science on poetry, religion, morality, and philosophy Frauenstädt argued that natural science was in no way detrimental to any of them. If there were some fundamental incompatibility between science and these other disciplines, then science would not be able to flourish alongside new developments in philosophy, religion, etc. Only when modern science was compared with historical forms, e.g., historical religion, did a clash appear. But, Frauenstädt maintained, no given historical form expressed the essence of religion.[26] Natural science, of course, helped to destroy superstitution, but not belief.[27]

Friedrich Lange's great work was even more critical of nineteenth century materialism, though for reasons similar to Frauenstädt's. Lange was displeased with Feuerbach's lack of understanding of Kant. Because Feuerbach had been a student of Hegel, he shared the delusion that the entire manifestation of essence could be apprehended in phenomena. Feuerbach, however, had gone a step beyond Hegel, and had explained reality simply as sensibility, "and this it is that brings him near to the materialists."[28]

Lange complained that the materialists by-passed the all important matter of the limits of knowledge. The illegitimacy of materialism could only be demonstrated by means of the theory of knowledge. But an understanding of that required the further development of philosophical culture; hence the materialists, with their appeal to natural science, would carry the day for the present.[29] This kind of aloofness never failed to anger the materialists, especially Büchner. Few things perturbed Büchner more than the condescending attitude explicit in the claim that materialism had been disproved, and that only the ignorant were not yet aware of it. Büchner for one took the popularity he enjoyed as evidence of the lasting truth of his position, and he repeatedly and vindictively castigated Lange for his air of superiority.[30]

There were other philosophical critics. Julius Schaller of Halle, like Feuerbach, had been a follower of Hegel early in his career, but, again like Feuerbach, had moved away from him in the forties. Schaller's defection, however, was not as radical as Feuerbach's, ending closer to the position taken by Whitehead in this century. Schaller appreciated the utility of mechanical explantion as an alternative to vital force, but he emphasized that regardless whether one chose atoms or a special soul-substance as a starting point inconsistencies and logical contradictions would crop up. It might be practical to begin with atoms, but at their base they prove to be mythical entities that give rise to intellectual difficulties. Schaller's advice was realistic: "Instead of despairing of knowledge we rather have to go back to the premises from which our perplexity springs."[31] Schaller's solution, like Whitehead's, was to try the other alternative, to view matter not as dead and inert, but alive and imbued with process.[32]

Obviously Schaller's persuasion was antithetical to that of the materialists. Schaller denounced the superficiality and inconsistency of materialists, especially their totally inadequate derivation of mental phenomena from nonmental entities, and their inability to view man as a moral being because of their outspoken denial of free will.[33] But since the materialists simply drowned out such objections with the cry: "we only know the facts," and since they did allow themselves to draw conclusions of a moral nature, Schaller explained that they were as much idealists as anyone. Materialism, he said, "as soon as it makes its appearance as knowledge (*Wissenschaft*), possesses already an idealistic interest through which it remains in close relationship with all other ideal tendencies of intellectual life."[34]

The final example of the philosophical responses to materialism comes from Catholic philosophy. Jakob Frohschammer, educated at Munich in theology and *Privatdozent* there in 1850, was promoted to professor of philosophy by King Maximilian II as the result of his articles against Vogt in the *Allgemeine Zeitung*, published separately as *Menschenseele und Physiologie* in 1855. By the time the book came out Vogt's reply to Frohschammer's articles had already appeared in the preface to the fourth edition of Vogt's *Köhlerglaube und Wissenschaft*, and with this yet another of Vogt's many polemical squabbles was under way.

Vogt dismissed Frohschammer as a professional theologian who was fighting to preserve the existence of his profession.[35] In this Vogt was wholly unfair, as subsequent developments showed. Frohschammer's defense in 1861 of the freedom of natural science from external restriction, his careful exposition of Darwin's work, his printed discussion of the Syllabus of Errors of 1864, and his opposition to papal infallibility all contributed to his eventual excommunication in spite of his life long protestation against materialism.[36]

Vogt argued that his own conclusions rested on perceivable facts while Frohschammer's faith was based on nothing at all.[37] To this Frohschammer replied with a question: How did Vogt reason about the facts of sensation? It could not be by means of other sensed facts. His point was that the scientific materialists simply did not recognize that they had to depend upon more than sense experience if they were to infer anything at all.[38]

Frohschammer's purpose was mainly to deny physiologists the right to decide the issues of the soul and vital force, which two entities Frohschammer felt to be identical since the soul was also the life principle.[39] If natural scientists could not clearly recognize the suprasensual in the sensual, the supernatural in the natural, then at least they could not deny the possibility of such, still less should they want to.[40] Frohschammer resented the elitest attitude of the natural scientists vis-à-vis the other sciences. Natural scientists acted as if they alone based conclusions on facts.

But scholars representing the other sciences also had come to realize that facts must be their point of departure as opposed to *a priori* constructions. And what about the nature of facts? Frohschammer quoted Liebig's *Chemische Briefe* to say that facts gained their real value from the idea to which they gave rise.[41]

Finally, the Catholic thinker rejected out of hand the relativity of morals preached by the materialists. His major concern with materialism was the same as that of other professional philosophers; namely, to show that one could not simply declare that the age-old philosophical problem of the relationship between mind and body had been solved by an advance in natural scientific knowledge, that to do so only confused an issue whose scope extended beyond the purview of facts.

For their part the natural scientists were not silent. In spite of the fact that it was properly undignified for an established German scientist to engage in popular debate, a few well known men did speak out. Liebig, although he had taken a solid position for experimentation over against speculation in the *Chemische Briefe*, had never supported the tenets of materialism.[42] In a speech delivered before King Maximilian II in 1856, Liebig, with his critic Moleschott in mind, referred to the materialists as dilettantes, as "those who stroll at the edge of natural science," and as "children in knowledge and natural law."[43]

The most famous encounter of the decade occurred in 1854 when Rudoph Wagner, a highly respected physiologist at Göttingen, took advantage of his position as host at the meeting of the Gesellschaft Deutscher Naturforscher und Ärzte by delivering a general lecture condemning Vogt's stance on the question of the soul and on the origin of man. When Wagner then published this lecture, and issued in addition another pamphlet on the topic, Vogt could remain silent no longer, and the highly polemical *Köhlerglaube und Wissenschaft* was born.[44] Wagner, who had suffered criticism from his colleagues for stooping to such messy business, wanted to have an end to it, and did not reply except to say that he would not reply.[45]

Rudolph Wagner's cause was taken up by Andreas Wagner, no relation to Rudolph, in *Naturwissenschaft und Bibel*. He had written against Vogt three time previously in reviews of books by Vogt, gradually moving from general approval to violent dissatisfaction. By the time Vogt's *Bilder aus dem Thierleben* came out in 1852, Wagner reported that Vogt's goal was merely to spread atheism and materialism, and that Rudolph Wagner's anti-Vogt

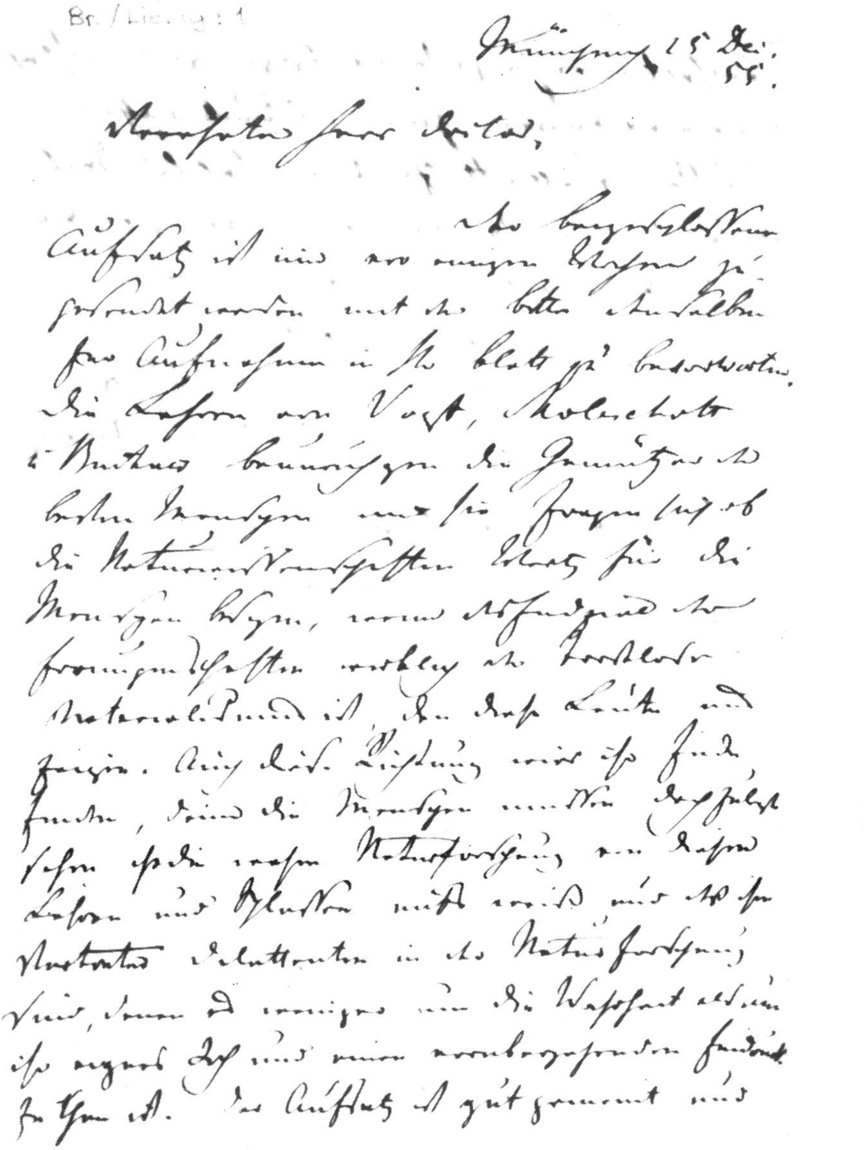

München 15 Dez. 55.

Verehrter Herr Doktor,

Der beigeschlossene Aufsatz ist mir von einigen Lehrern zugesandt worden und sie bitten denselben zur Aufnahme in Ihr Blatt zu befürworten. Die Lehren von Vogt, Moleschott u. Büchner beunruhigen die Gemüther der besten Menschen und sie fragen sich ob die Naturwissenschaften [illegible] für die Menschen [illegible], wenn [illegible] Materialismus ist, [illegible] Leute [illegible] zeigen. Aus dieser Richtung wie ich finde, [illegible] die Menschen [illegible] Naturforschung [illegible] Lehren und Schriften nicht weiß, [illegible] Dilettanten in der Naturforschung sind, denn die wenigsten [illegible] die Wahrheit [illegible] ich eigentlich [illegible] Der Aufsatz ist gut gemeint und

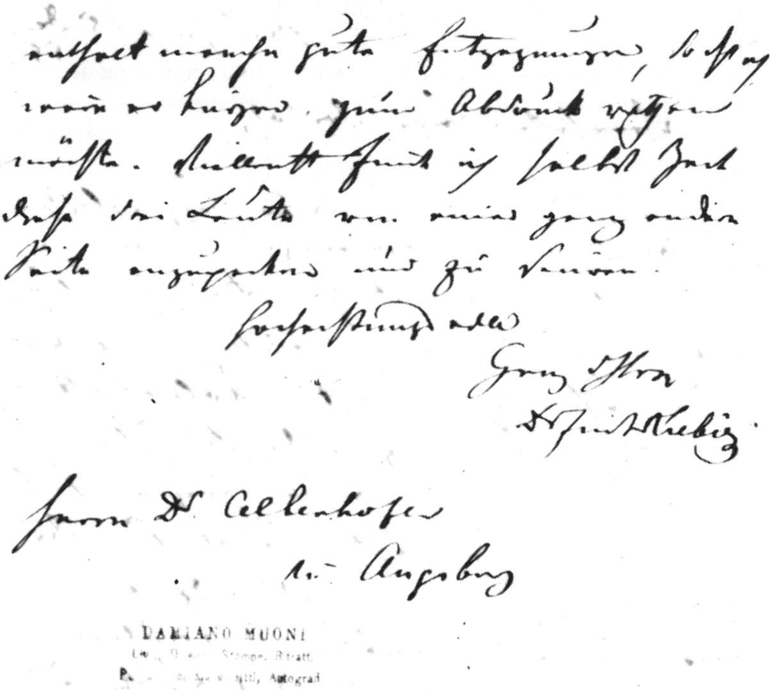
Hochachtungsvoll

Ganz Ihr

Dr. Justus Liebig

Herrn Dr. Altenhofer

in Augsburg

DAMIANO MUONI

Fig. 1. Letter on materialism written by Justus Liebig on Christmas Day, 1855 to A. J. Altenhoefer. Liebig was forwarding an article written by an unnamed author on the teachings of Vogt, Moleschott, and Büchner.

articles in the *Allgemeine Zeitung* had demolished all basis to Vogt's conclusions.[46]

Andreas Wagner held that Vogt was nothing but a conniving deceiver to imply that natural science led to a denial of the suprasensual. The history of science showed that such need not be the case at all. The greatest of scientists had been Christians, and the usual result of involvment in natural science was in no way atheism and materialism, but deeper knowledge. In contrast to what Feuerbach had proclaimed, Wagner declared that natural science was not to blame for radicalism any more than any other science. Materialism was the product of alienation (*Entfremdung*) from Christianity.

> It belongs to the diseased phenomena of our time which have resulted from the displacement of the Christian standpoint and have gripped all classes of civil society like an influenza, whence it is carried over into the scientific realm.[47]

Wagner continued by attacking Vogt in his most vulnerable position. Vogt, he said, taught that humanity was made up of several species (*Arten*), and that on this account the assumption that mankind originated from one pair was inadmissable. The obvious rejoinder to such nonsense was to point out that species were defined in terms of an ability of offspring to reproduce, and that all the allegedly individual human species were known to be cross fertile. Mankind, therefore, was not made up of several species, but one. Wagner went on to say that Vogt believed in the transmutation of one species into another. This was an inaccurate charge, for Vogt did not alter his acceptance of the fixity of species until after Darwin's *Origin*. The mistake on Wagner's part was Vogt's own fault, however, for throughout his life Vogt confused the ideas of species and race.[48]

It is interesting that Vogt was seen here as a defender of the descension theory, and that he was opposed for alleged ignorance of the facts. Everyone knew that only the offspring of parents of the same species were fertile, explained Wagner. Vogt's definition of species, which Wagner quoted as "an abstraction relying on the observation of similar individuals," did not rule out the possibility that two clearly different species might produce fertile offspring.[49]

Vogt's reply in the *Vorwort* to the fourth edition of the *Köhlerglaube und Wissenschaft* revealed that the definition of species was in fact bothering him. In the first place he denounced Wagner in typical Vogt *ad hominem* style as "a real Eckhardt of science," congratulating him for having been able to conclude that consciousness of God was somehow a "happy" inconsistency.[50] When he did get around to respond to the species matter, he simply declared that Wagner's definition of species, which implied that a tame dog or horse had to be of the same species as a wild dog or horse, was "incorrect and false."[51] As far as Vogt was concerned, he was not at all convinced that Wagner had successfully protected the truths of revelation as claimed.

Later in the decade August Böhner, identified only as a member of the Swiss Natural Scientific Society on the title page of his book, wrote *Naturforschung und Culturleben in ihren neuesten Ergebnissen zur Beleuchtung der grossen Fragen der Gegenwart über Christentum und Materialismus.*[52] Böhner's tone was identical to that of the Wagners, his work being a conservative defense of religion over against the horrors of godless materialism. Böhner was quick to point out the contrast between the humble spirit of the heroes of science of the past and the haughtiness of Vogt, Moleschott, Büchner, Czolbe, and Feuerbach, which latter he labelled "the principal leader of the most recent materialism."[53] Böhner noted that materialism could be linked historically to revolution. Materialism was in vogue not only prior to 1848, but also prior to the English and French revolutions of past centuries. To him this association was explained because social order, peace, cultural progress and human rights depended on the Christian view of man which materialism was determined to overthrow.[54]

Böhner was not above quoting scripture at the materialists. In fact, in chapter one of his book he listed the positions of Christianity and materialism in parallel columns. Nor was he at all discouraged by the present state of things. He declared that biblical criticism and the school of F. C. Baur had been defeated by the works of the theologian Adolph Harless and others, all of whom had proved the Bible true. He listed statistics to demonstrate how rapidly the Church was growing, even in Vogt's Geneva.[55] To those caught up with evangelism, materialism would not long prevail.

Not all scientists reacted with as much protestation towards materialism as did the Wagners and Böhner. The military doctor Hermann Klencke, who turned to writing popular science in the 1850's, took a more moderate position on the influence natural science was having. Klencke explained that developments in science had caused many to reflect on the traditional doctrine of God's providence, and that in this way science had in fact led to atheism for some, especially for the uneducated or morally confused.[56] But Klencke himself felt that the strict lawlike behavior of nature contained lessons for a reasonable manner of life and morality, for the man who uncovered the wonders of nature's laws learned to respect and revere the Creator.[57] Still, it was not orthodox religion he was defending, for he did not appear to be threatened by the great age of the earth, by the replacement of vital force with natural processes, or by the determining influence of diet on life.[58]

In fact, if one of Klencke's purposes in writing on science and its development in the previous fifty years was to suggest a corrective to the materialistic conclusions that were being drawn from the latter part of that period, he had several other goals in mind as well. One was to point out the positive practical effects natural science was having on the lives of the common man, a result he laid directly to the turn away from Hegel's approach to the more modern appreciation of facts and the real.[59] Another was to oppose the antiscientific attitude he sensed among many, mostly classicists and theologians, who condemned natural science because it was irreligious. Klencke replied that natural science itself was neither religious nor irreligious, it was a science of the senses and therefore could not be used directly to draw either theological or atheistic conclusions. In a bold move, he declared that it was not miracle as such that scientists opposed, since physical explanations did have their limits. Rather, it was the habit of making into a miracle anything that could not be immediately explained that perturbed men of science.[60] Yet scientists could lead men to God even if they could not prove God's existence. Klencke wanted to have it both ways. Scientists were to be neutral in religious matters, yet be able to lead men to God.[61] It was this attitude that scientific materialists like Vogt could neither understand nor tolerate.

There were a few other scientists who spoke out. The chemist and industrialist Karl von Reichenbach, most famous for his writings on the mysterious force called Od, opposed the materialists' position because he thought it meant that only the material world was capable of being sensed. To this claim he opposed the sensitive power of Od, which allegedly affected only some by way of the senses.[62] For the materialists, however, there was nothing about Reichenbach's Od that was qualitatively different from the problems involved in sensing the effects of the so-called imponderables (light, heat, electricity, magnetism, etc.), and they were well aware of these.

Another minor figure, the botanist and medical professor Heinrich G. L. Reichenbach of Dresden, published a highly speculative scheme in 1856 in which life was viewed as a progressive development from an unconscious state of darkness to the mature light of day. Beginning with birth, each movement up the scale of life was precipitated by a specific traumatic experience. Reichenbach associated the organic world with love and the inorganic world with egoism in a manner such that materialism clearly came out opposed to a humanistic world view, and was therefore condemnable.

On the basis of opposition to materialism by the scientists here cited, a fairly predictable result presents itself. Scientists holding a university position tended not to condone or accept materialism. While they were often willing to criticize official religion, they would not go so far as to question religion itself. Fed up with religious dogmatism, they had equally little use for the secular dogmatism of materialism, and when they did offer suggestions on how to modernize religion, it was in vague and often romantic terms.

The response to materialism from the clergy provides some indication of the degree to which scientific materialism had permeated all levels of German society after 1848. Materialism became the subject of innumerable religious tracts and sermons. By denouncing materialism from the pulpit, clerics were guaranteeing that public attention would be drawn to it. A parishioner may not bother to read an antimaterialistic pamphlet, but he would listen to an antimaterialistic sermon.

Some clergymen used the rise of materialism as an occasion to denounce the modern emphasis on reason and to sound a general call to repentence. Two pastors in particular viewed scientific materialism as simply an outgrowth of the godless idealism of Schelling and Hegel. No normal individual could possibly be expected to revere and honor, let alone comprehend the abstruse God of these pantheists. Materialism, on the other hand, was far easier to grasp. In reality it was a logical outgrowth of rationalistic idealism.[63]

Other pastors tried to use science in defense of their cause, arguing that the Christian faith was reasonable, and pointing to the glories of nature. Wolfgang Menzel, for example, declared that nature had to be viewed today just as it had been by Moses and the Psalmist – as the creation of a wise God and Father.[64]

By and large, however, most claims assumed a stance between these extremes. While openly critical of the materialistic conclusions which were being drawn from science, most pastors did not intend to criticize the enterprise of science itself. In 1855 a Catholic priest named Friedrich Michelis wrote an open letter to Matthias Schleiden in which he complimented the great scientist for his antimaterialistic position, but in which he expressed his regret that Schleiden had succumbed to the dangerous practice of popularizing natural science.[65] This popularization of science, he said, had blossomed in the wake of Humboldt's *Kosmos*, though Michelis was certain that Humboldt had not intended it to take on a materialistic flavor.[66]

Michelis's statement of the position of the Catholic Church was that natural scientists should move beyond a philosophy of resignation to a more positive recognition that the Christian standpoint is the same as that of science. In both science and Christian doctrine investigation is carried out in humility with the full recognition that finite, created being lies beneath infinite being.[67] Michelis characterized the Catholic position as the assertion that science and religion were separate realms, that science dealt with finite material being and that the claims of materialism therefore were not those of natural science, but heresies that some thinkers tried to pass off as scientific.[68] From Schleiden's popular lectures it was clear that he was not persuaded of such a view.[69] If Michelis identified with Schleiden it was only to oppose materialism, against which he had founded the Catholic journal *Natur und Offenbarung*.[70]

The Protestant theologian Adolph Harless, one-time professor of theology at Leipzig and in the fifties head of a Leipzig Christian mission to India, wrote a highly entertaining antimaterialistic work in 1856 in the form of a play. Harless, besides being a leading figure of Lutheran orthodoxy, had another reason for taking up the pen against materialism: he was the brother-in-law of Rudolph Wagner, the opponent of Vogt. Writing under the pseudonym of Dr. Mantis,[71] Harless attempted to reach the people through a satirical drama that brought Goethe back from the dead to see the disastrous level to which German culture had sunk. The *dramatis personae* revealed all. There were German professors named Moleküleschott and Hahnebüchner, a guard called Blauvogt, and Mr. Firebachlor (Feuerbach), the German Lord Byron. Even the *Theologe* of the play was a materialist, a fact that reflected Harless's continual real life campaign against rationalist theologians.[72]

The play was written in verse, and repeatedly brought Goethe to the point of despair and anger over the views with which the materialists were infecting the minds of the people. At the climax of the action Goethe, eavesdropping as the theologian taught a young girl a materialistic catechism, heard the girl instructed to deny all authority, to believe only in herself and nature, and, in good Feuerbachian fashion, to consider the species always above herself. Reference was made to a practice which Moleschott had urged in the *Kreislauf des Lebens*; namely, to use human corpses as fertilizer in the fields. At this Goethe could stand it no more, and he attacked the theologian physically. At the end of the story Goethe declared his sympathy for religion and invoked God to destroy materialism with the cry "To hell rather than to earth!"

Harless's use of drama is instructive. It reveals that opponents of materialism could be clever in the way they chose to combat the pervasive and persuasive simplicity of the materialists' works. The materialists were able to appeal to natural science to lend credence to the veracity of their conclusions. Nonmaterialists could not depend on astute philosophical reasoning to combat a popular movement like materialism, but they too could use the polemical tactics of satire and wit, and, in addition, draw on the image of a national symbol such as Goethe in their attempt to persuade the reading public that the issue was not as cut and dried as the materialists proclaimed.

Among the most interesting of the religious works against materialism is Friedrich Fabri's *Briefe gegen den Materialismus.* Fabri, a Protestant missions inspector, responded to a request to write against materialism because too many people were remaining silent in the face of this serious threat to humanity. He wrote his work as a warning against the one-sided popularization of natural science that had led some to replace religion with half-truths from science.[73] What Fabri found threatening was the apparent inevitability of the victory of materialism. Not scientific proofs, but titillating formulae enticed the masses into disbelief. Fabri took Feuerbach as an example. The secret of his effectiveness, Fabri explained, lay in his conviction that he spoke what already lived in the hearts of thousands.

> This consciousness so intoxicated with victory is what gives him that powerful way with words which so mightily affects so many. He is sure of his public, he knows that the more audacious the claim that flows from his mouth, spewn forth as it is in so dazzling a form, the more certainly it will be taken up as gospel by thousands. What more scientific proof does such authorship need?[74]

Materialism was much more a sign of the times than it was a philosophical development, said Fabri, for there were plenty of philosophical refutations extant, but they were all swept aside by Feuerbach's cry, "Speculation is inebriated philosophy."[75]

Fabri was unambiguous about where it had all started: Feuerbach.

> One searches in vain throughout the whole extent of our most recent materialistic literature for a thought which adds but one essentially new factor to the basic propositions of the Feuerbachian doctrine which has already been imparted.[76]

Feuerbach was then the "founder of today's sensualism and materialism," and Karl Vogt, Moleschott, and Büchner were all "the celebrities and knights of matter," who had shown themselves to be "devout and docile blind adherents of Feuerbachian dogmatics."[77] Interestingly, Fabri explained what Vogt's dependence on Feuerbach meant by saying that he was opposed to all types of authority, "in whatever form they are."[78]

What had been the cause for the rise of materialism? Fabri provided three reasons. First of all, the cardinal dogma in philosophy since Descartes had been the autonomy of the human mind, and with it a turn away from that which was higher than man. Secondly, natural science had increased greatly, and its results were being used to support atheism, though illegitimately in Fabri's view. Finally Fabri pointed to the worship of material interests in contemporary society. He was not so foolish as to condemn progress outright; but he did explain that the potential misuse of material advancement, the egoism that made means into ends, was easier the more technology moved ahead.[79]

Modern materialism was not something new to history, declared Fabri. The downfall of ancient culture following Greek materialism, and the deterioration of French social, religious, and political life following eighteenth century materialism had revealed that culture stood or fell with the freedom and independence of the human spirit, that complete nihilism and a despotism no less revolutionary than anarchism resulted when that spirit was denied.[80] Yet it was not simply a matter of the return of an old philosophy. There was a crucial difference: the 1850's was on the after-side of the French Revolution, and men should know full well what would happen if such thinking were not checked.[81]

Fabri brought up the common criticisms of crass materialism; e.g., its dependence on atomism, and the assumptions that change could be explained by the changeless, sensation by the sensationless, purpose by blind causation, and mind by the nonmental.[82] But the most interesting emphasis of the book was the critique of Rudolph Wagner's double-entry approach to the relation between science and religion. Like Lotze and Büchner, Fabri could not stomach the pure dualism that lingered behind Wagner's position, but his reasons were altogether conservative. In the long run Fabri was out to defend the one truth of Christianity, a truth that could not be broken up or tampered with.[83] He related that there was a heritage of a unity between faith and knowledge that came from the Reformation. Unfortunately the reformers referred this conviction almost exclusively to an ethical pole, with the result that its *theologia naturalis* was only a propaedeutic where nature was concerned. Fabri spoke of the general departure from biblical realism and of the rise of rationalism. Eventually reason invaded the realm of faith so thoroughly that the idea arose that reason and faith were incontrovertibly opposed. Fabri was clear about what was to be done and why.

It must be shown that all knowledge and understanding presuppose belief, that in the apology for Christian truth it is not a matter of a struggle between knowledge and faith, but between religious faith and irreligious faith, or a struggle between religious knowledge and irreligious knowledge.[84]

In fact, in Fabri's scheme faith held a position of priority over reason because faith sprang from immediate experience, both sensual and suprasensual. Faith operated not only where religious knowledge was concerned, for the suprasensual also existed quite openly in natural science. Whenever a scientist proclaimed that natural laws were unchangeable, whenever he maintained that matter and space were infinitely extended he was overstepping scientific thinking (*naturwissenschaftliches Denken*); these conclusions were not the result of sense experience.[85]

Nor was experience what the materialists thought it was. In addition to sense experience there was also the experience of self-awareness and of God and revelation. These to Fabri were as real as the testimony of the senses.

Whoever does not know within himself the suprasensual as a certain fact does not have or can never acquire a notion of it. Reason can never replace the reality of a thing, nor prove its reality or unreality.[86]

The picture here created contrasted with that Wagner had drawn. Fabri consciously rejected Wagner's diagram of two separate circles in favor of his own, in which two concentric circles, science the smaller and religion the larger, showed that science had boundaries beyond which lay religion. He held that religion had the same logical right to appeal to the authority of the facts and relations of suprasensual experience as science did with sense experience.[87] For this missions inspector not knowledge but faith determined the choice of *Weltanschauung*:

But neither believing or knowing are isolated activities of men ... *There is a constant influence of our will on the production of our thoughts.* Such is also a fact of empirical observation, even of careful self-observation. Therefore man is also morally responsible for his thinking; and the doctrine of the materialists of a mechanical necessity in thinking is basically nothing but a figleaf for their lack of free will, which denies moral responsibility.[88]

One might be tempted to read modern ideas into Fabri's emphasis on the inevitable influence of disposition, or will as he put it, on theoretical conclusions. Yet Fabri was completely a man of his own century in all of this. He was fighting for the truth, the Christian truth, which for him meant that one and only one description matched the world "as it really is." What Fabri admired about Czolbe's program of eliminating the suprasensual was its ruthless consistency and the fact that Czolbe openly acknowledged his presuppositions. Fabri was not only a naive realist, as most of the figures dealt with in this study were, but a naive realist with Christianity thrown in. This had the effect, if one were scrupulously consistent, of making his stance on scientific issues a matter of moral choice. Fabri could dismiss the places where science seemed to contradict doctrine by simply identifying the scientific evidence as a theory, as the result of thinking, not of experience.[89]

Thinking had for him an air of relativity. It did not tell one about reality itself. Only experience did that; more specifically, only the experience of Christianity. Yet not all people had had the Christian revelation of the believer, so the problem of truth in science was reduced to the matter of salvation, an appropriate conclusion for a missionary.

There were a number of anonymous publications in the 1850's which expressed general dissatisfaction with materialism. These fell into the category of lay productions by nonprofessional persons who took up their pens against the upsurge of radical thinking. Typical of such works was *Der Humor in Kraft und Stoff, oder die exacten Ungereimtheiten der modernen Realphilosophie*, published in Büchner's home town of Darmstadt in 1856. This was the second time the anonymous author had attacked Büchner, the first being in an earlier pamphlet entitled *Dr. Büchner's Kraft und Stoff, oder die Kunst, Gold zu machen aus Nichts*, which Büchner dismissed as worthless.[90] Büchner identified the author as an old pensioned *Hauptmann* of Darmstadt, an exposure that prompted the author to reply that he was proud of his grey hair and his former profession.[91]

The booklet was not about the humor in Büchner's work. The only reference to Büchner's use of humor was to charge Büchner with the appropriation of nonscientific means to cover up a lack of factual basis for his claims. Büchner, in other words, was a charlatan.[92] Much of the work was written in the form of satirical poetry. Büchner was denounced as a pretentious young upstart, a danger to religion and state. In his 'Invocation' at the beginning of the booklet the author lamented that materialists were confusing the people about God, spirit, holiness, and eternal justice, about the true faith that ought to prevail in the home and on the throne.[93] But our *Hauptmann* was also no orthodox fanatic. He criticized Büchner, for example, for the blind faith which made him no better than other religious extremists.[94] He held to a moderate attitude in religious and political matters, and he was an optimist where science was concerned. He believed in the positive influence and the noble value of science for the life of humanity.[95]

Still others were ready to come to the rescue of society. An author named Drossbach declared that men were unhappy because they no longer believed in immortality. If men could somehow regain their belief, they would be satisfied with the inequalities of life. Drossbach claimed that he had found the way to bring men back to this belief; namely, through the natural sciences. How? By recognizing that atoms were not simply dead and inert pieces of matter, but that they possessed consciousness. Fortlage, in his review of Drossbach, commented that this schema reminded him of the famous hunting horn of Freiherr von Münchhausen, which emitted no sound when blown during the hunt, only to sound out loud and clear when the chase was over.[96]

Others wrote of better worlds where the stains of evil thinking did not exist.[97] One writer openly admitted that his belief in God and miracle was irrational, and that he held to it in spite of having to acknowledge that Büchner's opposition in *Kraft und Stoff* to revelation, inborn ideas, and innate moral postulates proved these things to be unacceptable. This position was so clearly arbitrary that Fortlage suspected its author of not being a believer at all, but a wolf in sheep's clothing.[98]

A more serious effort against scientific materialism was the pamphlet by Wilhelm Schulz-Bodmer. Once again from Darmstadt, which seemed to be the location of so much of this kind of activity, Schulz-Bodmer had been imprisoned by a Hessian court for illegal political activities in the 1830's, but escaped to Alsace. Eventually on reaching Switzerland, he settled down to writing left-wing political tracts, and joined the Swiss fight for freedom in the Sonderbund War. In 1848 he returned to Darmstadt and took a seat at Frankfurt. Back in Switzerland after the collapse of the Revolution, Schulz-Bodmer continued in the main to fight against the thing he hated most, the military state.

In 1856 Schulz-Bodmer penned a polemical tract against materialism in which he satirically depicted the railings of Büchner, Vogt and their opponents as so much *Kinderei*, the whole affair being a battle between frogs and mice.[99] Although a republican, he was clearly the kind of liberal who had no sympathy for religious radicalism. Vogt's unmanly polemics with Rudolph Wagner found no favor in his eyes, but more tragic, according to Schulz-Bodmer, was that Vogt did not understand religiosity rightly so-called. It was hard to tell which he resented more, Vogt's impertinence or "the spiritual police state which usually accompanies the secular police and military state."[100] As far as he was concerned, genuine religiosity was thoroughly democratic, since it had its source in the highest rights of personality. With regard to the issue of the soul, Schulz-Bodmer declared simply that physiology could not decide this question since physiologists used the methods of natural science.

Most interesting of all the lay reactions to materialism was the open letter to Jacob Moleschott from a female admirer, one Mathilde Reichardt. Reichardt was an enthusiastic disciple of Moleschott for whom the beauty of the eternal *Kreislauf*, in which nothing could depart from natural law, had become elevated into a new religion. Yet she had a complaint to lodge against the Dutch materialist, for Moleschott had allocated a place for sin in his *Weltanschauung*. This ruined everything. Moleschott, she indicated, had associated sin with the unnatural, but how could there be any ugliness if beauty was determined by natural law?

> Sin is said to lie in that which is unnatural – but where is there unnaturalness in a world in which each effect corresponds with strict logical consequence to an endless series of causes, all of which themselves are based on natural necessity? There is nothing unnatural. The concept of the word "sin" circumscribes purely and simply the violation of the laws of beauty.[101]

This kind of criticism, coming from an admirer as it did, proved to be directed at one of the weaknesses of the materialists' position. It seemed to reinforce the claim that the materialists had no right indeed to make reference to ethics or morality in the justification of their assertions. But, as will become abundantly clear, the scientific materialists themselves did not at all conceive of their endeavor as amoral. To them the very future of the human race was tied intimately to the world view they advocated, and to whose formulation we now turn.

PART 2

THE SCIENTIFIC MATERIALISTS AND THEIR WORKS

CHAPTER III

KARL VOGT: SOUNDING THE ALARM

The Grand Duchy of Hesse was the birthplace of two major scientific materialists, both of whom pursued their studies in science at Giessen. Karl Vogt and Ludwig Büchner were not, however, students there together. In fact Büchner was just completing his medical education in 1847 when Vogt was called to Giessen from Switzerland as a professor of zoology. Vogt was the first and in many ways the most independent of the scientific materialists. He began writing just prior to the Revolution of 1848, and his radical position at Frankfurt was already the culmination of an eventful and an unusual career. Consequently he cannot be seen solely as a product of the reactionary years following 1848, though some of his contemporaries accounted for the rise of scientific materialism in this way.

It is simply impossible to ignore the contrast between the two family lines that came together in Karl Vogt.[1] His ancestors on his father's side had been traditionally butchers, though his grandfather had broken the pattern by becoming a pastor, and his father, a medical man, had traded the cleaver for a scalpel. When Karl's only uncle on his father's side became a pastor, and when three of his four aunts married pastors, the young Vogt could not help but notice what he later called "the numerous ministerial ties in our father's family."[2] Later he developed a vitriolic hatred for all things ecclesiastical, but when it came to his own clerical relatives he spoke fondly of them and was careful to note their virtues. Karl characterized the Vogts as conservative and solid in their opinions, analytically minded, less given to feelings than to reasons, critical, with little sense for poetry and even less for art. The Vogts were logical, they were interested in natural science and mathematics, and they liked to work with their hands. It was, for example, Vogt's paternal Aunt Luise who had first interested the young Karl in natural science.[3]

Vorgt's portrait of his father was that of a liberal German professor. As a member of the medical faculty at Giessen, he stood to the left of the so-called *Biedermänner*, the liberal opposition to the reactionary *Westfalen Koterie*. Among the virtues Karl recalled in his father were his openmind-edness, his opposition to anti-Semitism, and his academic excellence. Vogt noted that his parents had resisted class discrimination in Giessen, that his father, as a doctor, had won the trust of the people so completely that they even allowed their children to mix with his, something not ordinarily done between professor and townsman. Although the senior Vogt was elected several times as a delegate to the local governing body, he was repeatedly unable to take his seat, since he could never obtain the required leave of absence from the university. Those in power, knowing he would be a member of the opposition, blocked him by putting pressure on the university, a procedure that had the effect of making him all the more politically independent and all the more popular.[4]

If the Vogts had tended to be analytical and conservative in their approach to life, the same could not be said for the family on his mother's side. The Follenius name meant nothing but trouble to the police of the Holy Alliance. A defender of the Jews long before emancipation, Grandfather Follenius had three sons who inherited his dissenting and fiery temperament. The two eldest of Karl's uncles dropped the 'ius' from their name, and chose to be known as Adolf and Karl Follen. Adolf was one of the students who, on October 18, 1817, ended the Reformation festival with a revolutionary trial by fire. Eventually he was imprisoned in Berlin, and when finally released had to flee to Switzerland where Vogt later got to know him.[5] Karl Follen, member of the 'Giessener Schwarzen,' dedicated himself along with his friend Karl Sand to freedom through the murder of tyrants. After sharing with Sand in the plot to murder Kotzebue, he too fled to Switzerland. When pursued by the Holy Alliance he went on to America, where he founded the chair of Germanic Philology at Harvard.[6] The youngest of the trio, Paul Follenius, participated in the revolutionary movement of the 1830's, then settled down in a legal profession. Small wonder that Alexander Herzen described the Follenius family as one of those "blessed old Germanic families."[7]

One can only guess what impact these events had on the daughter of the family, Karl Vogt's mother. Vogt's general characterization of his mother's side of the family conformed well with what one might have expected. He described them as lively, romantic and poetic, possessing an inclination to the extraordinary and a tendency to dominate. The Follens preferred the artistic to the logical, they were a temperamental family with a cutting wit.[8] Between the Vogts and the Follens Karl inherited about as many diverse attributes as could be rolled into one person, an especially significant fact in light of the strong family ties he was taught to respect as a child.

The eldest of nine children, Karl Vogt was born on July 5, 1817, in Giessen. At this time Giessen was a town of 6,000 inhabitants with a university population approximately one-tenth that amount. Life there was simple; most inhabitants owned some pigs, sometimes even a few cattle. Besides the anti-Semitism, which was common enough in the Germany of that day, the identifying feature of town life was the rift between the university and the citizenry, largely due to class differences. Nothing delighted the townspeople more than to mock the professors by applying their noble titles to each other.[9]

The eroding power of secularization had had its impact on the religious life of this small town. Mysticism and pietism were, according to Vogt, hardly known in Giessen. His description of the students' mockery of the senior theological faculty, and of the disrepute in which any believing student was held support such a judgment. Men of the educated classes went to church only for the confirmation of their children. Other events that might have dragged them there, such as baptisms and marriages, were often held in the home. Students did go to church on Sunday, but only to watch the girls as they came and went to hear the sermon.[10]

Vogt would have us believe that all Hesse, with the exception of the Catholics from Westfalen, was a place of great religious tolerance, even indifference. According to Vogt these Catholics, through a concerted and cooperative effort, were able to infiltrate both the court at Darmstadt and the university at Giessen, in general stirring up an otherwise peaceful situation. Were it not that such a claim fits so well the argument that all local ills can be explained as a result of the influence of subversive outsiders, and had it not been a Catholic who had later tried to block Liebig's nomination of Vogt to fill the post in zoology at Giessen, we might be tempted to entertain Vogt's charge. One should realize, however, that Vogt's autobiography, on which much of our knowledge depends for this period of his life, was written at the end of a long and polemical career. The fiery events in the interim could not but have colored his memory.

As a youth Karl attended the *Gymnasium* in Giessen. His descriptions of student life there provide the historian with an extremely revealing commentary on the nature of German secondary education in this period. Vogt labeled Giessen's *Gymnasium* as in every way the worst in all Hesse, being to him only "a continual battle with the instructors."[11] All the lectures were on Latin and Greek, while history, geography, mathematics, and modern languages were done strictly on the side, and only in the last semester was there as much as a hint of the natural sciences in the curriculum. Time up to that point was often enough spent in memorizing trivia as absurd as the codex numbers of the variant manuscript readings in the Vatican library![12] The classroom was a notoriously unruly place in which pupils, who had scandalously little respect for their teachers, delighted in mocking them when their backs were turned, and in general in creating an atmosphere Vogt identified as "a bitter war between teachers and students."[13]

Karl went on to the university in the fall of 1833. Giessen was just far enough south to be considered part of southern Germany. Indeed, it was true that life in the Duchy of Hesse contrasted with the more formal atmosphere of Electoral Hesse farther north. Giessen students, for example, looked south to Heidelberg while those of Marburg, just north of Giessen, felt more affinity with Göttingen.[14] Karl's first year at the university was spent doing all the things student life in Germany was famous for, "dueling, drinking, boasting, brawling and arguing amidst the usual tobacco smoke."[15] When his family moved into a new house, his father promptly erected an exercise building "for his children," for gymnastics was forbidden to university students as dangerous to the state. Needless to say, students were frequent visitors to the Vogt household.

By the end of a year Karl had tired of wild carryings on, and began to settle down to study. His father had never pushed him in any particular direction, but when Karl declared that he wanted to take up medicine the elder Vogt privately arranged for a course in anatomy for his son and two other Giessen students. It was in this context that Karl came into contact with a very unusual student of the natural sciences, one Georg Büchner by name, who became known to the world subsequently not for his abilities in science, but for his literary work and for his pre-Marxian revolutionary activity. Vogt's impressions of Ludwig Büchner's brother convey an image of an aloof, withdrawn, independent and bright student who took the brunt of a great deal of mockery because of his nonconformism.[16]

In the summer of 1834 Vogt became involved in the course in experimental chemistry taught by Justus Liebig. Liebig had studied in Paris with Gay-Lussac, and had returned to Giessen in 1824 out of respect to the Hessian sovereign who had paid for his education. In 1826 he was made full professor in the philosophy faculty, and he proceeded to establish a chemical laboratory in Giessen second to none in Europe. Imbued with the methods of the French, Liebig eventually went beyond his teachers as an experimentalist, and gave Giessen an international reputation. Obviously his great emphasis on experimentation was not the usual approach of a professor of philosophy, and natural tensions between Liebig, with his "barbarian cohorts" as they were called, and others on the philosophy faculty soon appeared.[17]

Vogt became captivated both by Liebig and chemistry. As for Liebig, he soon recognized the analytical talents of the young Vogt, and determined that he should be a chemist. By the fall of 1834 Liebig and Vogt's father agreed that Karl should put aside all else and work with Liebig in the laboratory, where he would labor alongside the Frenchman Regnault.[18] From morning to night every day but Saturday afternoon and Sundays, Vogt worked under Liebig's direction, and at the end of one year Liebig wrote a glowing assessment of Karl's work.

It eventually became impossible for Vogt's father, with his free thinking views, to remain at Giessen. Both he and Paul Follenius had been confidants in the unsuccessful seizure of the Federal Diet in Frankfurt in 1833, and he no doubt became highly suspect when those arrested in the case demanded that he be the doctor to treat them.[19] Relieved of his teaching post in 1834, he accepted a call from the medical faculty at Bern in Switzerland. It was Liebig who persuaded him that Karl, still a teen-ager, should stay behind and continue his studies in chemistry. But it was not long before Karl too was forced to retrace the path his uncles Follen had taken.

Many of the students at Giessen, including Vogt, belonged to the student corps known as Palatia. Karl had not actively participated at first, thinking it wiser to ground himself thoroughly in chemistry. As he continued to pour over Berzelius in the summer of 1835, it became evident that those of his friends who had become active in political affairs were disappearing one by one. Up to this point Karl had accumulated three black marks on his record for brawling, seconding in a duel, and teasing a student.[20] but later in 1835 he was drawn into political intrigue without intention, for news of how he daringly aided a student escape from police arrest soon reached the ears of the university officials. Karl became the next target. Vogt's narration of his flight from Giessen to Daurenheim to Darmstadt to Jugendheim and eventually to Strassburg is a first-hand account of a fugitive student's life in reactionary Germany. In Strassburg Vogt met other fugitives, mostly those from the Hambach festival of 1832. After a time he escaped to Basel and walked to his family in Bern.[21]

In the 1830's Switzerland had but one university, Basel, but a movement was afoot to set up another either at Zürich or Bern. The choice went to Zürich, whose *Hochschule* was begun in 1833, and which would be the site of the peasant uprisings against the appointment of D. F. Strauss some six years later. Rejected by the government in favor of Zürich, Bern decided to erect a *Hochschule* of its own. It was this that had occasioned the call to a number of foreigners to come to Bern as teachers, including large numbers of Germans who were not always viewed in a favorable light by the Swiss.[22] Vogt's father quickly identified with the radical party in Bern, as was to be expected, and before long the Vogt household, with the complete sanction of Vogt's mother, became a haven for many of the student fugitives of the thirties.

The younger Vogt also enjoyed the overwhelming contrast of Bern's atmosphere of freedom to the confining and restricted student life at Giessen. Here the janitors were not policemen – one could travel without a pass. There was no special university court, gymnastics was openly permitted, and students were even expected to engage in politics.

Karl soon found that it was impossible to continue his studies in chemistry because of the lack both of equipment and of a professor with Liebig's facilities. G. Valentin, a professor of physiology called from Breslau, turned out to be Vogt's savior, since it was he who successfully engineered the transfer of Karl's studies from organic chemistry to anatomy, physiology, and zoology. Valentin taught Vogt how to use a microscope, how to prepare insects for study, etc. In turn Karl assisted his teacher in lecture demonstrations. In the summer of 1839 Vogt successfully completed his doctoral examination in the medical faculty, but never having intended to practice medicine, he now began to cast about in search of a career.[23]

It was just prior to this point that a young fugitive from Giessen happened to arrive at the Vogt home. Eduard Desor, one of the unfortunates of the Hambach Festival, had gone first to Paris, where he had used his gift for languages to support himself through translating. There he had forsaken his interrupted education in jurisprudence for a newly found fascination with geology and paleontology. Out of work, he had come to Switzerland in search of a job. When Louis Agassiz passed through Bern on his way to his laboratory in Neuchâtel, he happened to mention to Vogt's father his need for an assistant who had a gift for languages. The older Vogt not only suggested that Desor was exactly the man Agassiz was looking for, but made arrangements for Karl too to work with him upon the completion of his medical degree.

Arriving in Neuchâtel in 1839, Vogt spent the next five years with Agassiz. Vogt's relationship with the Swiss naturalist at Neuchâtel is difficult to assess, the question being to what extent Vogt's views were affected by the older man who, respected as a scientist, was also revered by many as an orthodox believer in matters of religion. The autobiography does on occasion speak highly of Agassiz, but the general impression given of him is overwhelmingly negative, a flavor which is intensified in Vogt's son's biography. Others who have looked into this phase of Vogt's life have left open the contrary possibility; namely, that Vogt at Neuchâtel was proud to be associated with Agassiz's name.[24]

Trouble began with the first tasks Agassiz gave to Vogt, which were concerned with the anatomical description of fossil fish, the history of the development of fossil fish, and the redaction of the German edition of Agassiz's book on glaciers.[25] Agassiz's ideas about publishing were exactly contrary to what Vogt had experienced under Liebig. In Giessen procedure demanded that whatever a student wrote up by himself was to be signed only by him when published. Liebig regarded the desire to publish as one of the strongest motivations a student could have. Agassiz went to the other extreme. He considered it a favor on his part to allow young scholars to work under him, and he was not beyond using the results of their research as if they were his own. Such a situation was bound to create a problem with the strong-willed young Vogt, whose name had already appeared in print for work done before coming to Neuchâtel.[26]

The issue seemed to be the lack of credit given Vogt and Desor for their work on the later volumes of Agassiz's monumental *Recherches sur les poisson fossiles*, published between 1833 and 1844. Vogt complained that Agassiz insisted on signing his own name to original work he, Vogt, had done. While it is surely going too far to declare, as did Vogt's son, that Agassiz had not written more than seven pages of all the work that appeared under his name during Vogt's years at Neuchâtel, it apparently was true that later Agassiz did spread the word in America that he was the author of volume one of the *Histoire naturelle des poissons d'eau douce de l'Europe centrale*, but that he had graciously allowed Vogt's name to appear as the author. Such a state of affairs is at least probable in light of the legal suit that eventually erupted between Agassiz and Desor over similar claims.[27]

If there was friction between them it took time to develop to the breaking point. Vogt had acted as Agassiz's spokesman for his glacial theories at an Erlangen meeting of scientists in 1840, prompting the celebrated geologist Leopold von Buch to declare that he would refute Vogt's claims one by one at the next annual meeting in Mainz. Agassiz had no desire either to renounce his own ideas or to do battle with an associate of his friend Alexander von Humboldt, hence Vogt was once more commissioned for the task. This was Vogt's first exposure to debate, but by no means his last; even Buch remarked that the scientific world would hear more from this young man.[28]

Nor did Agassiz insist on signing his name to everything Vogt wrote. In 1842 Vogt published his first book, *Im Gebirge und auf den Gletschern*, travel accounts of the expeditions of the Neuchâtel group to the Swiss glaciers. This work reveals the romantic enthusiasm of the young scientist for nature. The long periods spent away from the rest of the world are described in highly poetic language. The return to Neuchâtel was pictured as an effort, a return to the burden of etiquette and the distasteful presence of railroads and steam engines.[29]

These expeditions, in themselves costly, were even dearer because of the guests and scientists attracted to Neuchâtel by them, so that Agassiz was continually in trouble financially. Vogt and Desor were paid for their assistance only when they requested funds. When, therefore, the opportunity arose for Agassiz to go to America, the situation had become bad enough that Agassiz had no choice but to go.

Vogt's decision not to accompany Agassiz to Harvard in Vogt's account is made the result of worsening relations with the Swiss naturalist. For one thing, Agassiz allegedly was using Vogt's work more and more to cover his own deficiency, and for another he seemed to be becoming increasingly religious. Vogt related that when Agassiz's domineering wife, who lived most of the time in Karlsruhe, came to Neuchâtel for a visit, Agassiz would become nervous, he would work less, and he would try to relieve the situation by engaging in theoretical discussions on the essence of things and on the wisdom of the Creator as seen in nature.[30] These discussions had no appeal whatsoever to Vogt, who disliked philosophy in general. Son Wilhelm once commented: "There are few examples in the history of science of a teacher and a pupil remaining so dissimilar, so separate in their views."[31] Given this it is hardly believable that Agassiz, as Vogt recalled it, begged him with tears in his eyes to accompany him to America.[32]

Misteli's argument that Vogt's relationship to Agassiz was far from that represented in both the autobiography and the biography is much more likely. For one thing, every complaint about Agassiz came after the great upheaval in Vogt's life in Paris. In fact, references to Agassiz by Vogt from the period immediately following Neuchâtel are generous indeed,[33] though one must remember that at this time Vogt was a young man without steady employment and would have needed Agassiz's name.

With Agassiz set on going to America, Vogt left in 1844 for Paris, where he spent what he called the most interesting three years of his life. If for no other reason these years must have been exciting simply because of the people with whom he associated. Arriving in Paris, he went to the hotel of the Jardin du Roi, where he arranged to share a fourth floor room with the Belgian Adolphe Quetelet. Not long thereafter he was summoned by a Russian gentlemen, who anxiously requested that he come and help a friend who was dying of cholera. Vogt diagnosed the "cholera" as a case of indigestion, but the whole affair established a lasting friendship with the sick man, Michael Bakunin, the Russian anarchist.[34]

Vogt supported himself while in Paris mainly by acting as a science correspondent. He reported on the meetings of the Academie des Sciences for the *Allgemeine Zeitung* from January, 1845 to September, 1846.[35] His time during these years was divided among attending sessions at the Academie des Sciences, listening to the lectures of Parisian scientists (mainly Elie de Beaumont on geology at the École des Mines), and periodic research trips outside the city with friends made since his arrival. While the association with the *Allgemeine Zeitung* provided the occasion for publishing his well known *Physiologische Briefe* of 1845–1847, the lectures at the École des Mines served as the basis of his textbook on geology published in 1846, and the scientific research excursions resulted in his two volume *Ocean und Mittelmeer* of 1848. Each of these works reveals the radical changes Vogt underwent in this period.

We are fortunate indeed to have a book length study of the crucial and productive Paris years of Karl Vogt. Hermann Misteli argues that Vogt came to Paris from Neuchâtel as a romantic enthusiast of natural science, but that he was by no means either a political revolutionary or a materialist. Exposed to Bakunin, Herzen, Herwegh, and others in Paris, he left that city with an altered perspective that showed itself both in his political radicalism and in his scientific materialism. Misteli is concerned to uncover this gradual radicalization through an analysis of two of the works Vogt wrote during this period, the *Physiologische Briefe* of 1847 and the *Ocean und Mittelmeer* of 1848. Vogt's first book of the Paris years, however, appeared in 1846, and it is necessary to examine it before evaluating Mistelli's claim.

In the first edition of the *Lehrbuch der Geologie und Petrafactenkunde* Vogt's only deliberate departure from de Beaumont's geological views came with the glacier question, concerning which Vogt defended the Neuchâtel position.[36] But the predominant concern in geological circles of Paris in 1845 was catastrophism, pitting the traditions of Cuvier, represented essentially by de Beaumont and Brongniart, both of whom Vogt had heard lecture, against the notions of geological development that could be inferred from the ideas of Geoffrey St. Hilaire on the variation of species. More than once Vogt's articles for the *Allgemeine Zeitung* examined both sides of the issue for the benefit of his readers, though Vogt himself always ended by opting for the catastrophist position.[37] Vogt's catastrophism, and his belief in the fixity of species later became points of embarrassment for him, since he was to do an about face shortly after the publication of Darwin's *Origin*.[38]

If Agassiz had influenced Vogt in favor of geological catastrophism, he had not successfully persuaded Vogt of its religious implications. Vogt's experiences in Paris could not help but reinforce his irreligiosity, and if the *Lehrbuch* is any indication, they did precisely that. Vogt rejected in this work all notions of a Creator in favor of a pantheistic and romantic catastrophism. This was equivalent to Agassiz's brand of catastrophism minus the Creator, and shows, according to Misteli, that Agassiz had more of an impact on Vogt than Vogt's own account allows. All this does not suggest that Vogt never spoke of some kind of creative action, for this he plainly did.[39] But he was careful to spell out his antipathy to the idea that a supernatural process was at work. As Feuerbach would explain some three years later at Heidelberg, man was born and kept alive in a natural way; hence theological explanations explain nothing.[40] Vogt put it as follows:

> It is said that paleontology above all points unmistakably to an ordering spirit, a personal Creator, who has called individual creatures into life according to a reasonably devised plan and has engineered its gradual completion. It seems to me that, on the contrary, precisely this gradual perfection of organization, this successive destruction of imperfect types and appearance of more perfect ones provides the strongest proof against the existence of such a personal Creator. We subscribe to the view that organic matter as such has in it, as a characteristic, a capacity for gradual development which is not separated from matter. . . . We see . . . here the moments of independent development which the earth goes through as a material body and to whose ultimate goal it certainly has not as yet attained.[41]

Vogt's attitude here is noticeably more tolerant of the religious believer than it will later be, for he says regarding the presumption of a Creator: "Let him who wishes adopt such an assumption."[42]

Though the *Physiologische Briefe* was not published until a year after the *Lehrbuch* had made its appearance, Vogt had made the agreement to do the work in December of 1843. Originally the intention was that Vogt would write a series of articles on physiology for the *Beilage* to the *Allgemeine Zeitung*, but after submitting the first eight installments in 1845, the decision was made that a book was the more appropriate medium.[43] The letters were submitted in three sets, one set each in the years 1845–47. Mistelli leans heavily on this fact to support his claim that Vogt gradually grew towards full scale materialism as the book progressed, and that in 1847 he had not yet adopted materialism as a world view.

The point may seem trivial, but Misteli is addressing a real issue, for to adopt materialism as a world view, to hold that all is matter in motion, leaves one with a deterministic framework. But the years 1845–1847 were filled with talk of challenge and unheard of opportunity. It was hardly a time to insist that the world was determined by the laws of the motion of matter! In his examination of Vogt's *Physiologische Briefe* (1847) and his *Ocean und Mittelmeer* (1848), both compiled between the years 1845–1847, Misteli concludes that Vogt displays in these works characteristics inherited from both family lines. As a romantic, pantheistic geologist he is drawn to the political left under the influence of Bakunin, Herwegh, Alexander Herzen, and others both in Paris and on the expeditions described in the *Ocean und Mittelmeer.* As the young scientist reporting on the meetings of the Academie des Sciences, he finds his own atheistic position reinforced through exposure to the French materialistic tradition of Cabanis, *et al*, and reveals this attraction to materialism more and more openly in the *Physiologische Briefe*.

Misteli's interpretation has much to support it, as we shall see in our examination of the works mentioned here. The unanswered question, however, is how these two sides of Vogt, viz., the romantic and the materialistic, can exist alongside each other in one person. The answer lies in Vogt's ability to deceive himself, even if he did not deceive all those around him. If revolutionary politics and materialism share anything in common it is their radical nature. Both views are extreme positions that violently oppose accepted canons, yet they are in a sense mutually exclusive. Where, for instance, would the deterministic materialist find the motivation to defend the barricades? It is the dilemma Marx tried to solve with dialectical materialism as opposed to scientific, or in his terms, vulgar materialism. Here in 1847, and indeed for the rest of his life, Vogt the romantic won out over Vogt the rationalist. Through an appeal to revolutionary rhetoric in both cases, he made it sound as if one unified position was being represented when such was not in fact the situation. It is no wonder that Misteli thinks he catches, in spite of the apparent materialism of the *Physiologische Briefe*, "a hidden voice. . . , which feels otherwise than is evident from the outward tone."[44]

How then does the materialism of the *Physiologische Briefe* show itself? At the outset of the task in 1845 Vogt did not reveal his materialism. In that year letters one through eight were composed, and only twice was there a

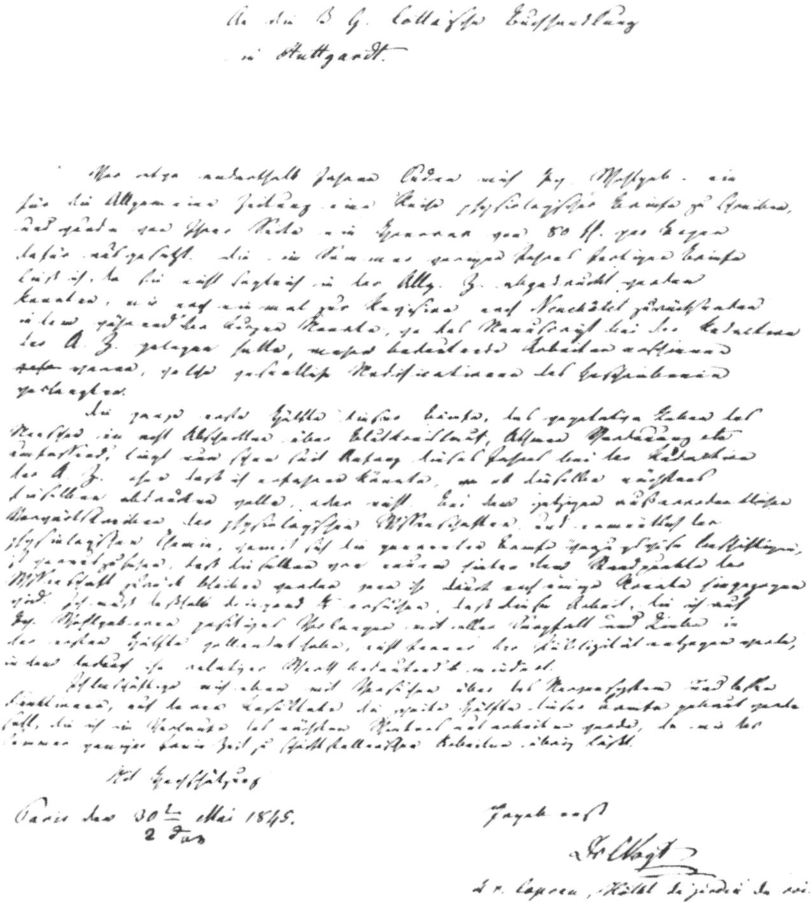
An die J. G. Cotta'sche Buchhandlung
in Stuttgardt.

Paris den 30ten Mai 1845.

Dr. Vogt

Fig. 2. Vogt's first contract with the Cotta Publishing Company for the series of articles to be entitled 'Physiologische Briefe.'

Fig. 3. Pencil sketch of Karl Vogt in middle age.

hint of materialism. Once he somewhat gingerly conjectured that it was characteristic of organic life not to develop of itself completely new forces, but only to make use of those already present in inorganic matter in a special way ("if I do not deceive myself"). Later he more boldly denounced belief in a vital force as "the refuge of lazy minds . . . who do not take the trouble to look into something they find incomprehensible, but are satisfied to stand amazed before the apparent miracle."[45]

The next set of letters, written in 1846, became progressively more daring, and contained the famous analogy that would be repeated *ad nauseum* by Vogt's critics. Modifying a proposition of the Frenchman Cabanis, whom along with Voltaire, Diderot, Holbach, and Condillac he mentioned in one of his reports to the *Allgemeine Zeitung*,[46] Vogt brashly proclaimed:

> Every natural scientist who thinks with any degree of consistency at all will, I think, come to the view that all those capacities that we understand by the phrase psychic activities (*Seelenthätigkeiten*) are but functions of the brain substance; or, to express myself a bit crudely here, that thoughts stand in the same relation to the brain as gall does to the liver or urine to the kidneys. To assume a soul that makes use of the brain as an instrument with which it can work as it pleases is pure nonsense; one would then be forced to assume a special soul for every function of the body as well . . .[47]

Vogt would never live down this vulgarism; even Büchner attacked him for it. After the final thirteen letters were completed and the book appeared, the relative multitude of materialistic tenets of the last section from 1847 were not recognized by his critics as the consummation of a gradual radicalization. Little did they realize the changes that Vogt had undergone since starting his book.

Vogt's presentation in the *Physiologische Briefe* was not philosophical. His work was not intended as a close analysis, but as a popularization of physiology. He did not like philosophy at all, as is evident from an oft-cited incident involving Bakunin and Proudhon. One evening he left the company of these two radicals when they began to discuss Hegel, only to be amazed when he returned in the morning to find the two still at it. How anyone could talk philosophy the night through was incomprehensible to him.[48]

It is therefore quite possible and even likely that he himself had not read the works of the philosopher-theologian Ludwig Feuerbach. But it would have been almost impossible for him not to have become acquainted with Feuerbach second hand, since he spent so much time on research trips with Georg Herwegh, a graduate of the Tübinger Stift and a one-time student of F.C. Baur. Besides, the Paris Russians, especially Herzen, were enthusiastic about Feuerbach.[49]

Vogt's world was natural science, and his audience was the people. He took his stand, as he put it, "before the jury of public opinion."[50] And it was first in the *Physiologische Briefe* that the people got to know him. In addition to the notoriety that the letters achieved in Germany, their fame spread throughout Europe, prompting the French churchman Dupanloup later to say of them: "These shameful and fatal doctrines are the foundation for the politics that the Commune has made live again in Paris."[51]

It was the common man that Vogt had in view, though sometimes to heed Vogt's advice could be dangerous. For example, Vogt decided after deliberation to include a chapter on sex education because of the widespread ignorance and misconception regarding human reproduction, and in order "to provide a way to limit the increase of the family without being forced to deny the enjoyments marriage offers."[52] But woe to the couple who listened to the latest from science *à la* Vogt, for after frank talk about menstruation and the process of conception, he assured the reader that conception during the last fourteen days of the menstrual cycle was impossible, because the fertile period occurred immediately following menstruation.[53] Vogt's materialism was not the only way he made enemies!

During these years Vogt organized and conducted two research trips. The first excursion, a month long, lasted from the end of August to the end of September, 1845. Vogt took Bakunin and Herwegh with him; in fact, it was this trip Feuerbach had envied so much in his letter to Herwegh of 1845.[54] The party travelled to the Breton and Norman coasts of France where, led by Vogt, they investigated the world of the lower sea organisms, which up to that time had been examined by only a few, mainly by Milne-Edwards and Quatrefages.[55] Back in Paris by the beginning of October, Vogt's next journey from Paris was not related to science, but took him home for a visit to Bern. There he witnessed some of the political unrest that led to Switzerland's Sonderbund War in the following year. It was here in Bern that Vogt for the first time carried a weapon in the fight against aristocracy.[56]

With Herwegh as a companion, the second scientific research excursion got under way during the first days of December, 1846, the goal this time being the Mediterranean. In Geneva the two met James Fazy, the democratic leader with whom Vogt later was to work closely, then on to Turin and finally Nice. The diversity of sea life presented to them at this juncture of north and south so delighted them that they remained caught up in observing and describing it until the beginning of February of the new year. Detouring on their return to Paris, they passed through Genoa and stayed in Rome for most of the month of February. During the trip Vogt had received a letter from Liebig informing him that he was invited to fill the newly created post in zoology at Giessen.[57] By March 2 the party was back in Paris, and in April Vogt returned to the city and the university from which he had fled some twelve years earlier.

Before leaving the *Ocean und Mittelmeer*, a few words regarding its general significance are in order. Meant for laymen, the work takes the form of letters to an unnamed person.[58] In general the book is not particularly materialistic, the closest allusion to heresy coming when Vogt boldly challenged the idea that man and animal are different.[59] More typical of the work is its emphasis on what Vogt called the new science which was replacing the old. As early as 1842, referring to the work of Schwann, Vogt had noted: " We stand at a turning point in physiology in every respect."[60] Here in the *Ocean und Mittelmeer* he was more specific. What was needed now was to join chemistry and physics to physiology.[61]

Showing little patience for the members of the scientific community who resented his youthful enthusiasm for innovation, Vogt denounced all older men who used their high positions of influence to hold back younger thinkers whose work might have left behind cherished theories. Trying to cover his own bitterness at being left outside the professional ranks, Vogt dismissed the charge that he had been insolent to his seniors by noting that he was an independent scientist, tied to no institution. Particularly angry because of his inability to penetrate the world of French academe, he lashed out at the deferential character of the French scientific community.[62] In nature too, Vogt wrote, the old became slow and unable to move. In nature this disturbs no one. Only when one showed the same thing at work in man did cries of opposition arise and charges of impropriety abound. Paraphrasing a line from one of Herwegh's poems, Vogt declared that he had been humble too long and that he wished finally to be proud.[63]

The use of nature as a basis for reasoning about man was exploited for political purposes to an even greater extent. A certain progression in Vogt's willingness to use nature in this fashion is clear, especially seeing that the two volumes of *Ocean und Mittelmeer* were written between 1845 and 1847, when political activity in Europe was escalating towards its peak in 1848. There is a parallel here with Vogt's development as a materialist in the *Physiologische Briefe*, written roughly over the same unusual period of time. Here in the *Ocean und Mittelmeer* there is also a gradual radicalization, this time political.

In the first volume, with a tongue-in-cheek attitude, Vogt reported on

Bakunin's discoveries of proofs that communism was founded in the natural order of things.[64] His own use of the term 'proletarian', in reference to the moonfish, was to point up its laziness. By the second volume, perhaps because of long association with radical thinkers such as Bakunin, Vogt depicted nature's ways as attributes of the Establishment, referring to himself as a "free proletarian of science."[65]

Between the end of the second volume and the writing of the preface of the *Ocean und Mittelmeer* in the fall of 1847, Vogt delivered his inaugural lecture at Giessen. Nature as justification for revolutionary ideology now became explicit. In this noteworthy speech Vogt sang the praises of the new science, which he viewed as the best mirror of the spirit of the time, the sure foundation on which the future would build. As a zoologist he confessed that he was envious of chemistry and physics, for they had a more immediate effect "in the wheels of society," but, he assured his listeners, he did not aspire to science for science's sake.[66] The natural sciences, organic and inorganic, had one thing in common – their goal of investigating matter and its laws of combination and transformation. The common point of all sciences, without which they could not exist, was the eternity of matter. All positive knowledge and exact science depended on this, and the time was near when natural science would draw the obvious materialistic conclusions.[67]

Natural science had an obvious material effect on man, Vogt continued, but it also was beneficial intellectually.

> Every new gain in trade, industry and agriculture is also tied to intellectual progress, to the growing perfection of the free spirit which natural science carries within itself.[68]

A free spirit was what natural science required. Vogt denounced the school of *Naturphilosophie* that had flourished earlier in the century. Modern science was based on facts. Yet Vogt did not want to be recognized as a strict empiricist; hence he explained the rise of *Naturphilosophie* as a natural reaction to the dry cataloguing of facts of the encyclopedists. He refused the title *Naturforscher* to those who had not gone beyond mere classification.[69] This is not the place to ask for a thorough unpacking of Vogt's position on the nature of scientific theory, but we may note that his works demonstrate the tightrope walked by the scientific materialists. Just how did the creation of a scientific theory differ in kind from speculative and hypothetical thinking?[70]

The conclusion of the speech reveals Vogt's political radicalization, and it is here that he attained the full scale exploitation of nature only approximated in the second volume of the *Ocean und Mittelmeer*. Citing Buffon, Cuvier, von Buch, and de Beaumont, Vogt argued, on the basis of the catastrophism these great names represented, that progress from one level to another came by means of revolution. "The principle of revolution is common to every development of inorganic and organic nature."[71] Nature's progress was like the conqueror who burned his bridges behind him, resting only to collect new forces for new revolutions. Those who investigated nature were themselves but links of a chain buried in the earth's history. Someday, after new revolutions, modern scientists would also be links in that chain. Though Vogt did not explicitly draw the political object lesson, it was clear enough. No doubt many in the audience that day became convinced that calling Vogt to Giessen definitely had been a mistake.

That was April. In the fall of 1847 Vogt wrote the preface to the *Ocean und Mittelmeer* in the form of a letter to Herwegh. The latter had written Vogt about the publication of the account of their excursions, and had taken the opportunity to ask how things were going in Giessen. In the reply Vogt revealed that he considered himself a materialist; further, that materialism was an unavoidable and in some ways unfortunate result of physiology.

> That is the misfortune, that the crude materialism I mentioned is so deeply grounded in physiology that one cannot occupy oneself with the latter without carrying on one's body and soul the offensive blemishes of the former.[72]

That Paris had a profound impact on Vogt's life is without question. He came to that city a likely candidate for the radicalization he experienced in his political views and with regard to his world view. By the eve of 1848 Vogt had formulated a brand of materialism he thought resulted from the tenets of natural science, and he had adopted an attitude toward revolutionary political change that he believed could be supported from science and expressed in scientific terms. Misteli has argued that each of these aspects of the young Karl Vogt derives from the two sides of his heritage, the political radicalism from the Follens and the scientific materialism from the Vogts. In addition Misteli has concluded that the two are mutually contradictory, that Vogt had to subordinate the determinism of materialism to the passion of his romantic temperament. In support of this he has argued that Vogt's materialism developed only gradually, that even the last sections of the *Physiologische Briefe* do not represent Vogt's complete adoption of materialism, that they only opened up the sluice gates through which his complete materialism would later pour forth. Only in the

1847 preface to the *Ocean und Mittelmeer* did Vogt openly confess to being a materialist. According to Misteli, as he went to Frankfurt he suppressed his deterministic materialism in favor of an idealism of revolutionary politics. At no time prior to 1852, claims Misteli, did Vogt interpret matter as the ultimate ground of all things. Since an exclusively materialistic system would run counter to revolutionary catastrophism, Vogt juxtaposed the two in the Giessen lecture, allowing them to run parallel without clashing.

In all this Misteli had demanded too much of materialism, and for that matter, of Vogt. He has assumed that materialism is a logical system of thought. Because Vogt does not spell out his materialism in a logical and finalized form, Misteli concludes that Vogt is not yet a materialist. But materialism has always been more slippery than that. In spite of Misteli's attempt to spell out in twelve propositions what materialism entails,[73] and to measure Vogt by this standard, Vogt has eluded him. Vogt *was* a materialist by the end of the *Physiologische Briefe*; it is not that he merely tended towards that position. Had Vogt been a philosopher, Misteli's perspective would have been more legitimate, for he could then demand more in the way of logical rigor from him. Vogt was a natural scientist who disliked philosophy, and it is primarily because he saw himself in this light that he remained unconcerned whether his materialism was accepted as philosophically respectable. The results which the philosophers detested as the fruit of shoddy thinking Vogt saw as the unavoidable testimony of the facts. This difference in perspective is why it is important to remember that Vogt was a scientist, and that as such he distinguished himself from the philosophers.

If not popular among the Giessen officials, Vogt did find favor among the students, at least among the most active. Giessen had not changed as far as the individual freedom of the students was concerned. From the account of Alexander Büchner, Ludwig Büchner's younger brother, who along with Ludwig was a student there at the time, the university was like the Prussian army – everything was forbidden that was not expressly permitted.[74]

When news of the Revolution came to Giessen, it was Karl Vogt to whom the students went. They asked him to call together an assembly and to preside. Vogt's response was cool; he would speak, but not preside. As the days passed and the elections to the Frankfurt Parliament loomed in the offing, Vogt, who had become the leader of Giessen's citizen corps, ran as a candidate for the district's delegate. As a popular professor he enjoyed the support and enthusiasm of many of the students.[75] Successful in the elections, Vogt went off to Frankfurt, first to the Vorparlament, and then to the Nationalversammlung, where he was one of the most notable and active of the Assembly's members. With the failure of the Parliament at Frankfurt, Vogt refused to give in, moving to Stuttgart as one of the leaders of the Rumpparlament, which also eventually collapsed.

Forced to leave Germany anew, and once again out of a job, Vogt returned to his father's house in Bern, where he worked on an analysis of the prospects for the political left in the coming future.[76] This situation was hardly satisfactory for a man of thirty-two, however, and he returned in the fall of 1850 to the Mediterranean coast at Nice for research. If his bitterness at being unaccepted and out of work had come out in his books during the Paris years, it had been mild compared to the venom he called forth during his second stay at Nice from 1850 to 1852. In addition to the political tract mentioned above, he produced during this period the *Untersuchungen uber die Thierstaaten*, a highly unusual work in which Vogt denounced the middle class, religion, and all forms of government. He called for the establishment of anarchy, by which he meant the lack of an official governmental organization, as the solution to the problem of individual freedom.[77]

Particularly caustic were his words about the German university system that had shut him out. The universities now represented to him the place within which the bureaucracy, with the help of the German professors, made free men into servants of the state.[78] Students, once enthusiastic for the idea of freedom, in the main had disclaimed on their heritage. The work teemed with materialistic tenets. Vogt was no longer cautious about determinism; now he openly proclaimed it. With the revolution quashed, the contradiction between determinism and the call to action was not nearly so bothersome.

While still at Bern, Vogt had seen through the press the translation of Chambers' *Vestiges of the Natural History of Creation*, which he had finished before the March Revolution. It is difficult to know exactly why Vogt chose to do this translation. Was it because the author, as yet unknown to Vogt, had discussed ideas Vogt held? Was it the author's emphasis on geology and paleontology (about one quarter of the book)? Or, what seems most probable, was it the unconventional theological position, which Vogt praised highly in the preface to his translation?[79] It was not the ideas Chambers had talked about in the *Vestiges* that attracted Vogt, for he did not agree with the emphasis on descension and the transformation of species. Rather it was the furor the book had created in England on all sides that caught his eye, including the denunciation from theologians and official scientists. Vogt, who had been rejected by the Establishment, had something in common with this unknown author who was drawing forth the scorn of many.

With his time his own, Vogt returned to research, and discovered among other things how certain cuttle fish succeed in mating. A polyp arm separates from the rest of the body and swims free in the sea to the female.[80] But not all of his time was given over to this kind of work, for he produced during this second stay in Nice two more long popularizations of biological science, each well stocked with allusions to materialism, and a long article on physiology in *Die Gegenwart*.

In the article Vogt declared that physiological knowledge gradually was becoming the common property of educated people of all classes as opposed to the secret possession of the doctors. Mocking those who still opposed research in the name of God or the church, and deploring the approach to physiology of the vitalists, Vogt described how modern physiology had been able to free itself from these superstitious perspectives with help from the physical sciences, and specifically through the realization that physiology, as an applied science, sought the general laws of matter in their special modifications within the body.[81] As in the *Thierstaaten*, so here Vogt pointed to the recent progress in the science of nutrition, and the possibilities that had been opened up for that subject because of the materialistic perspective. It is uncertain whether he borrowed from Moleschott, or Moleschott from him, but the parallelisms are clear, for both cover the same foods in their respective discussions of the subject. In addition, both notice the demoralizing effect that potatoes have on a people if not supplemented with other foods, and both suggest beans as a substitute.[82]

One year later the two long volumes of *Zoologische Briefe* arrived on the scene. Like the *Physiologische Briefe* this work was addressed to "the educated of all classes." For the most part the work was purely descriptive, with but 107 pages of the 1359 total devoted to topics such as the history of zoology, the question of species and its significance, reproduction, geographical distribution of animals and extinction. Vogt saw his task to be the coordination of new knowledge of comparative anatomy, comparative physiology, natural history, paleontology and zoological geography. He was convinced that the relations displayed among the different groups would make clear the history of each individual one, "from its origin to its dissolution into the elements."[83] In contrast to the *Naturphilosophen*, modern zoologists depended on observable reality. Since political uprisings had been known to spill over into science, Vogt added, 1848 might well speed up the new direction in zoology just as the events of the 1790's aided the approach of Cuvier.[84] Nor could Vogt resist taking one more swipe at the German universities, this time thanking the liberal ministers for not allowing him to become part of such a despicable institution.[85]

The *Bilder aus dem Thierleben* of 1852 was much less formal than the *Zoologische Briefe* had been. In it Vogt leaned heavily on the research carried out on the organisms of the Mediterranean at Nice. Although he intended to impart information of a scientific nature to his reader, he also had another purpose in mind which had nothing in common either with that or with popularization. It was to bring the world of science to its senses, to help avoid the tendency to succumb to authority – the shibboleth of the reactionary church and state.

> German scholars sink more and more into the bog, and it is time that something *solid* be tossed into this mutual admiration society, to which natural science also is being elevated.[86]

Dotted with the narrative of personal experiences, the *Bilder* described subjects such as the habits of tuna fish, the various kinds of reproduction in sea organisms, and extinct animals of the sea. In the process Vogt was able to incorporate what had become standard diatribes against the concept of creation, which for him always meant the theological *creatio ex nihilo.*

> One can understand by the expression "creation" nothing but the construction of new forms from existent matter. Whether these be inorganic forms . . . or organic beings that come forth from a process of creation, one must always adhere to the basic propostion that from nothing also can nothing come, . . . and that the creation of matter is therefore palpable nonsense.[87]

Vogt singled out Rudolph Wagner as the paradigm representative of those superstitious scientists who still believed in a Creator, and declared that such people have been forced by science to abandon divine creation in certain instances of parthenogenesis, if that were in fact what really occurred, and to account for the first human pair.[88] Wagner was also the major object of attack in the last section, entitled 'Thierseelen.' As a reply to the articles Wagner had been writing in the *Allgemeine Zeitung*, this polemical section of the *Bilder* closed round one of the battle between Vogt and Wagner, and intimated much of what was yet ahead. One of Vogt's favorite tactics against Wagner was to declare that his blind faith (*Köhlerglaube*) was the result not of his views as a scientist, but as a believer.[89] In contrast to this, Vogt would pass off his own position as the fruit of scientific research and stamp it with the mark of discovered truth. Near the end of the *Bilder* he summed up the work this way:

> We have achieved our goal. We have proven that the same law of ascending material development rules everywhere organic matter treads.[90]

It was as if he had said, let him who dares oppose the truth of science!

Vogt's reputation was beginning to match his confidence. While at Nice Vogt received an offer to come to Geneva as professor of botany, but he turned it down; botany was simply too far from his expertise. The university persisted, however, and eventually he received another offer, this time in geology and paleontology. Vogt accepted, and after a trip home to Bern, he was off to the city of John Calvin,[91] where he became a naturalized Genevan and spent the rest of his life.

Even after Vogt's departure from Nice to Geneva the articles against materialism, penned by the Göttingen *Hofrat* Wagner, continued to appear in the *Allgemeine Zeitung.* Vogt later said that he had written much of the *Bilder* against Wagner's letters, since their quality was so appalling that somebody had to do something.[92] Wagner in turn directed his attention solely at Vogt soon after the *Bilder* appeared. He pointed to Vogt's political views as proof of his errors, and referred both to his own 'Physiologische Briefe,' number six, and to his Göttingen colleague Lotze's recent *Medicinische Physiologie* for a rebuttal to Vogt's nonsensical, materialistic view of consciousness. Wagner denounced Vogt's ideas concerning the Creator, the soul, suprasensual knowledge and the basis of morality, but denied that he wished Vogt's work suppressed. Better that it be spread around so all could see its superficiality. In highminded fashion Wagner announced that he would not reply further to Vogt because Vogt was not worthy of it.[93] Vogt's reply from Geneva to the *Allgemeine Zeitung* was printed only in part, and things quieted down for a while.

At the Göttingen meetings of the Gesellschaft Deutscher Naturforscher und Ärzte in the fall of 1854, Wagner took it upon himself to give a lecture in which he addressed himself to the two subjects indicated in its title, 'The Creation of Man and the Substance of the Soul.' In it he took the stance that man had descended from one pair and that man's soul was immortal, arguing that such assertions were neither provable nor refutable by natural science.[94] He concluded his speech, however, with a statement that betrayed his conviction that there was a link between science and these issues.

I cannot believe that in a serious encounter with the subject one would come to results which necessarily bring the natural sciences under suspicion and completely destroy the moral foundations of social order. Only because we support and preserve these do we fulfill a duty to the nation. Our successors will hold us accountable for it.[95]

Vogt's account of the episode in the *Köhlerglaube und Wissenschaft* related that Wagner had purposefully chosen a time and place when Vogt was unable to be present to give his lecture. Wagner, explained Vogt, invited his opponents to a disputation the following day, but when the physiologist Ludwig of Zürich postponed his planned departure in order to take part in it, Wagner was taken ill suddenly, and sent word to the waiting assembly that the debate would be held three days later. Wagner did not pass up the opportunity, however, to invite those who had come for the disputation to a lecture he would deliver the following day at the Institute of Physiology in Göttingen, where Wagner was director.[96] In his own earlier account Wagner did admit to a "sudden cold," though he did not indicate that he had wished to debate anyone, only to discuss certain things concerning the latest knowledge of the central part of the nervous system.[97]

A month after all this happened Wagner published a second pamphlet, a response to a communication Vogt had sent to the *Allgemeine Zeitung* just prior to the Göttingen meetings. Vogt there had reviewed the past differences between Wagner and himself and had quoted a passage from Wagner's *Neurologische Untersuchungen* of 1854, which he considered to be self-incriminating: "It is not physiology that necessitates my assumption of a soul, but the idea of a moral world order which is imminent in me and inseparable from me."[98] Wagner's purpose in writing this second pamphlet was to respond not only to Vogt, but as well to other critics of the *Neurologische Untersuchungen.* Few persons, conservative or liberal, appreciated the 'double book entry' approach Wagner took in this work.[99] Wagner did not, however, back down in the face of the criticism, but used the second little work to defend his position.

Vogt could take no more. In January of 1855 the first edition of the *Köhlerglaube und Wissenschaft* saw the light, followed quickly by three more before June. Much of the first section of the work was an open and viciously *ad hominem* attack on Wagner's incompetence. He was lazy, but clever enough to act as an editor of scientific works, thereby receiving the credit due mostly to others. When his incompetence as an anatomist finally could be hidden no longer, said Vogt, Henle had been called to Göttingen to take over. In general he was despicable in Vogt's eyes.[100]

When Vogt finally did get down to business, he opposed the two major issues Wagner had brought up, monogenism and the existence of an immortal soul. Once again Vogt's strong antireligious bent, and his natural rebellion against anything beyond man's ability to know by natural as opposed to supernatural means, caused him to run rough shod over the epistemological difficulties of a crass materialistic solution to the problem of consciousness.

Although the debate itself was not carried out in a respectful atmosphere, it proved to be extremely important for several reasons. First of all, Vogt had succeeded in drawing out a prestigious opponent. The works he had published up to then might have gone unnoticed by many who would, however, take note with interest of the well reported debate that Vogt's materialism had called forth. Secondly, the publication of the *Köhlerglaube und Wissenschaft* came at the beginning of the same year that produced Czolbe's *Neue Darstellung des Sensualismus* and Büchner's *Kraft und Stoff*. Coupled with Moleschott's attack on Liebig in the *Kreislauf des Lebens* of 1852, it appeared as part of a virtual frontal attack by the forces of scientific materialism. Thirdly, the age-old problem of the relation between mind and body had been raised in such a way that it could not be ignored or simply answered by dogma. Materialism now appeared to have claimed the forces of modern science for its own. Somehow the church and idealistic philosophy had to explain, in language as simple to understand as that used by the materialists, their side of this and various other complicated issues. Wagner had forced the issue of the relation between science and religion. By declaring that the two had nothing to do with each other he actually supported the position materialists like Moleschott had already proclaimed. It did not take Vogt long to point out that on his own grounds Wagner had no right to appeal to an analogy between the substance of the soul and the ether, but that Wagner's position was solely dependent on his particular moral and political values. Wagner was criticized from the left and center. Theologians were unhappy with the ease with which he had discarded natural theology and the unfragmented truth of religion, and philosophers were uneasy about his simple explanation of the relationship between fact and value. By the end of 1855 the fight that had been simmering came to a boil.

As during the prime of his career, disputes marked Vogt's last years as well. In 1859 the *Allgemeine Zeitung* had picked up a report from London that Vogt was being paid by Prince Jerome Napoleon to aid France in her position vis-à-vis Austria. When this was published Vogt took the journal to court on a libel charge, and in the ensuing trial it was exposed that the report had come from the party of the exiled Karl Marx. In neither Vogt's account of the affair in his *Mein Prozess gegen die Allgemeine Zeitung*, nor in Marx's rejoinder of the following year, *Herr Vogt*, was the matter settled once and for all. That would have to wait until the 1870's. The story of this encounter is instructive for a comparision between scientific and Marxist materialism.[101]

The early 1860's were otherwise productive for Vogt. After a trip north along the Norwegian coast to the arctic in 1861,[102] Vogt returned to Geneva to work on both scientific and nonscientific projects. One such project had occurred to him while still at Nice, but never came to fruition in spite of repeated attempts by Vogt to interest others in it. This was the founding of a maritime zoological station, a permanent installation on the sea coast equipped with laboratories for scientific research.[103]

In general Vogt devoted the 1860's to paleontology and anthropology, largely because of the stir Darwin's book had made. In addition to his last major popular work, the *Vorlesungen über den Menschen* of 1863, he authored numerous articles on the topic.[104] Before their appearance in book form, his *Vorlesungen* had been the subject of public lectures to a group in Neuchâtel. Although his acceptance of Darwin was qualified, like that of the other German scientific materialists who rocketed to fame in the fifties, Darwin *had* changed his mind about the fixity of species and the relative importance of catastrophes. Hence it is important to note that Vogt's *Vorlesungen* was one of the earliest pro-Darwin statements in Germany, and also one of the first works anywhere to apply Darwin's ideas specifically to man, earlier even than Haeckel, who later became the German Darwin.[105]

The *Vorlesungen* preserved Vogt's image as a scientific materialist throughout the sixties, but, except for his political opposition to the rise of Prussian militarism, his life seemed to have left its high-point behind. Loyal Germans everywhere, especially *Grossdeutsche*, learned to hate Vogt not only as a materialist, but also as a traitor.[106] Although he despised Austria, and had in 1859 urged Prussian neutrality in any Franco-Austrian dispute, he also had no great love for the Prussians, in spite of the fact that he viewed the solution of the *Kleindeutschen* as better than no unification at all.

Prussian militarism disgusted him thoroughly. Concerning the Prusso–Danish war of 1864 he said:

> It is a time of unprecedented barbarism. Thousands are slaughtered . . . out of vile lust for murder and the search for adventure only to prove what one can do and how strong one is. . . . The Schleswig–Holstein question is not a question of freedom.[107]

A similar stance vis-à-vis Bismarck's triumph over France in 1870 produced outrage among Germans. First of all, he opposed the annexation of Alsace.[108] In addition he declared that France had not been the aggressor, he repeated his disgust for war in general, and he recorded his expectation of resurgence of a "throne and altar" reaction in Germany. He painted a very black picture at a time when most Germans were filled with pride.[109] When it was uncovered that the real author of certain satirical verses mocking the German military was not one Christoph Veital, but Karl Vogt, Austrian journals reproduced them along with an expose of the author. This development led to the breakup of several of Vogt's friendships back in Germany.[110]

Nor were the sixties without disputes with other scientists. Vogt became embroiled in minor affairs with two great names in German science, Rudolph Virchow and Julius Robert Mayer. In the *Vorlesungen* Vogt had taken the position that the simian characteristics of human idiots were a result of an arrested development of the fetus, "therefore a transitionary level between ape and man which is produced by the development grounded in the laws of formation of the human species."[111] Developing his ideas further at a meeting in Copenhagen in 1868, Vogt found himself opposed by Virchow, who insisted on accounting for microcephalism as a pathological, not an atavistic condition. As a result of this and of the work of the Tübingen anatomist Luschka, Vogt abandoned his theory in 1872, admitting that he had never personally investigated the brain anatomy of a microcephalus.[112]

Then in September of 1869, at the Innsbruck meetings of the Gesellschaft Deutscher Naturforscher und Ärzte, Robert Mayer presented a lecture entitled 'Über notwendige Konsequenzen und Inkonsequenzen der Wärmemechanik,' in which he referred to the work of G. A. Hirn, whom he regarded as an independent discoverer of the mechanical equivalent of heat along with himself, Joule, Colding, Holtzmann and Helmholtz. Hirn, according to Mayer, had established that there were three categories of existence: matter, force (*Kraft*), and soul.[113] Mayer conceded that the molecular processes of the brain were closely tied to the mental states of an individual, but it was, he said, an error to identify them. Mind was not part of the sensed world, and therefore was not an object of investigation for physicists and anatomists. Mayer concluded his speech with the following:

> Whatever is correctly thought subjectively is also objectively true. Without this eternal harmony between the subjective and the objective worlds, pre-established by God, all our thinking would be unfruitful. . . . A real philosophy must not and cannot be other than a propadeutic for the Christian religion.[114]

In spite of a letter Mayer wrote to his wife in which he mentioned that his lecture had been well received,[115] there was general agreement later among several who had been present that Mayer's concluding remarks produced a murmur of dissatisfaction in the audience.[116] Vogt reviewed the session for the *Kölnische Zeitung* four days after the event. Following a favorable evaluation of Helmholtz's remarks, which had preceded Mayer's and in which Helmholtz had praised Mayer and conceded priority to his enunciation of the principle of conservation of energy, Vogt criticized Mayer's lecture for its lack of clarity and logic, reporting that Mayer had astonished the assembly with his views, and commenting that whoever had invited Mayer to speak should have known better.[117] Clearly Vogt was more impressed with the religious implications of a scientist's accomplishment than he was with his reputation as a scientist.

During the last two and one-half decades of his life Vogt openly became the German liberal and humanitarian he had really been all along. Active in Swiss politics from early on, Vogt was a member of the Grand Conseil du Canton de Genève three times, from 1856–1862, 1870–1876, and 1878–1880; the Conseil des États Suisses from 1856–1861 and 1870–1871; and of the Conseil National from 1878–1881. In the past he had never claimed affinity with the radical party of the communists. As an anarchist his was a peculiar ideological stance. It did not conform exactly to that of the German Liberal Party, but eventually it was more at home there than with any other group.

Besides antimilitarism Vogt gave vent in his later years to vocal opposition to Treitschke and anti-Semitism.[118] As a humanitarian he couched his criticisms of society in the context of the world of science. He emphasized, for example, the neutrality of science where nationalistic disputes were concerned,[119] and he acted as a peacemaker at the Bologna scientific meetings of 1871, where a considerable problem had presented itself.[120] He wrote against the revival of antivivisectionism in England in the 1870's,[121] and mocked the growing romanticism among German university youth, calling for more emphasis on science in education and less on the classics.[122] Finally, on the personal level, he wrote polite "no thank-you's" to several individuals who had written him about the state of his soul, urging him to repent.[123]

In the 1880's Vogt travelled, wrote articles, translated, and in collaboration with others continued to write scientific books.[124] In 1882 he became the last Chevalier de la Legion d'Honneur that Gambetta commissioned. At the age of 70 he began to criticize the superficiality of German society through the novel, completing three of them before his death.[125] In his very last years Vogt returned to the scientific research with which he had begun: "With the fish I began, with fish I will also end."[126]

Throughout his life Vogt had been the correspondent of many of the great names in science in the nineteenth century, most of whom are mentioned in Vogt's own writings. But surprisingly there is almost no mention at all of Vogt's immediate family. That a son wrote a biography is known, but only once is there a reference to Vogt's family in it, and that is simply an acknowledgement that he left everything to his wife in his will.[127]

If after his death on May 5, 1895 he had honored his wife for her loyalty, he had not done so while she yet lived, at least she seems to have played no part in his life. The explanation of this riddle is solved in a letter from Engels to Marx of January, 1868. As it turned out Vogt, friend of the worker and enemy of all aristocracy, had not been able to overcome the class difference between himself and his wife. Engels tells us what official publications would never mention, for in an argument with a friend of Vogt's, Engels was asked to go easy on Vogt because of Vogt's difficult personal situation. He was told, and he in turn passed it on to Marx, "his wife was a highland peasant maid whom he had virtuously married after he had gotten her pregnant."[128] Intrigue and difficult situations had followed him everywhere in his long life, even to his own doorstep.

CHAPTER IV

JACOB MOLESCHOTT: 'FÜR DAS VOLK'

We are almost completely dependent on Moleschott's autobiography, *Für meine Freunde* (1894), for information concerning his life. Unlike Karl Vogt, who became the subject of several studies, there is no major secondary work dedicated to the examination of the writings of this notorious Dutchman. Jacob Moleschott, eldest of the Moleschott children, early showed signs of the streak of stubborn independence that characterized so many of the scientific materialists. Reprimanded as an eight-year old at school by a high ranking visitor for a mistake in calculation, Moleschott, who knew he was not in error, stubbornly and insolently refused to be corrected, prompting the visitor to leave the room.[1] For the eventual scientist and materialist certainty would always be among the highest values in life.

Born five years after Vogt in 1822, Moleschott, like Vogt, grew up under the strong paternal influence of a very liberally minded medical doctor. Unlike Vogt, and more typical of the age, the mother of the family did not share the radical views of her husband; rather, she chose to influence her children in other than purely intellectual ways. From his mother Moleschott acquired a great love for music, and it was from her he learned that he was ultimately responsible for his own views, and that he must believe only what he could honestly accept.

Moleschott's father had been born Catholic, but a rather dramatic incident early in his life turned him altogether away from Catholicism. When the family home and business were destroyed in a fire in Leyden, Moleschott's grandfather, a member of the city council, received aid not from the church of his faith, but from a Protestant citizen who helped him to rebuild his apothecary. This made a deep impression on both the grandfather and his son, so much so that Moleschott's father grew up favorably disposed to Protestant thought in general, and to the Protestant critical tradition in particular. This he passed on to young Jacob, along with a hearty enthusiasm for literature and poetry of all stripes, and an interest in Dutch political affairs.

With his next younger brother Fritz, Moleschott left his native Herzogenbusch to attend a Dutch grammar school for children of upstanding families. The boys were never happy there, and when forced to go home temporarily because of a rash of measles, they did not return, but enrolled in a French school. Moleschott on his own kept up the study of English begun at the first school, and now in the new setting he soon became as fluent in French as his father.

During these early adolescent years his favorite pastime was literature. Often he would go to the open fields and read drama aloud to the breeze. Less exciting to him were his classical studies, in particular Latin and Greek, which he considered rote instruction, but not teaching.[2]

The elder Moleschott knew as well as anyone the drawbacks of an overemphasis on the classics, and thus he determined to give his sons the benefits he had enjoyed as a youth at one of the best schools in Leyden. When the first attempt failed due to the director's hesitation to permit the Catholic Moleschott to room with Protestant boarders, the decision was reached to send the two boys to the *Gymnasium* at Cleves. Moleschott remembered the four and one-half years spent at Cleves fondly. Mostly he recalled the faculty's dedication to teaching, a value he himself would try to emulate later.

No better contrast can be found between the personalities of Vogt and Moleschott than that of their respective descriptions of their *Gymnasium* years. Where Vogt had been highly critical of his early education, Moleschott was not only satisfied, but genuinely impressed. Being a foreigner Moleschott had reason to be uncritical. He was up against a new language once again, and because German was difficult for him at the beginning, his confidence was eroded by the demanding atmosphere of this north German school. But even though Vogt noted, as we have seen, the difference between northern Germany and Giessen, the discrepancy between the accounts of *Gymnasium* life is simply too great to attribute to regional differences alone.

The classics at Cleves ceased being the mechanical exercise they had been for Moleschott up to then, becoming, especially under the Hegelian Moritz Fleischer, a means of communicating the spirit of the old heroes as they grappled with the meaning of life. According to Moleschott Fleischer was a master teacher, deeply concerned that his students learn what education could do for them personally. He had a way of instilling confidence in Moleschott, feeding the spark of independence he sensed. On walks with Moleschott he passed on his own values: "If someone has not banged his head in a fall but has contemplated something seriously, then he is always justified in giving an opinion."[3] It was Fleischer who imparted to Moleschott his life-long respect for Hegel's philosophy, and taught him the democratic value that he was no better or worse than any man.[4] And it was not just Fleischer; all the teachers at Cleves were educators, all of them were concerned about the students.

Obviously the contrasting portraits of *Gymnasium* life are also affected by the different outlooks of the two men. Vogt, the ruthless and fiery polemicist, had no use for philosophy, especially for Hegel; but Moleschott, who argued as stubbornly and as vehemently as anyone for materialism, always respected the idealistic tradition, even if he chose to follow Feuerbach in modifying it.

It was during these formative years at Cleves that Moleschott finally rejected the Christian beliefs of his mother. In one of the inevitable discussion/arguments between father and student son home on vacation, Moleschott collided head on with his own doubts. When his father mentioned that no thinking man any longer believed that Jesus was the Son of God, he had gone beyond his son's position. Moleschott has recorded for us his personal religious crisis on that occasion.[5] Back at Cleves, the crisis resurfaced when on a final examination in religion he wrote on the concept of Christian hope. He was determined not to write something he did not believe, and he had come to realize that he could no longer accept much of the traditional Christian doctrine. His solution was to answer the question using as a foundation the idea of Christian love, thereby avoiding the issue of God's providence altogether. His teacher did not approve of Moleschott's answer, but, in conversation with him about it, he agreed to accept it. On his oral examinations before graduation, however, he did not escape quite so easily. Reprimanded by an important examiner for presumptuousness, and advised to cultivate his character before entering the university, Moleschott was merely strengthened in his conviction, and once more expressed his stubborn pride, this time by boycotting a dinner given in honor of the examiner.[6]

From Cleves Moleschott went back to Holland for a vacation before going off to the university. He had toyed with the idea of going to Berlin, "primarily because I hoped to find Hegel discussed there," and also because one of his teachers at Cleves had recommended the medical program there.[7] But his father advised beginning with a smaller school, and the two agreed on Heidelberg where Tiedemann, Gmelin, Nägeli, and Bischoff all taught science, and philosophy was cultivated by the Hegelian Moritz Carriere. Moleschott's father had encouraged him to go into law, but in the end Jacob decided to follow in his father's medical footsteps.

Natural science had not been a large part of the curriculum at Cleves, so Moleschott did not know what to expect at Heidelberg. Arriving in 1842, he was at first distressed to find that his studies were all pieces of patchwork. Nowhere did anyone try to integrate the many approaches to knowledge pursued separately throughout the university. Moleschott found refuge in renewing his study of Hegel. At this point he was far from the materialistic Moleschott of 1850, rejecting Carriere's philosophical position because it was too dangerous. Carriere gave too much importance to Spinoza and Erasmus.[8]

His father had already convinced him that natural science, and physiology in particular, was the real basis of medicine, and that medicine should be studied as an experimental science in contrast to the speculative approach of the *Naturphilosophen.*[9] Moleschott gradually became convinced, much like Feuerbach, that natural science provided a sound basis for philosophy and that philosophy in turn enriched his perspective for medicine. Under such men as W. Delffs in chemistry, Tiedemann in physiology, Gmelin in physiological chemistry, and Bischoff in anatomy, Moleschott learned well the role of experiment and observation.

Philosophically he began to entertain and study the new brand of Hegelianism represented in the *Hallesche Jahrbücher* of Ruge and Echtermeyer. He learned of the young Hegelians D. F. Strauss and F. Vischer, and even translated some of Strauss into Dutch in 1843. One year earlier, during his first vacation at Heidelberg, he had travelled to Württemberg where he met Strauss and Vischer. Another journey, to Switzerland in 1843, was undertaken in order to meet scientists such as Vogt's teacher Valentin in Bern, and Henle, Oken and Kölliker at Zürich. Moleschott's attitude towards Oken's *Naturphilosophie* was not the satirical one of Vogt or Büchner. He was aware of the contemporary feeling of the many scientists who held Oken and his school in open derision, yet he could and did sympathize with Oken's goals:

> The imaginative contemplation of nature (*die denkende Naturbetrachtung*) is able to exhaust the particular only when it seeks, with the help of reliably determined facts, to raise itself to a general concept.[10]

Moleschott's university days also contrast with Vogt's. On his arrival he decided he would avoid the beer hall, while Vogt had wasted his first year with such activities. Moleschott consoled himself when jilted by a girlfriend by reading Goethe, Feuerbach and Spinoza and by playing Beethoven on the piano, means which Vogt never would have used. Moleschott's contact with politics was informal, moderate, and carried on at a distance with several of his friends,[11] again not the extreme and direct involvement of Karl Vogt.

In 1844, still a graduate student, Moleschott submitted an essay to a scientific society in Holland, in which he discussed Liebig's theory of plant nutrition with the particular goal of finding out which of Liebig's claims rested on confirmed facts and which did not. It turned out to be the preliminary round in the continuing battle between Liebig and Moleschott, but here at the beginning things remained friendly. Moleschott had been quick to point out in the essay that Liebig did not always base his argument on experimental results.[12] After hearing in December of 1844 that his had been the prize essay, he sent Liebig a copy of it, and received in reply a letter in which Liebig expressed his admiration for Moleschott's gentlemanly criticism and help. Later Moleschott learned from Alexander von Humboldt that Berzelius had spoken to him of his article. Surely he was in league with the masters![13]

Earlier in 1844 Moleschott suddenly experienced the urge to finish his degree and to choose a profession. Henle, who was called to Heidelberg from Zürich in 1844, gave him the choice of working in his dissertation on the anatomy of either the liver or the lungs. Moleschott chose the latter, and attempted in his investigation of their histological structure to demonstrate that the small bronchoids ended in vesicles and not canalicules. He showed that these vesicles can move themselves by means of their own smooth muscles, that they increase in size with age, and that in emphysema they atrophy but do not hyperatrophy.[14] By translating the first part of the Dutch chemist Mulder's work on physiological chemistry he was able to earn enough money to pay for the cost of the promotion himself, and to surprise his father with the completion of his degree on January 22, 1845.

Returning to Holland, Moleschott went to Utrecht to work in Mulder's physiological workshop, having met Mulder and the subsequently famous opthomologist F. C. Donders during the fall of the previous year.[15] Moleschott admired Mulder for breaking the path, along with Liebig, in physiological chemistry. It was in Mulder's laboratory that Moleschott learned the methods of scientific research.[16] Assuming that he had paid attention to what he had translated, Moleschott must have learned the importance of factual as opposed to speculative knowledge. Only the indisputably true, only that which is brought to light by objective observation had lasting value in natural science according to Mulder.[17] Further, Mulder's position on vitalism was thoroughly progressive. He declared it his purpose to explain vital phenomena by means of a knowledge of inorganic matter, and labelled it a 'false idea' that organic and inorganic matter did not obey the same laws.[18]

The tutelage under Mulder was not to last very long, for Moleschott decided to break his promise to translate Mulder's reply to Liebig on the so-called protein question when he saw the nature of the charges and countercharges that were being exchanged.[19] Still, Moleschott did translate into German Mulder's *Die Ernährung in ihrem Zusammenhange mit dem Volkgeist* in 1847. One cannot help but notice the points of potential influence of Mulder on Moleschott. His antivitalism, his strongly experimental approach, his concern with physiological chemistry and *Stoffwechsel*, and even his application of the study of diet to its effects on the national mind all had become positions or concerns of Moleschott by 1852.[20]

With a medical degree from Heidelberg, Moleschott had merely to pass an examination required by his homeland before he could set up practice. Even though he left the Mulder laboratory to practice privately, Moleschott did not break contact with young Franciscus Donders, nor with another doctor he had met, I. van Deen, who also had studied under Bischoff at Heidelberg. The three formed a close friendship that lasted throughout the century, and here, during the period from 1845 to 1847, they joined in editing the *Holländische Beiträge zu den anatomistischen und physiologischen Wissenschaften*, which appeared for three volumes from 1846 to 1848. In this short-lived publication all three men contributed material from their own research, and attracted articles from other Dutch scientists, including Mulder.

Moleschott spoke of his Utrecht days with mixed emotions. He continued to read widely in literature, especially Goethe and, after a visit from his old *Gymnasium* teacher Fleischer, Feuerbach.[21] He was involved in scientific research, which he enjoyed, and continued to use his proficiency for language in translating. With Donders he was especially close, and later considered his almost daily contact with this young and gifted scientist the best part of the Utrecht experience.[22]

But there were reasons why he was unhappy to be back in Holland, the greatest of which was that he missed the excitement of the German university setting. In addition, none of the three young Dutch scientists was able to capture the vacated position in anatomy and physiology at Leyden, though all three had applied, and all were convinced that one of them would prevail. Moleschott eventually was offered a position in legal medicine at the *Hochschule* in Utrecht. It was too late, for he had decided to leave Holland and return to Heidelberg as a *Privatdozent.*

In the summer of 1847 he gave his first lecture in physiological chemistry to only a few students, but eventually he began to distract students from Henle's lectures by his repeated use of experimentation and demonstration in his lectures. Tiedemann soon asked for Moleschott's collaboration in the nutritional section of the physiological handbook he was preparing, and prospects began to improve.

Far more comfortable back in the whirl of academe, Moleschott was carried away with enthusiasm for the German approach to education. When van Deen was turned down for a teaching post in Leyden, Moleschott vented his pent-up hostility against what he considered the backwardness of Dutch universities by denouncing his native country. In an article written in *Griesinger's Archiv für physiologische Heilkunde* he complained that the Netherlands knew nothing of *Privatdozenten*, and that professors were promoted there without regard to their work. He exposed and denounced the corrupt system by which candidates were selected, one in which private influence, favoritism, or in van Deen's case, anti-Semitism, took priority over talent. In his fury he wrote things during the heat of 1848 for which he later apologized: "No Prussia, no Austria, no Schleswig, no Holland, but Germany! – That is the only solution from which the Netherlands can expect its salvation."[23]

Moleschott was completely caught up in the events of 1848. During the winter of 1847–1848 Moleschott gathered regularly over lunch to discuss politics with a small group of young men whose ranks included Berthold Auerbach, Gottfried Keller, Hermann Hettner, and others. Hettner in particular was a favorite of Moleschott's. Approximately the same age, they soon realized they had a great deal in common, in spite of the fact that Hettner was primarily interested in art and Moleschott in science. Both of them had worked through Hegel to Feuerbach, and both used the other to broaden his perspective in *Wissenschaft*, each being convinced that his particular approach represented one branch of a common philosophical root. Hettner took Moleschott to Mannheim where he instructed him in architecture and opened for him a completely new world.[24]

The group frequently assembled socially in the house of Christian Kapp, where they met, among others, Kapp's friend Ludwig Feuerbach. On their first meeting Moleschott politely did not engage Feuerbach in conversation, but one evening when Feuerbach found his views being challenged on physiological grounds, Moleschott, who was quietly sitting across from Feuerbach, came to his rescue and demonstrated that the facts of physiology supported Feuerbach's position.[25] Feuerbach later told one of the Kapp daughters that he would like to see more of Moleschott, and proved his interest and respect a few years later by giving his name to the radical position Moleschott had taken in his *Die Lehre der Nahrungsmittel.*

Baden was perhaps the most liberal atmosphere in Germany in the *Vormärzliche Zeit*, and Moleschott's group knew it. But that did not mean that Baden was free from conservative elements. In fact, even more than the forces of the Right, the group detested, as did Vogt, the Gothaer party, which in Heidelberg represented to Moleschott the paradigm of the impotence of German liberalism.[26] The news of the revolution was a source of great excitement and expectation for these young enthusiasts of freedom. Moleschott expressed his excitement, as we have seen, by means of his incautious and total commitment to German unification. He eagerly read the works of Georg Forster, the eighteenth-century science consultant on one of Captain James Cook's voyages and an enthusiast for the French Revolution. When Feuerbach came to Heidelberg to lecture in December of 1848, it could only raise to a higher peak the unsettled atmosphere of such abnormal times.

In the midst of all this fevered activity Moleschott fell in love. Of Fräulein Sophie Strecker, whom he had met at Auerbach's house, he wrote to van Deen: "During the Whitsuntide break I met a girl in Mainz who, through her own development in religion, life and politics, completely shares my views."[27] Daughter of a free-minded doctor, Sophie demanded of a suitor that he be not wholly unknown, a republican and a free thinker. Engaged in the wild spring of 1848, they married the following March, and lived a happy life together of mutual respect.[28]

When the dust of the Revolution had settled, the yoke of political reaction did not pass by Heidelberg unnoticed. Nevertheless Moleschott had not lost students, and now, with nowhere else to turn, he settled down to research and writing. From 1847 on he had been working at Tiedemann's urging on a systematic treatment of dietary questions.[29] Here in 1850 *Die Physiologie der Nahrungsmittel*, Moleschott's first book-length scientific work, finally came out. The first edition contained Tiedemann's name on its title page along with Moleschott's, for, as Moleschott explained in the preface to the second, revised and enlarged edition (which no longer listed Tiedemann as an author), the first edition had depended largely on the third part of Tiedemann's *Physiologie der Menschen.*[30]

The format and style of the work was somewhat like Vogt's *Zoologische Briefe*, which came out in the following year. It was an extremely fomal and detailed reference work, full of information on the nutritional composition of different foods, on digestion, and on the nutritional requirements of the body, all well stocked with 254 pages of tables.[31] Hints of his later theme of a *Kreislauf* taking place within the body were not absent in *Die Physiologie der Nahrungsmittel*. Food was pictured as the raw building blocks of the body, and oxygen as the building master that prepared and altered the blocks. When oxygen oxidized the protein of the blood, the building matter was formed into cells, nerves or formless intermediate matter. But oxygen destroyed as well as shaped; it ate its own children. The cells and nerves into which oxygen transformed the blood, according to Moleschott, had an equal inclination to unite with oxygen, so that they gradually reduced to simpler compounds and ceased being organic.

What oxygen helped to develop it also must undo, for the law governing its affinity is realized only when protein and fat have been transformed beyond the highpoint of tissue back into carbonic acid, water, ammonia, and nitrogen.[32]

Moleschott and Vogt were good examples of those who believed in the "physiological hegemony of the blood,"[33] for emphasis on the interaction between the blood and the oxygen of the air caused him to conclude that the blood was of primary concern in overall respiratory activity.

In the same year (1850) Moleschott had put out a popular version of the results he had come to in his research. No doubt urged on by his reading of Vogt's books,[34] and by his study of Feuerbachian materialism, Moleschott pulled out all the stops in his *Die Lehre der Nahrungsmittel: Für das Volk*, which was translated into Dutch, English, French, Italian, Spanish, and Russian.[35] In one sense Moleschott wrote the work because of the lesson he had learned from Fleischer and from his examination of the material composition of the human body; viz., that one man was of no more intrinsic worth than another. "I sought to write for everyone because I believed that everyone was capable of the genuinely human,"[36] he wrote in the foreword. In the text he said:

Only because differences in circumstances alter the matter and force of our tools, only because of this can we be different. We are all similarly dependent on air and soil, on men and animals, on plants and stones.[37]

To account for the obvious differences in mental abilities, Moleschott appealed to an idea which Büchner, admittedly borrowing from Moleschott, made his central theme: the unity of *Kraft* and *Stoff*. It would seem, explained Moleschott, that we lack the transitional steps of the proof that differences in force are only differences in matter, but in thousands of instances this has been shown to be true. Force is an inseparable characteristic of the body, conditioned by the particular composition of its matter.[38] Moleschott provided separate diets for artisans and thinkers, for, he declared, it was an error to believe that mental activity did not require the consumption of matter.[39]

Moleschott had little use for those who insisted that the essence of things was something external to them, hovering over them. Since Spinoza, he assured his readers, we have known that the total of all the qualities of a body makes up its essence. Such a separation of matter from force would lead to ascribing an immortal spirit (*Geist*) also to steel or amber.[40] Life itself was defined in terms of changing forms of matter:

> There is no activity without a constant metamorphosis of composition, without an eternal genesis and passing of forms. Therefore I have been able to derive all life from the bonding and breaking up of the matter of our body. Life is an exchange of matter (*Stoffwechsel*).[41]

The tone of *Die Lehre der Nahrungsmittel* was as condescending as that of a doctor advising his patient. His concern for the poor, whose material and mental condition he attributed to the diet they were forced to consume, was open and direct. Rather then deplore the poor for turning to alcohol, Moleschott, unlike virtually everyone else,[42] explained that alcohol helped burn up the food matter in the blood more thoroughly, thus one could eat less and yet receive more nourishment. Of course he opposed inebriation, but he felt that to ban alcohol because some men are drunkards put mankind back into the Middle Ages.[43]

Like a concerned parent Moleschott outlined the requirements of a balanced diet, and often provided housewives with inside information on what to give their families and how to cook it.[44] He even gave instructions to expectant mothers on what to eat and what to avoid.[45] In general he advised the poor, who rarely enjoyed the nourishment of meat, to eat peas and beans as the best way to make up for the nutriments they were missing.

Moleschott acknowledged that while many forces combined to influence man's life, here he was talking about diet. As a modern day physiologist he felt that he was breaking down the barriers that had been erected against the material explanation of life.[46] Spices, for example, could in excess lead to restless passion and insidious jealousy. Had they never entered Europe, the Spanish, Portuguese and Dutch would be able to erase a bloody page of their history.[47] Hunger too was clearly a motive force in the making of revolutions in the past, more so than the causes usually given. Hunger desolated the head and heart and accentuated all other pressures of life. "And therefore it is not *presumptuous desire* which has awakened the belief in a right to work and in a respect for the most conscious creature on earth."[48]

Moleschott deliberately circumvented the contradiction between the call to action and a deterministic world view, most likely because he had not yet become a determinist. On the one hand he did champion the universal and necessary law of attraction by which the delicate balance of nutritional material was controlled.[49] Departure from this balance, he explained, soon showed the injurious consequences that necessarily resulted from ignoring the supremacy of natural law.[50] But on the other hand he did not declare that deliberate change was impossible; rather, he urged that man work in cooperation with natural law. It was no wonder that the poor of all lands were in a position of subjection – look at their diet!

> [The potato] can overload the blood and the tissues with fat, but because it furnishes the blood only sparsely with protein, it can supply no fibrin and no power to the muscles, and to the brain it provides neither protein nor phosphorous fat.[51]

He had explained the idleness of the Italian Lazarone and the mysticism of the Hindus by means of their diet. Now he lamented over the plight of Ireland vis-à-vis England in a passage delicately omitted from the English translation:

> Sluggish potato blood, is *it* supposed to impart the power for labor to the muscles, and the enlivening verve of hope to the brain? Poor Ireland, whose poverty breeds poverty. You cannot win in the struggle against your proud neighbor. . . . You cannot win! For your diet awakens powerless despair, not enthusiasm, and only enthusiasm is able to blow over the giant through whose veins courses the energy of rich blood.[52]

Moleschott had tested *Die Lehre der Nahrungsmittel* on his wife and his friend Hettner, reading it to them section by section to see if it actually had been written "for the people." He later said it was an error to think that it was easier to write for the people than for scholars, for one really had to know the material to be successful at popular writing. The response to the work was favorable. A new edition was called for in 1853, and another in 1858. Alexander von Humboldt wrote the budding scientist a letter of congratulations, adding that Moleschott's work had an emphasis rare in Germany. It had demonstrated that one need not be totally broad in approach to write a popular work.[53]

But the most important review by far was that of Ludwig Feuerbach, which appeared in 1850 in the *Blätter für die literarische Unterhaltung* of November 8. Moleschott had sent a copy of *Die Lehre der Nahrungsmittel* to Feuerbach with the request that he publicly discuss "the relation of my work to general ethical questions."

For I am honest enough to express to you my hope that you will allow my work to serve as one of the blossoms into which the bud of your principle . . . has developed.[54]

Das Wesen des Christentums, Moleschott continued, had been negative criticism, and it would find more general acceptance when replaced with "positive knowledge . . . , which, like all true knowledge, sets one free at the same time it satisfies the demands of feeling."[55]

Feuerbach responded just as Moleschott wished him to – in a scathing and polemical attack against the re-established forces of the Reaction. It was ostensibly a review of *Die Lehre der Nahrungsmittel*, but it served more broadly as an occasion for Feuerbach to denounce the Establishment. 'Die Naturwissenschaft und die Revolution' began with the warning that the censor was making a huge error to exempt natural science from the list of those suspicious disciplines where ideas contrary to official doctrines arose.[56] True, the natural scientist did not deal directly with political matters, but no one could treat natural science as a separate class of ideas which had no effect on the whole. Whoever became convinced that he could no longer believe in natural miracle, declared Feuerbach, would soon not believe in religious miracles either. Copernicus, therefore, was the first revolutionary of modern times.[57] The German government failed to understand that it was not godless philosophy, but Copernicus who had sapped the Bible of its charming power.

Feuerbach referred to Moleschott's work as a timely proof of the revolutionary significance of natural science.[58] It might have seemed harmless, but the way in which it dealt with eating and drinking demonstrated that nutrition was the bond between body and soul. At this point Feuerbach began to carry Moleschott's materialistic emphasis to the extreme, making statements like "Being is one with food (*Essen*). Being is food.", and "The matter of nutriments is the matter of thought." He repeated the cryptic phrase "Without phosphorous, no thought," which Moleschott had coined in *Die Lehre der Nahrungsmittel*.[59] Crass materialism of this sort was not typical of Feuerbach; in fact, he had opposed precisely this unsophisticated philosophical position all his life.[60] Yet, liberally sprinkled with quotations from *Die Lehre der Nahrungsmittel*, Feuerbach seemed not only to have become a total materialist, but to have been carried away with the social and political implications that Moleschott had emphasized, particularly the caustic remarks about the Germans' dependency on potatoes, and the need to relocate this dependency to increase material and intellectual powers. His peroration was reached as he cried:

> We see here also of what important ethical as well as political significance the teaching of the means of nutrition is for the people. Food becomes blood, blood becomes heart and brain, thoughts and mental substances. . . . If you wish to improve the people then give them better food instead of declamations against sin. Man is what he eats. (*Der Mensch ist was er isst.*)[61]

If there had ever been doubt that Feuerbach's materialism and the materialism of the new natural science were partners in sin, there could be doubt no longer.[62]

During the year following the two works on nutrition Moleschott published another reference work with the same general theme as *Die Physiologie der Nahrungsmittel*, but this time broader in scope: *Die Physiologie des Stoffwechsels in Pflanzen und Thiere, Ein Handbuch für Naturforscher, Landwirthe und Ärzte.* The explosion, however, came with *Der Kreislauf des Lebens* of 1852.

Moleschott was not lacking in shrewdness when it came to making a name for himself. What better way to attract attention than to write a popular work against a well known, well respected and well established person? The subtitle of the *Kreislauf* read, *Physiologische Antworten auf Liebig's Chemische Briefe*. The book began with an open letter to the Giessen chemist in which Moleschott paid his respects, but served notice that his approach would dissent from Liebig's.[63] Moleschott had not been certain until the last minute that he wanted to oppose Liebig in the work. In addition to the exchange of letters with Liebig described earlier, Moleschott had met him personally in the house of his father-in-law in Mainz, where they had come to a friendly disagreement regarding the work of Jolly. At one point Liebig had even supported Moleschott as a candidate for a position at Giessen, though for unexplained reasons nothing ever came of it. With the publication of *Die Physiologie der Nahrungsmittel*, Liebig sent his thanks for the copy Moleschott had sent him, but unwittingly revealed that the two were drifting apart by complaining that Moleschott had sided far more with Mulder than with him in his treatment of proteins. When, therefore, the new edition of Liebig's *Chemische Briefe* came out in 1851, in which Liebig deliberately misrepresented what Moleschott had said in *Die*

Lehre der Nahrungsmittel about the presence of phosphorus in the brain,[64] Moleschott made up his mind to make the *Kreislauf* a reply to Liebig.[65] As a result, it was Moleschott whom Liebig had in mind in his condemnation of materialism before King Maximilian II in 1856.

The central theme of the book, written in the form of letters, was hinted at in the title, but it came out in ruthlessly explicit fashion in the text: life was one particular variation on a general theme, and that general theme was matter in motion.

> The motion of the elements, combination and separation, assimilation and excretion, that is the essence of all activity on earth. This activity is called life when a body preserves its form and its general composition in spite of the continuing alteration among the smallest material parts that compose it.[66]

This was the overture to the major symphonic theme yet to come, for the point was that the particular motion of matter that defined life was by nature cyclical, and cyclical in more ways than one. Weaved throughout the book was the cycle of the individual's life, the *"Kreislauf des Stoffes"* in which animal and plant cooperate for mutual benefit,[67] and the cycle of activity by which the species, or life in general was preserved.

To explain these cycles Moleschott appealed to the concepts of *Blutbildung* and *Rückbildung.* Plants, he said, transformed inorganic matter into protein, which, when consumed by animals, was used to form tissues. While the first part of the cycle was accomplished when plants removed oxygen from the raw materials of carbon dioxide and water, the latter phase was completed when oxygen was added to the protein in the blood, when the protein was oxidized. But oxidation did not stop here, it continued its action on the tissues themselves (*Rückbildung*), producing end products of carbon dioxide, water, and urea, thus constantly requiring the formation of new tissue in the animal organism. The difference between plant and animal was that the former robbed compounds of oxygen, while the latter caused oxygen to be added to the plant's product, protein.[68]

This individual cycle did not, however, last forever. There was also another cycle, one that involved the life and death of the individual. Tissues of every individual organism one day ceased replacing themselves, and when that happened a final oxidation of them occurred. The matter that made up the body was then broken down into its elemental forms, perhaps to be taken up into the constant exchange of matter of some other individual organism. Put this way, the life of the individual did not seem so very important.

It seems silly to me, if not vacuous, to find it miraculous that the carbon of our heart, the nitrogen of our brain once belonged to an Egyptian or a negro. The miracle lies in the eternality of matter throughout the change of form. ... For the grandeur of creation ... is that every individual being falls as a sacrifice to the species, that death itself is nothing but the immortality of the cycle.[69]

For page after page Moleschott corrected Liebig's errors. By the end of the book it did not seem possible that one man could have been so wrong. Moleschott now clearly opposed Liebig's view of proteins as the sole building material of tissue,[70] and he criticized his division of nutritional matter into *Athemmittel* and *Baustoffe*. Moleschott denounced this classification because, he thought, it was based on the view that oxygen was a hostile power against which the body must fight; i.e., it involved a teleological perspective.[71]

In addition to specific attacks on Liebig, Moleschott attacked Liebig's defense of vital force and the existence of the soul. The *Kreislauf* had opened with a chapter entitled 'Offenbarung und Naturgesetz,' in which Liebig was soundly castigated for having said in the *Chemische Briefe* that knowledge of the laws of nature points to a Being "for whose contempation and knowledge the senses no longer suffice."[72] For Moleschott this was nonsense. The fifth edition, in which incidentally he dropped the anti-Liebig format, added a clarification which summed up well his own perception of the issue.

Here is the point where humanity is divided into two classes. For the one the power of nature is not only unexplained, but it is inexplicable; for the other doubt rules more than belief, and, by unceasingly eavesdropping on what is in process, it is deemed possible to understand what has already come to be.[73]

The following chapter, 'Erkenntnisquellen der Menschen,' explained more clearly why he thought Liebig's defense of revelation was a basic error. "There is nothing in our understanding which could not have entered through the gate of our senses."[74] Revelation differed from knowledge, declared Moleschott, in separating the cause from the effect by thousands of unknown transitional steps, so that different religions have come up with different remote causes, all driven by the feeling of dependence Schleiermacher and Feuerbach had exposed to be the basis of all religion.

It is like carrying coals to Newcastle if, in the land where Ludwig Feuerbach wrote his immortal critique of the essence of Christianity, one wishes to pile up examples in order to debate the insoluble contradiction between the omnipotence of a Creator and natural law.[75]

Remote causes, therefore, were unreal, and the prime test of their unreality was their inability to be perceived through the senses. Moleschott approached this theme one final time by arguing against the separation of matter and force, for forces could only be expressed by motion in space and time.[76] Since motion had to be motion of something, Moleschott pronounced it the motion of matter. Force was a quality of matter which enabled it to move, and all the properties of matter were the result of combining forces. Most importantly, never did a property arise apart from matter, nor was there ever an "essence" of matter that eluded sense perception.[77] Moleschott was careful to acknowledge that phenomena such as heat, light and electricity were neither forces separated from matter, nor a kind of matter themselves. They were conditions (*Zustände*) of matter which corresponded to different states of motion.[78] He went on to oppose vital force as an unknown power violating the restriction that matter and force could not be separated.

> Life is not the emanation of a very special force, it is much more a condition of matter, grounded in its inalienable qualities, governed by characteristic phenomenal movements which call forth to matter heat and light, water and air, electricity and mechanical vibration.[79]

That one must not separate force from matter, that in doing so one only created something like a childhood wish for oneself, this had become plain in modern times, proclaimed Moleschott, and the Herculean deed had been performed by none other than Ludwig Feuerbach. He had made clear the human basis for all intuition and thinking.[80] Moleschott had sent Feuerbach a copy of the *Kreislauf* in June of 1852, again declaring his debt to him: "A glance at the table of contents will show you how deeply I have ventured into a field which, with your critique, had become arable for the first time."[81]

Elsewhere in the book Moleschott had attacked the ideas of free will and of consciousness as something separate from matter. But his own brand of materialism, like Feuerbach's, was not without religious emphasis. His version of the Golden Rule was taken not from the Sermon on the Mount, but from Madame de Staël: "To understand all is to forgive all."[82] Unlike the bitterly anticlerical Vogt, but like his master Feuerbach, Moleschott found an historic value in Christianity with its emphasis on love, and he denied that the perversions of that value would ever be able to destroy it.[83] He defended his view of life as noble and beautiful. It was a moral obligation to act first and foremost as a member of the species rather than as an individual.

Like Feuerbach before him he transformed the transcendence of traditional religion into a transcendence of the species over the individual. In particular he stressed the generalized concept of matter in motion over the unique form of matter in motion we call life. At the end of *Die Lehre der Nahrungsmittel* he talked of his religion in terms of man's dependence on matter in motion: "It is fitting for the philosopher to recognize this dependence, and it is genuine piety to cherish joyously the feeling of relationship one has with the universe."[84] Now there was an almost ascetic quality about his materialistic attitude towards death.

> There is death in life and life in death. This death is not black and horrible. For swaying in the air and resting in the mould are the eternally smouldering seeds of blossoms.[85]

When he suggested, as a practical application of this, that corpses be used to fertilize fields and that every meal be seen as the eucharist, he ran into great difficulty with the administration at Heidelberg, trouble that eventually led to his departure.

Nor was his book without a heavily socialistic emphasis. All modern social movements, he naively suggested, strove toward proper distribution of resources, though each had pursued its goals in the dark. Not communism, but socialism was the view of the future, declared Moleschott, and the ignorance of the present concerning the distribution of force and matter was being dispelled by the scientist, who held the key to the solution in his hand. Moleschott concluded the *Kreislauf* on this note:

> The natural scientists are the active cultivators of the social question. . . . Its solution lies in the scientist's hand, which is guided with certainty by the experience of the senses. . . . Knowledge is insurmountable power, it is the power of tranquility. Knowledge is not merely the greatest prize, it is also the broadest foundation for a life worthy of human beings.[87]

Moleschott's name was everywhere after 1852. The *Kreislauf* brought him new friends and admirers like D. F. Strauss[88] and Emil Rossmässler. Largely through the efforts of Rossmässler and Otto Ule, editors of the new journal *Die Natur*, the book reached a wide audience. Students from Erlangen, for example, wrote Moleschott that they had formed a group to discuss his book.[89]

Yet in the midst of the fame Moleschott was clearly in trouble. He received a warning in July of 1854 that if he did not cease infecting the minds of the youth with his immoral and frivolous doctrines, his right to teach would be withdrawn. Immediately the young instructor, once more demonstrating his streak of independent pride, resigned from his post rather than be a part of an institution where *Lehrfreiheit* meant nothing. This action rallied many of the Heidelberg students to his cause,[90] but to no avail.

Out of a job, and with children to support, Moleschott lived from the proceeds from his notorious books. He continued to live and engage in research in Heidelberg until 1856, when he accepted a call to succeed Karl Ludwig at Zürich as anatomist and physiologist. He had been chosen in the face of stiff opposition over DuBois-Reymond, winning the post largely because of the advice given the Zürich officials by Rudolph Virchow.[91]

During the turmoil of 1848 and 1849 Moleschott had discovered the figure of Georg Forster, German natural philosopher and botanist who with his father had accompanied Captain James Cook on his second voyage from 1772 to 1775. What attracted Moleschott to Forster, he said, was his love of humanity, but surely one must not overlook the similarity in the religious experience of the two men. Forster, like Moleschott, had come to the world of science from a strict religious background, which in Forster's case was Rosicrucian. Neither ever lost sympathy for religion in spite of an eventual radical religious position and a political sympathy for revolution.[92] In addition, Forster, like Moleschott, had an extremely high regard for anthropology.

Moleschott's portrait of Forster, issued on Forster's hundredth birthday in 1854, bristled with hero worship, but in so doing Moleschott revealed himself as the humanist that he was. What Moleschott respected in Forster was his honesty, his respect and sympathy for others, his ultimate confidence in reason, his rejection of faith, his dedication to freedom, and his Feuerbachian love of humanity.[93] Forster never became an atheist, nor was he opposed to the idea of religion. His religion was more like Moleschott's, and it may well be that Moleschott borrowed from Forster the notion of *Kreislauf* to express his religious attitude towards life. In his *Georg Forster* he quoted Forster to have said: "Today we are nothing more in body and soul than we were yesterday, tomorrow no more than today. Everything is a cycle (*Kreislauf*), all is change."[94] Forster was portrayed as a sensualist and a materialist in Moleschott's biography,[95] as a scientist, humanist, and philosopher, but above all he was what Moleschott himself wanted to be, a "natural scientist of the people" (*Naturforscher des Volkes*).

Moleschott's inaugural lecture at Zürich, 'Licht und Leben,' was the first in a series of lectures he delivered throughout the remainder of his life. With the exception of three nonscientific books, the one on Forster, a later one on Hermann Hettner, and his autobiography, Moleschott wrote no book-length material on science after the *Kreislauf.*[96] His major concerns for the rest of his life were two-fold: teaching, and the writing of an unfinished and never published work on anthropology.

Anthropology he deemed to be the most all-embracing approach to knowledge, again a position directly attributable to Ludwig Feuerbach's influence.[97] Feuerbach himself had once referred to anthropology as "the crown of the natural sciences," and Moleschott indicated that he had taken Feuerbach's man-centred viewpoint to heart in a letter of 1854 in which he talked of his own projected magnum opus: "Such a work will necessarily be carried out from the anthropological standpoint of which you have brought our century to full consciousness."[98] Here in 1854 he told Feuerbach that it would be a long time before his *Anthropologie* would see the light, and again fourteen years later he mentioned that he was still at work on it.[99]

The emphasis on teaching he underscored as he assumed his duties at Zürich.

> And you, honored students, . . . I have little to say to you; for I see teaching as an art in the highest sense, and the artist should put deed before word. . . . Let us increase the examples of master and students living with one another as comrades, comrades in free and independent research.[100]

Moleschott remained in Zürich from 1856 to 1861. During his stay he conducted scientific research on the nerves of the heart, on the influence of heat on respiration and other subjects. Like thc work in his last year and a half at Heidelberg, he often published his results in his own *Untersuchungen.* He spent much of his spare time with Gottfried Keller, Georg Herwegh and Franz List, and he lived for several years in the same building with Rudolf Clausius, to whom he dedicated the French translation of *Die Lehre der Nahrungsmittel.* By this time Moleschott had established an international reputation great enough to attract visitors such as G. H. Lewes, George Eliot and others, while he himself frequented what had come to be a center for scientists at Combe Varin, an estate owned by Vogt's Neuchâtel colleague Eduard Desor.

The move to Zürich had its drawbacks. This was the third country in which he had lived, and since he was apparently reluctant to take Swiss citizenship, he could not vote. Then again, he was not without colleagues who opposed what he stood for. Nor was he part of a genuinely Swiss educational institution, since so many Germans, who in Moleschott's eyes tended to be cliquish when in foreign lands, had been called to fill Swiss academic posts. But the Italian influence at Zürich was also notable, and Moleschott, always interested in new cultures and languages, decided both to study the Italian language and to attend lectures on Italian literature given by Francesco de Sanctis.[101] It was de Sanctis who later as an Italian educational administrator was responsible for calling Moleschott to Turin.

Soon after arriving in Italy Moleschott was made a member of the commission on Italy's higher education.[102] After years of wandering, Moleschott had finally found a fatherland that he was willing to call his own. When he left Heidelberg he had already tired of polemics, but at Zürich he had never been accepted in spite of his willingness to forsake polemical writing and to dedicate himself to teaching and research.[103] In 1866 he became an Italian citizen, and a decade later he was made a senator by King Victor Emmanuel, whereupon in 1879 he left Turin for Rome and remained there until his death in 1893. In his last years his main concern became less and less laboratory research and more and more politics, yet on the celebration of his seventieth birthday in 1892 he received congratulations from great names of nineteenth-century science such as DuBois-Reymond, Helmholtz, and others.[104]

Moleschott died a humanist. By temperament he was unlike his German counterpart Karl Vogt, who enjoyed vicious debate. "To humiliate an opponent was not his style," notes Walter Moser.[105] Moleschott had had his day in the sun in the early 1850's, but soon thereafter he redirected his energies back to teaching and research. And even in his heyday Moleschott betrayed the same religious attitude that Marx detested in Feuerbach. Both Feuerbach and Moleschott vehemently opposed religion as it was practiced in Germany and elsewhere, but neither viewed religion as such to be a pathological condition. His socialism might have been anti-Establishment, but it was far from revolutionary.

CHAPTER V

LUDWIG BÜCHNER: SUMMARIZER AND SPOKESMAN

As far back as the early eighteenth century the Büchners had been medical men.[1] Ludwig's father studied medicine in Holland, where two of his uncles had established themselves as doctors. From there he entered the Dutch and then the French army as a regimental surgeon. After accompanying Napoleon's troops throughout Europe Ernst Büchner studied for a while in Paris before taking up a short-lived Dutch civil service position. In Germany he entered the service of the Grand Duke in Reinheim, and in 1816 became a medical councilor in Darmstadt. Much remains unclear about his life, especially concerning his medical studies outside Germany. We do know that he finished his degree at Giessen in 1811, and that he was married a year later to Caroline Reuss, a cousin of Eduard Reuss, the noted theologian and orientalist. Of the eight children of the marriage, four earned reputations far greater than that of their father.[2]

Eldest of the children was Georg, the only Büchner whose fame has survived into the twentieth century. Although he was almost unknown throughout the nineteenth century, he is recognized today as the author of *Woyzeck* and other dramatic works. As a young medical student at Giessen in the 1830's he became involved in political intrigue and revolutionary activities. His father, a hard working doctor dedicated to liberal causes, was tolerant enough of religious criticism, but not of politics. As Anton Büchner put it:

> The father was open-minded, and, if conscience demanded it, ready to protest. Science (*Wissenschaft*) had made him into a sceptic (perhaps more!) in religious questions, but he was of an obstinate temper when it came to attacks on the existing order of state.[3]

One of the earliest communists in Germany, Georg had to keep his political activities a secret, at least from everyone except his next younger brother Wilhelm.

Unlike either the homelife of Vogt, whose parents were more or less like-minded in most things, or Moleschott, where there was a sharp difference of conviction between mother and father, the Büchner parents represented neither extreme. The father was sympathetic to liberal ideals where religion was concerned, but differed from the elder Vogt, for example, in his political conservatism. Frau Büchner, though not religiously orthodox like Moleschott's mother, did share her sensitivity, compassion, and respect for the integrity and independence of her children. The Büchner father and mother were alike in their religious scepticism, but very different when it came to tolerating their childrens' political radicalism.

Not everything Georg was doing could be kept from his father. In the winter of 1834 he had to flee to Strassburg because of the publication at Giessen of his revolutionary flyleaf 'Der hessische Landbote,' and because of his founding there of the secret Gesellschaft der Menschenrechte.[4] Once in Strassburg Georg turned away from politics to writing, and in 1836 he was named a *Privatdozent* in comparative anatomy in Zürich, only to fall victim to a fatal bout with typhus in February of 1837.[5]

That all this had a profound impact on Ludwig is clear from an autobiographical sketch he has left for us. Georg's activities created an unbearable tension within the Büchner home, one that took its toll on the eleven-year old Ludwig. He later recalled the trauma that had interrupted his otherwise carefree youth:

> I soon came to learn the seriousness of the situation in terrifying fashion, for participation in the political intrigues of that time brought sorrow and worry into the family. How often our father sat at the table angry and bitter, while our mother sat or stood beside him weeping. We children, who understood nothing of it all, would stand silently around our parents. And when finally, after a long time of grieving and worry, our good parents thought they had a ray of happiness in view, death suddenly stepped in between fear and hope and spoke his horror-filled No. Who can describe the distress? Never will that dismal evening go from my memory when the letter which daily brought the report of the progress of the terrible sickness finally brought the news that all was over. Our father was serious and quiet, our poor mother near desperation. The scenes which I must here relive cast the first, though deep and abiding shadow into my youthful heart, which up to then had been filled only with cheerfulness.[6]

Alexander Büchner agreed that this scene was one of the most vivid recollections from the youth of the Büchner sons.[7]

About this time Ludwig enrolled in the *Gymnasium* at Darmstadt, distinguishing himself as an excellent student. While there he showed a certain predilection for philosophy and literature, and later he might have continued in this area had not his *Gymnasium* instructor been such a poor teacher.[8] Ludwig, however, did not know what vocation to take up. He toyed for a while with philology, then spent a year in Darmstadt's *Gewerbeschule*, where he studied mathematics and physics.

When in 1843 he enrolled in the university at Giessen, he was still attracted to philosophy and literature, particularly that taught by Moritz Carriere. But his decision concerning the first priority in his studies had been made before he came to Giessen: he would follow in his father's medical footsteps. In an autobiographical sketch Büchner claimed that his decision to go into natural science was due to his desire to learn more about the Creator of nature. As if this were not enough of a contrast to the youthful attitudes of Vogt and Moleschott, Büchner further called attention to his personal piety during the time he was studying for confirmation in the church.[9]

During a year spent in Strassburg in 1844, most likely at his father's wish, he became disillusioned with the July Monarchy of Louis Philippe, and returned to Giessen where his younger brother Alexander, the constant companion of his youth, was now also a student. Both brothers became active in the student society Alemannia, which Ludwig, Wilhelm Liebknecht and a few others had founded.[10] When the Revolution of 1848 broke out, Ludwig and Alexander worked together with Georg's old friend August Becker on a new journal. Becker had returned from exile in Switzerland, where he had lived since his release from a German prison in 1840. Full of enthusiasm for the revolution, Ludwig had travelled with Becker to Frankfurt as a correspondent for their new periodical, *Der jüngste Tag*, which was meeting with great success.[11] He became a leader in the newly organized volunteer corps of which Karl Vogt was head, and gave himself willingly to the task of getting Vogt elected to the Frankfurt Parliament.

Somewhere in the midst of all this activity Büchner found time to complete his medical degree. With a dissertation entitled *Beiträge zur Hall'schen Lehre von einem excito-motorischen Nerven System*, he took up a discussion of the work of the English physiologist Marshall Hall, which had appeared in the previous decade. His study criticized Hall for neglecting to note that excitomotor activity was not exclusively the province of the spinal cord. The brain too could act as a central organ for reflex action.[12] In general he used Hall's work as an occasion to emphasize that explanations of reflex action should be based on the material substratum. According to Büchner, all reference to consciousness in such an explanation merely confused the issue. If reflex motion was purposeful at all, it was unlike the motions that resulted from the orders of the will, and solely due to the physiological structure of the organs involved. In that case the physiological structures and processes were but a means to an end. The reflex nerve system had nothing to do with the activities of the psyche, nor could it have, for it had an independent life, purely material and operating according to material laws.[13] Like Vogt, who, as a teacher at Giessen in 1847–1848 may well have been an influence on his thought, Büchner looked to a "new physiology" which had done away with all theories "of the force, energy (*Energie*), and activity of the nerves as such."[14]

The nerves for Büchner formed a passive substratum, and that was the message he wanted to get across. The best men of science, he claimed, had long banished into folklore those explanations of reflex action based on a spinal soul (*Rückenmarkpsyche*).[15] When he defended the nine theses of his dissertation the morning of September 9, 1848 this was made absolutely explicit, as was the more general claim that a personal soul was unthinkable without a material substratum. Such statements prompted the presiding official to append a note to the printed version of the dissertation to the effect that the subject had been chosen by the student independently, and that the views expressed in it were not necessarily those of the examiners.[16] Already in his first publication Büchner's conclusions produced only qualified acceptance by his superiors. While the dissertation was no systematic statement of materialism, it certainly did reinforce the association that many people of the day were making; namely, that modern scientific explanation rejected nonmaterial categories however traditional they might be.

Done with the examination, and with his medical degree in hand, Ludwig, or Louis as he was also known, returned to Darmstadt by way of Frankfurt, where he witnessed the tumult of September 18 first hand. Alexander reports that Ludwig's revolutionary fervor fell off rapidly after the failure of the Frankfurt Parliament, especially as he settled down to the routine of a medical practice in his home town. He turned his attention briefly to further education, travelling to Würzburg to hear Virchow for a while, and then on to Vienna. On a different note, he undertook the editing of his brother Georg's works in a volume entitled *Nachgelassene Schriften von Georg Büchner*. This effort did little for his brother's reputation as a writer,[17] though it did bring Georg's name to the public's attention.

In 1852 Büchner landed a post at Tübingen as assistant to Professor Rapp at the medical clinic there. In addition to his regular duties he was allowed to give lectures; yet this period of his life was particularly depressing for him. From his brother's description of his personality we learn that Ludwig was a very emotional person, far more like his mother than his father.[18] He wanted to be optimistic, but everywhere he felt this attitude challenged – in Georg's pessimistic works, in the sickness he encountered daily as a doctor, and even in the materialistic and deterministic philosophical implications he considered the foundation of his vocation as a man of science. His lectures were going well, yet his letters home betrayed an overriding dissatisfaction and vacillation, "and one saw in them," commented his brother Alexander, "that he felt totally alone in that Swabian context."[19] His mother, sensing the depression, decided to come to his rescue by sending his younger brother for company.

Alexander Büchner had majored in jurisprudence at Giessen, and had entered into state service after 1848 in Langen, which lay between Frankfurt and Darmstadt. Because the campaign for democracy was still alive in his heart, Alexander decided to take advantage of London's Great Exhibition of 1851 to meet and talk with some of the German political fugitives who had settled there. On his return he found himself charged with treason and was forced to leave Hesse. For a short time he became a *Privatdozent* at Zürich, but at his mother's request he joined Ludwig in Tübingen to bring him out of his depressed state.[20]

Ludwig began to improve. He busied himself with a work on odic force,[21] that hypothetical power popularized at that time by Freiherr Karl von Reichenbach. Reichenbach believed that od pervaded all nature and could be found not only in selected persons of sensitive temperament, but was exhibited in magnets, crystals, heat, light, and chemical action. He had collected and republished as a book his articles in the *Allgemeine Zeitung* under the title *Odisch-magnetische Briefe* in 1852, and Büchner was no doubt concerned to evaluate this German version of mesmerism as a critic from modern science.[22]

In addition he was enthusiastically occupied with the thirtieth annual meeting of the Gesellschaft Deutscher Naturforscher und Ärzte, held in Tübingen in 1853. It was the inspiration these meetings imparted to him, plus his reading of Moleschott's *Der Kreislauf des Lebens*, that moved him to follow the lead of his own brothers and sister and write down his ideas on paper. Alexander's account of the birth of one of the most popular books in Germany in the nineteenth century emphasized the personal ambition of its author and the significance of the political context of the times.

> Thus passed two highly enjoyable years, in which only an occasional lack of funds was of concern. I had, like my older sister Luise, tried my hand at writing, and we expected that Ludwig would show himself as the family's esthete. But he remained as closed up as the grave. Then he came into my room one morning with a thick packet of papers under his arm, and informed me that he also had made an attempt at paper scratching. "Aha," I said, "have you also caught the family disease? *Il n'en etait que temps*! What kind of waste paper do you have there under your arm? Is it a historical novel like

Alexandre Dumas or Eugene Sue, or a blood and dung smelling drama such as Victor Hugo writes?" For anything else I could not expect from my brother, who up to then had shown his literary talent only. "No," he said, "they are studies in natural science," and assuming a completely "dogmatic" tone he continued: "This kind of thing has strong appeal these days. The public is demoralized by the recent defeat of national and liberal aspirations and is turning its preference to the powerfully unfolding researches of natural science, in which it sees a new kind of opposition against the triumphant Reaction. Look at Vogt, Rossmässler, and Moleschott, all of them are finding good publishers. Let me ask you to read this manuscript and then tell me whether you think I can get a publisher for it, and from which one I could hope for a little success." "What is the title?" I asked him. "Force and Matter," was the reply. "Force and Matter!" I cried, jumping up. "The title alone is worth gold, and every publisher will take the book and pay you for it sight unseen. Send it right away to young Meidinger in Frankfurt. He's the man for it," "You are as always too optimistic," he replied cautiously. "Read it first and then give me your opinion." "That I'll gladly do," I answered, "but I know already what the result will be."[23]

The author spoke his intent in the preface, and he spoke it as one who had decided to end a masquerade by courageously, if sometimes painfully, telling the truth. His dogmatism, if we may call it that, was based on a view of science which from the standpoint of the twentieth century is markedly overconfident.

If these pages dare to attribute a merit or character to themselves right at the outset, they do so in the resolve not to draw back prudishly from either simple or unavoidable consequences of an empirical-philosophical examination of nature, but to stand up for the truth in all its parts. One cannot make a thing other than it *is*, and nothing seems to us more wrongheaded than the attempts of respected scientists to introduce *orthodoxy* into the natural sciences.[24]

The *Kraft und Stoff* was a conscious attempt to bring together into one place and to make explicit all the implications of the materialistic tendencies others had been expressing. It was unlike the major works of Vogt and Moleschott in that it did not appeal primarily to one science like physiology, geology, or nutritional theory. Its aim was general and philosophical, and it was written in a way that did not put off the lay reader with its erudition. Due to its popularity it soon earned the reputation as the Bible of materialism. Here was a book in which one did not have to sift through a wealth of detail in order to get the writer's meaning. It was as if Büchner had already done the hard work for the reader, and now he was telling the significance of his findings in plain talk. No wonder the work went through twelve editions in seventeen years (eventually reaching a total of twenty-one editions), and that it was translated into seventeen foreign languages.[25]

One of the primary targets was the old school of *Naturphilosophie*, which, he declared, had made the word *Naturphilosoph* into an opprobrious term. The approach of this group of thinkers was completely misguided, according to Büchner, for one could not construct nature out of thought.[26] It was not because these philosophers of nature were overly religious that Büchner turned against them, for many of them could be classified as atheists. Rather, it was their speculative disposition he disliked. By mid-century it had become commonplace to speak of speculation as the mistake of a former era which had been replaced by a realistic emphasis on facts.

> The failure of those older philosophical attempts can also serve as the clearest proof that the world is not the realization of some centralized creative Idea, but a complex of things and facts.[27]

"We have to take things as they really are, not as we imagine them to be," quoted Büchner from Virchow, though he might as well have taken it from Feuerbach. Indeed Feuerbach was not overlooked. Not only was he cited as an opponent of speculation repeatedly throughout the book, but the task Büchner had set for himself sounded like an application to natural science of Feuerbach's philosophy.

> Starting with the knowledge of the fixed relationship between force and matter as an indestructible foundation, the empirical-philosophical view of nature has to come to results that decisively ban every kind of supranaturalism and idealism from the explanation of natural events. Its explanations must be conceived completely independently of the assistance of any external power existing outside things.[28]

This in general *was* the program of the scientific materialists. They of course did not think that such an undertaking was biased in any way, for objectivity was their watchword. We shall see, however, that there were several issues connected with natural science for which the Feuerbachian perspective clearly determined their judgment.

Büchner relied heavily on his predecessors. The whole idea for the title and main theme of his book had come from a chapter in Moleschott's *Kreislauf*, and by page four he had cited Moleschott, Mulder, Czolbe, Vogt, DuBois-Reymond, and several others. His message in the first chapter was an elaboration of the presupposition of his whole program: no force without matter and no matter without force.[29] Force in Büchner's eyes was the entity above all others which had been exploited in order to establish the existence of supernatural powers. If he could show from natural science that force could never be separated from a material substratum, he would have relocated this traditionally transcendent category within nature, thereby denying its allegedly supernatural quality.

This was not to say that Büchner refused all reference to things other than matter. Force was not *identical* to matter, as Büchner acknowledged both here and in his next book.[30] But just as thought was not material, it was also a mistake to believe in a "general spiritual matter" or "psychic substance" (*Seelensubstanz*), as Rudolph Wagner put it. "Spiritual matter" made as little sense to Büchner as did the concept of "imponderable matter".[31] Such things were clearly self contradictions. The proper way to think of the soul was as "a special way of expressing the force of life, determined by the characteristic construction of the materiality of the brain. The same force that digests by means of the stomach, thinks through the brain."[32]

In other words, thought was the sum of the individual effects or properties of matter, produced in a fashion analogous to the steam engine. It was the resultant action of a peculiar combination of force-endowed materials. It was not the steam that was analogous to thought, as Vogt would say, but the total effect of the machine. Force was an inseparable *property* of matter,[33] while the imponderables (heat, light, electricity, etc.) were no more or less than alterations (*Veränderungen*) in the aggregate state of matter.[34] Büchner here was walking a tightrope. On the one hand he wanted to deny that force had any existence separate from matter, while on the other he did not wish to say that there was no difference at all between the two. His way out of the problem, at least for the moment, was to declare that they were different things only with regard to our conception.[35]

The young author had addressed several scientific issues of the day, but he neatly sidestepped the polygenist-monogenist disagreement highlighted in Vogt's *Köhlerglaube und Wissenschaft* of the same year. He explained that the important fact was that nature produced man by natural forces, and he merely suggested that if nature could do so in one case, she could do so in several places.[36] Nor did he make a decision between catastrophism and uniformitarianism. He emphasized that gradual changes could accumulate deceptively when an immensity of time was involved, but that catastrophes, though they were not the rule but the exception, have occured throughout history and could happen again. Of the two sides Büchner favored the slow, gradual and cumulative factors, all the while making it clear that no unknown or unnatural forces were required to explain the changes the earth had undergone.[37]

The book had as a subtitle 'Empirisch-naturphilosophische Studien.' It was Feuerbach's *Sinnlichkeit* through and through:

> Any knowledge that stretches beyond the world around us, which is accessible to our senses, any such supernatural, absolute knowledge is impossible and not existent.[38]

Fig. 4. Jacob Moleschott.

Fig. 5. Ludwig Büchner.

Weltanschauung als ein schlagendes Beispiel nennen. Heilmittel in dem Sinne aber, daß sie bestimmte Krankheiten mit Sicherheit und unter allen Umständen vertreiben und so als für diese Krankheiten zum Voraus bestimmt angesehen werden könnten, giebt es gar nicht. Alle vernünftigen Aerzte leugnen heute die Existenz s. g. specifischer Mittel in dem angeführten Sinne und bekennen sich zu der Ansicht, daß die Wirkung der Arzneien nicht auf einer specifischen Neutralisation der Krankheiten beruhe, sondern in ganz anderen, meist zufälligen oder doch durch einen weitläufigen Causalnexus verbundenen Umständen ihre Erklärung finde. Daher muß auch die Ansicht verlassen werden, als habe die Natur gegen gewisse Krankheiten gewisse Kräuter wachsen lassen, eine Ansicht, welche dem Schöpfer eine baare Lächerlichkeit imputirt, indem sie es für möglich hält, daß derselbe ein Uebel zugleich mit seinem Gegenübel geschaffen habe, anstatt die Erschaffung Beider zu unterlassen. Solcher nutzlosen Spielereien würde sich eine absichtlich wirkende Schöpferkraft nicht schuldig gemacht haben. — Um noch einmal auf die Mißgeburten zurückzukommen, so wäre noch anzuführen, daß man künstliche Mißgeburten erzeugen kann, indem man dem Ey oder dem Fötus Verletzungen beibringt. Die Natur hat kein Mittel, diesem Eingriffe zu begegnen, den Schaden auszugleichen; im Gegentheil folgt sie dem zufällig erhaltenen Anstoß, bildet in der falsch ertheilten Richtung weiter und erzeugt — eine Mißgeburt. Kann das Verstandeslose und rein Mechanische in diesen Vorgängen von irgend Jemanden verkannt werden? Läßt sich die Idee eines bewußten und den Stoff nach Zweckbegriffen beherrschenden Schöpfers mit einer solchen Erscheinung vereinigen? Und wäre es möglich, daß sich die bildende Hand des Schöpfers durch den von Willkühr geleiteten Finger des Menschen in ihrer Thätigkeit aufhalten oder beirren ließe? Es kann hierbei nicht darauf ankommen, ob man das Wirken einer solchen Hand in eine frühere oder spätere Zeit versetzt, und es ist nichts damit geholfen, wenn man annimmt, die Natur habe nur den uranfänglichen Anstoß zu einem zweckmäßigen Wirken von Außen erhalten, vollbringe nun aber dieses Wirken weiter auf mechanische Weise. Denn der zweckmäßige Anstoß müßte ja nothwendig auch eine zweckmäßige Folge erzeugen. Ueberdem läßt sich nachweisen, daß auch schon in der allerfrühesten Anordnung irdischer Verhältnisse dieselben Fehler von der Natur begangen wurden. So hat dieselbe nicht einmal so viel Vorsicht besessen, die organischen Wesen jedesmal dahin zu versetzen, wo die äußeren Verhältnisse am besten für ihr Gedeihen sorgen. Im Alterthume gab es in Arabien, wo heute bekanntlich die edelste Raçe dieses Thieres erzeugt wird, keine Pferde; in Afrika, wo das Kameel, das s. g. Schiff

Fig. 6. Remarks on abortion from the first edition of the *Kraft und Stoff*. Büchner omitted the underlined section on abortion and its religious significance from the late editions of his book.

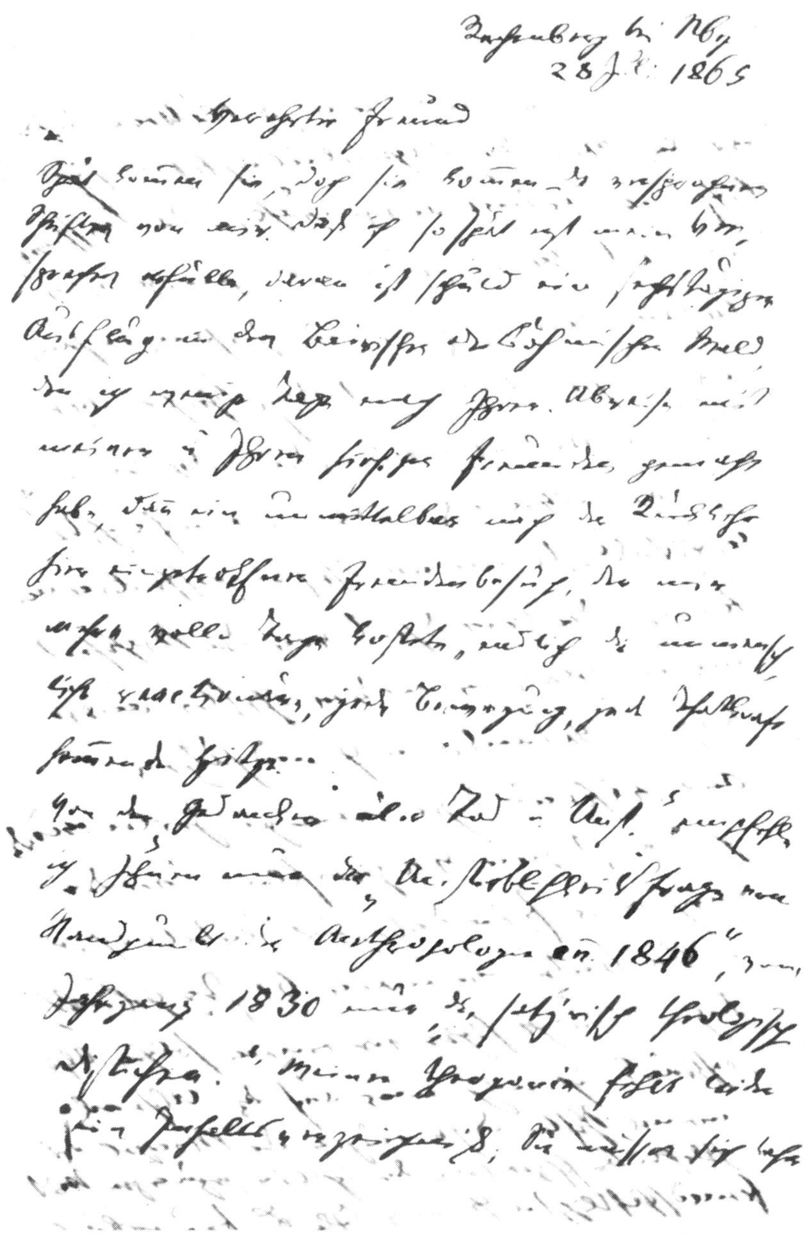

Fig. 7. Letter from Feuerbach to Büchner, written on Feuerbach's sixty-first birthday. On a recent visit Büchner had requested Feuerbach to provide him with material for lectures. In this letter Feuerbach recommends his *Theogonie* of 1860 as "the most easily understandable of my writings, written in the triumph of the certainty of the truth of the thoughts and propositions I have uttered in other works."

As a result of his strict empiricism, Büchner took a stand against metaphysics and the existence of all innate ideas. Even mathematical concepts had for him an origin rooted in experience.[39] Along with this belief in the empirical origin of ideas came a growing conviction, explicit by the fifth edition of 1858, that reason was a reliable and trustworthy tool, that the categories of thought corresponded to the real world because it was from our experience in the real world that such categories were born.

Since [the soul] knows nothing of so-called absolute, supernatural, immediate or transcendent ideas, but achieves all its thinking and knowledge only from observing the objective world around it, therefore being only a product of this world and of nature itself, it cannot but be that the laws of nature are mirrored or repeated in the human soul.[40]

This strong appeal to realism in science formed the basis for his pronouncement on epistemology near the end of the work.

Whoever discards empiricism as such, discards all human understanding whatever, and has not yet once seen that human knowledge and thought without real objects are nonentities. Thinking and being are just as inseparable as force and matter. . . .[41]

Büchner described the battle that was occurring between the forces of materialism and the forces of speculative philosophy. The latter, being dependent on hypothesis and abandoning sense perception as the source of human knowledge, would inevitably succumb. "Hypotheses, however, can never serve as a foundation for a scientific system."[42] Quoting from Virchow, Büchner declared that philosophy, in so far as it was the science of what was actual, could only proceed by the road of natural science.[43] He admitted that empiricism could never give positive answers to transcendental questions. But it could answer them negatively, it could put an end to them.

The conclusion of the work in the first edition was a resounding hymn to the inevitability of truth prevailing, but by the edition of 1858 this ebullient, overconfident attitude was dropped. He closed the fifth edition with a quotation from Cotta to the effect that natural science was dedicated to finding the truth whether it was beautiful or ugly, consoling or disturbing. The quotation was directed to those, "who might feel themselves grieved in their existing philosophical or religious convictions through one or another result of our studies."[44] What the reader would never have suspected was that these words were meant above all to ease the mental anguish of Büchner himself, for somewhere between the earlier editions of the *Kraft und Stoff* and the later ones, with their denial that materialism was an immoral position, Ludwig Büchner sank into a deeper depression than he had ever known.

Denunciations showered down thick and fast upon the young new author. Many of them were predictable, but some must have been unexpected.[45] Secondly, Büchner lost his position at Tübingen. In Darmstadt, with the private medical practice that would occupy him the remainder of his life, he became more and more despondent. He did not particularly enjoy taking the pulse of the sick, especially as a contrast to the relative excitement of academic life. Like Vogt before him, he soon learned how to find fault with the university system that had shut him out.[46] Alexander, looking back to these years, depicted Ludwig's plight as follows:

From then on for brother Ludwig the cleavage between the born ideologue and the materialistic philosopher was felt more and more. In the midst of the great success of his many writings he often complained to me that he had missed his calling, explaining that if he had it to do over, he would not again write his books.[47]

Although it did not appear in print until 1885, and even then under a pseudonym, Büchner wrote down about this time in his life a series of totally pessimistic pieces of *Weltschmerz* entitled *Der neue Hamlet: Poesie und Prosa aus den Papieren eines verstorbenen Pessimisten*. In poetry and prose he poured forth thoughts of misery, death, and injustice. Life itself was pictured as a meaningless cycle in which social injustice seemed to be almost deliberately intended. Büchner's hatred of mankind's apathy towards the poor found vent in biting verse.[48] But most pessimistic of all were his portraits of the vanity and meaninglessness of human existence itself.

We are like dogs on a treadmill. The glowing irons of life prod us to unceasing running on the eternally revolving, yet eternally unmovable wheel, to restless running without goal, until we fall dead from exhaustion in the grave we have made for ourselves.[49]

Included at the end of the work was an autobiographical sketch, which detailed the depths of his moroseness and described his emergence from it. That had been a time of great anxiety, Büchner explained, about which he no longer cared to think. For a while he had hated the world, man, and God; he had nowhere left to turn. He had wanted to penetrate the darkness around him, but could not. Only thoughts of his parents and brothers and sisters saved him from suicide.[50] Through it all Büchner found just one other person with similar despondency, and although he never named that individual, it was through discussions with him that Büchner came out of his depression and pessimism. He spoke of his salvation experience as if he had been a prisoner set free.

I looked clearly around myself. I stood on the basis of a satisfactory philosophy of life and had over me a – no, I had no image over me, only a feeling of the eternal, invisible Spirit of the All permeating me, the Spirit of which the Bible says: "God is a spirit, and they who worship him must worship him in spirit and in truth." . . . I had learned to understand the famous words of Schiller: "What religion do I confess? None of all those you name. - And why none? Because of religion!"[51]

By the time he wrote volume one of the *Physiologische Bilder* in 1861, probably even by his marriage in 1860, he had come out of it. Unfortunately no dates are mentioned, but in the intervening work, the *Natur und Geist* of 1857, his position had so greatly compromised the dogmatic tone of the *Kraft und Stoff* that fellow materialist Otto Ule wrote concerning it to Moses Hess: "This last book really commits the most pitiful betrayal of the unique concern of the author of *Kraft und Stoff*."[52]

Natur und Geist was dedicated to the memory of Georg, who was identified simply as the author of "Danton's Tod." It consisted of nine conversations about materialism between August, a materialist, and Wilhelm, an idealist. August did not emerge the clear victor in the dialogues, a fact that reflects the doubts Büchner was experiencing. What began in the first conversation as a fundamental difference in viewpoint had by the end of the chapter been acknowledged by both participants to have been an overly hasty disagreement. Wilhelm had not hesitated from the outset to make use of suprasensual categories, while August early made it known that an educated man of science could acknowledge nothing supernatural because his research brought him repeatedly into contact with the natural and necessary causal relationship of things.[53]

When in the first conversation the discussion turned to the development of matter into organic forms, Wilhelm declared that here was an example of mind guiding nature, not nature guiding mind. August replied that the crucial thing was not that matter developed into certain forms, but that the laws of development lay in matter itself, not outside it. Wilhelm would not agree that the law of development could be derived from reasonless matter alone. August, on the other hand, felt that to the extent nature conformed to reason it did so *unconsciously*; the categories of reason were themselves products of man's experience of reality. When the discussion ended the two conversers parted, each with more respect for the other's position. The unsettled question between them was the same one that distinguished Feuerbach from Hegel: to eliminate or to preserve transcendence in human experience. "You look for eternal reason more outside and prior to nature," said August as they parted, "I more in and through nature."[54]

The first conversation set the pattern for those that followed. In each of the remaining topics Wilhelm continued to expose the metaphysical nature of materialism by forcing August to the point of his Feuerbachian disposition. August, said Wilhelm, did not want to acknowledge the existence of suprasensual categories, and that was the reason he did not permit them. Materialists might think their position was ultimately logical and based solely on facts, but in the last analysis their argument led back to the presuppositions of the Feuerbachian predilection.

In these dialogues Büchner did not hide the difficulties in the materialistic position. On the contrary, he argued the idealist's point of view persuasively. His own vacillation was evident from his failure to resolve the differences between August and Wilhelm. Wilhelm, for example, suggested that he could just as easily make *Kraft* the primary entity from which *Stoff* was derivative, that in effect he could construct an entire dynamical system.[55] He confessed that he defended this claim not because he himself felt that *Kraft* was the primary basis of reality, but to point up the insufficiency in the opposite approach. The mistake, Wilhelm said, was to assume that the phenomenal world was equivalent to the essence of things. Force might be identical to matter on the level of phenomena, but one could not demand that in their essence they were two separate things. There was, declared Wilhelm, a basic dualism in the universe,[56] and August was ignoring it by inferring that no immaterial force existed anywhere simply because none existed in the phenomenal world.

When August finally saw the distinction Wilhelm was trying to draw, one that separated August's metaphysical assumptions from the epistemological categories Wilhelm had identified, he reluctantly agreed that there might be limits to knowledge, but revealed that he did not think Wilhelm's distinction was significant at all.

> But just as certainly also I know that in the things where we do have knowledge . . . it is an indispensable duty for those who love the truth to adjust their philosophical and religious consciousness accordingly.[57]

Truth was one thing, and philosophical niceties were another. Büchner would have us believe that he understood the epistemological objections that were being raised by the Neo-Kantians and others, but that he did not take them seriously.

August's attitude Wilhelm called the perspective of the natural scientist.

His eloquent rebuttal to it showed that Büchner knew well the arguments of the other side. The human spirit could not be confined to facts, said Wilhelm. One must learn where the limits of knowledge are located in order to distinguish factual knowledge from the unsuppressable musings and conjectures of faith and speculation.[58]

Throughout the rest of the book the emphasis shifted back and forth according to Büchner's indecision. On some occasions August recognized that his own view involved nondemonstrable assumptions of a metaphysical nature, while on others he blindly dismissed Wilhelm's analytical qualifications with a sweep of his pragmatic hand. In chapter three they fell to discussing atoms. Wilhelm pushed August to admit that atoms were conceptually an impossibility since one could always subdivide them in thought. August was forced to conclude that thought in this case was misleading, for in reality matter was obviously not infinitely divisible.

> To assume infinite divisibility is absurd; it means as much as assuming nothing, and it places the existence of matter itself in doubt – an existence which in the last analysis no unprejudiced person can deny successfully.[59]

The scientist did not care about such theoretical difficulties, said August. With incredible inconsistency he explained that matter must have a lower limit or the structure of crystals would be totally unthinkable![60] August's utilitarian criterion for truth finally came clearly out into the open.

> What our speculative philosophers have imagined up to now about the essence of matter is worth less than nothing, for they have been harmful instead of useful, and have veiled the truth instead of disclosing it.[61]

The limits of knowledge were evaluated further in discussions of motion, form, space and time, natural law, creation, and purpose, always ending in a draw. Wilhelm challenged August's sensualism by pointing out that such postulates as the eternity of matter or the unchangeability of natural law were inferences, not immediate results from sense experience.[62] Once again August was forced to appeal to emotion. He declared that Wilhelm knew natural science only from hearsay, not from first hand contact,[63] and that only an unreasonable man would see things the way Wilhelm did. Wilhelm made him laugh with his need to escape clear and empirical reasoning to preserve illusions.

At the end of the work the disagreement had not gone away. Wilhelm still refused August the right to his metaphysical move from the inconceivability of an idea such as immaterial force to its nonexistence or impossibility,[64] and August displayed the same impatience he had shown throughout with Wilhelm's cautious qualifications.

The only semblance of a resolution came when August granted that reason was active in the world, and Wilhelm conceded that its action was mechanical and unconscious. When Wilhelm then went on to suggest that reason might not always have been unconscious, but that the history of nature was an expression of the gradual re-emergence of the consciousness that had been lost, August reminded him that he was speculating again.[65]

With Büchner as well, the high point in his life was reached before Darwin's *Origin* of 1859. For the rest of his life he wrote, lectured, travelled, and practiced medicine. His reputation was made by the work of 1855; everywhere he went he was known as the author of *Kraft und Stoff*. Once he had overcome his doubts and completed his conversion to materialism he gained the new confidence that so often accompanies a conversion experience. He became willing to give himself to the cause of materialism in spite of intellectual difficulties. After 1860 he began to turn out popular books on science one after the other, so many that Marx said of him "He is clearly a 'bookmaker,' and he is probably called 'Büchner' because of it."[66]

His first literary product after his marriage in 1860 was the first volume of the *Physiologische Bilder*, in which his intent was primarily to disseminate information. Though not without a few lessons from science concerning the diet of the poor, etc., the book was more a reference tool in which the work of Virchow, Schleiden, Schwann, von Baer, Buckle, Darwin, and others was laid out in layman's terms. He did not miss the opportunity to side with Moleschott against Liebig,[67] nor to re-emphasize the opposition between supernaturalism and realism. Mainly, however, he instructed his reader about the workings of the heart, the lungs, the blood, etc.[68]

Here in the *Physiologische Bilder* Büchner began to talk about his philosophy in terms he had employed briefly in the *Natur und Geist*; viz., in terms of the opposition between monism and dualism. Each new discovery in natural science led nearer to the great truth that nature was an immense, single, indivisible whole extending to infinity in uninterrupted relationships of cause and effect.[69] Later, indeed by the second volume of the *Physiologische Bilder* in 1875, Büchner would prefer the label monist to materialist, and in this at least he was not unlike Ernst Haeckel.[70]

It is unclear whether Büchner was trying to make some future historian's work easier, or whether he was simply an egoist. In any event he had the habit of gathering together articles he had published and issuing the collection as a book. This he did, for example, in 1862 in a book entitled *Aus Natur und Wissenschaft*. Volume one included journal articles, mostly essay reviews of books, that appeared between 1856 and 1860. Volume two, which was not published until 1885, covered the interim period. These reviews are an invaluable source of information about Büchner, being a running account of his attitude towards many diverse topics of the day. One can find him here criticizing Comte's positivism, recalling the catalytic significance of Humboldt's *Kosmos*, or reviewing Moleschott's *Kreislauf*.

By the time of the publication of the second volume Büchner was 61, firmly settled into his role as the popular spokesman for materialism. Outside of academic life, he was merely tolerated by university professors, especially by philosophers who thought his position superficial and long ago discredited. In several of the pieces from volume two Büchner attacked the professional scholars for their attitude of superiority towards popular knowledge. He replied to charges that his program of general education was more harmful than good by denouncing the specialist who so lost himself in detail that he had no overview at all.[71] Büchner depicted the state of education in late nineteenth-century Germany as deplorable, and became involved in a project to found a free educational institution in Frankfurt to help correct the situation.[72]

Nor did scientists escape his criticism. Virchow was called to account for his views on the theory of descension and for his attitude regarding the teaching of this theory in the schools. Wilhelm Wundt came under judgment for what Büchner considered an inconsistent materialistic position. Wundt's ultimate acknowledgment of a special soul was something Büchner could neither understand nor sanction.[73]

During the last half of the decade of the 1860's Büchner dedicated himself patriotically to the service of his country as a doctor to the sick and wounded of war, and for this he received the only official recognition his fatherland ever granted him. As his reputation spread, he was often in demand as a public lecturer throughout Germany and even in the United States. During the winter of 1872–73 Büchner toured 32 American cities and delivered approximately 100 lectures on various topics relating to natural science.[74] Within Germany he travelled and lectured extensively, becoming an honorary member of no less than 15 scientific and free thought societies.

In 1881 Büchner founded the Deutscher Freidenkerbund, and became an active participant in the international free thought movement.[75] On July 13, 1886 he delivered a lecture at the dedication of a monument to Diderot in Paris in which he emphasized the internationality of free thought, and in which he called for the solidarity of all peoples in the struggle towards impartial truth. When he made reference to the *Völkervereinigung* that the future would bring in spite of the hatred sown by the barbarous practice of war, he called down upon himself criticism not unlike that Vogt had endured for his antimilitaristic views.[76]

Opposed to capital punishment, to inheritance laws, to the existence of ground rents, and to many other aspects of the capitalistic society of his day, Büchner remained a humanistic, liberal critic of the treatment given the poor by the wealthy. But he opposed with equal vigor the forced self-help tactics of communism and of social democracy. As co-founder of the Deutscher Bund fur Bödenreform, as first speaker of the Darmstadt Turngemeinde for 30 years, as a town councilor in Darmstadt for 9 years, and as a member of the Hessian Landtag for 6 years, he had become directly involved with the political process. Unlike his elder brother Wilhelm, a prosperous director of a dye industry, he never became a member of the Reichstag, mainly because he could not afford it.[77]

During the course of his life Büchner wrote some 19 full-length books and published numerous pamphlets, but most of his works after the 1850's repeated and elaborated on themes he had dealt with earlier. A notable exception was the *Liebe und Liebesleben in der Naturwelt* of 1879, an exaggerated hymn to the all-pervasive principle of love in the universe. In this book the idealistic monist clearly showed through, for the heights to which love was raised were rapturous. Büchner spoke of hydrogen "loving" chlorine, and even explained gravitational attraction in terms of love. All this was done as an application of his monistic belief that everything in the universe was related to everything else.[78]

Büchner wrote for the people, and that was the secret of his success. His work was, in his words, "not for scholars, but for the people or for the entire educated public."[79] He played a role similar to the one he had assigned his mentor Feuerbach, whom he visited in the latter's waning years. Feuerbach had emancipated humanity from the chains of superstition through his writing.[80] After his death it was said of him by Hugo Ganz, an enemy of all Büchner represented, that he had done more to popularize natural science than all the universities with their professional scholars.[81] Whatever one thought of him, there was no way to avoid the conclusion that he had had a great influence on the image science carried in nineteenth-century Germany.

On his gravestone was etched a verse taken from a late edition of the

Kraft und Stoff. It harked back to the turning point in his life – the period of crisis he had had to overcome. He had done it by coming to view his doubts as so many riddles, useless musings that no man could settle. As a scientist he was convinced that he could avoid the tantalizing realm beyond the limits of the senses. He would always remain amidst the true and the real, from whence all the needs of humanity would be met. The Why was unimportant – it was the How that counted.

Das Warum wird offenbar
Wenn die Toten auferstehen,
Doch das Wie ist sonnenklar
Wenn die Welt wir recht versteh'n.[82]

CHAPTER VI

HEINRICH CZOLBE: IRREFÜHRENDER MATERIALIST

While Vogt, Moleschott and Büchner were the most well known and the most prolific authors of materialism, the title "materialist" was frequently attached to others as well. Some of these were well known for their work in other connections, but their pronouncements on the significance of science for society were sufficient to earn them the epithet materialist. Such was the case with David Friedrich Strauss, the Tübingen Biblical critic, and Moses Hess, the Jewish socialist. Strauss, although he never forsook his Hegelian training,[1] found nevertheless that his *Der Alte und der Neue Glaube* of 1872 was denounced as a materialistic book. Hess turned aside after 1848 from his devotion to the socialist movement and began studying science. From 1852 to 1855 he attended lectures of Parisian scientists and read the works of Vogt, Moleschott and Büchner, whereupon he took up his pen and began writing about the materialistic implications of science in the popular journals.[2]

Others were simply minor figures who had in one way or another taken up the materialistic cause. Otto Ule and Emil Rossmässler[3] devoted themselves to the popularization of science and the propagation of materialism through journalism. The colorful Austrian *Bauernphilosoph* Konrad Deubler, on the other hand, made materialism more a personal than a public creed. Friend of Feuerbach, Vogt, Moleschott and Büchner, Deubler remained firmly convinced all of his life of the importance of scientific materialism for the education and betterment of mankind.[4]

One of the minor materialists is of special interest, however, because of his unusual understanding of scientific materialism. He is Heinrich Czolbe, often considered along with Vogt, Moleschott and Büchner as a major champion of scientific materialism in the 1850's.[5] From Vogt's, Moleschott's, and Büchner's point of view Czolbe was *ein irreführender Materialist*; i.e., he was responsible for confusing the issues and for leading people astray. As we shall see, Czolbe's conclusions were in fact strange and extreme, but Czolbe's work helps to remind us that scientific materialism could lead different individuals to very different results.

Born in 1819 near Leipzig, Czolbe was reared within the Protestant tradition. At thirteen he went off to the *Gymnasium* at Danzig where he fell under the influence of Theodor Mundt. Mundt's *Madonna* of 1835 proved to be a good preparation for Czolbe's later encounter with Feuerbach, for it "combined criticism of society and religion with a plea for humanism."[6] In 1840 Czolbe entered the university at Breslau, where during his three semester stay he entered a *Burschenschaft* and made friends with young Prince Esterhazy. With the Prince he travelled on foot from Breslau to Poland, Austria and Hungary.

In the summer term of 1842 he transferred to Heidelberg, where he began medical studies. No mention is made by either Czolbe or Moleschott of mutual acquaintance during the short overlap of their Heidelberg years, but since Czolbe soon moved on to Berlin to finish his degree in medicine, it is most likely that they did not meet at Heidelberg.

Under Johannes Müller Czolbe completed his dissertation at Berlin in 1844. Once again there is no record of contact between Czolbe and other of Müller's students who were in Berlin at the time, specifically Helmholtz, DuBois-Reymond and Brücke. There can be little doubt, as will become evident, of the influence the Berlin atmosphere had upon him.

Once graduated Czolbe elected to set up practice as a private physician. Then in 1848 he decided to join the military, beginning his service in Freiburg. Later in 1859 he was promoted to staff doctor and transferred to Spemberg. In 1860 Czolbe successfully completed the state examination for a position as head staff doctor in Königsberg, where he established his close friendship with the philosopher Friedrich Überweg.

All his life Czolbe retained an active interest in philosophy and the development of science, contributing himself to the field with books and articles from the early 1850's on. When in 1868 there was a reorganization of the army that threatened to interrupt his ability to pursue philosophy, he decided to retire.[7] Free to engage in what he liked to do best, Czolbe spent a year in Leipzig in interchange with philosophers and scientists. He almost never missed an opportunity to participate in such activity, and rarely was absent from the annual meetings of the Gesellschaft Deutscher Naturforscher und Ärzte. Throughout his life he remained unmarried, although he did become the guardian of Überweg's children after the latter's death in 1871.[8] He continued writing until his own demise four years later, leaving his last work uncompleted.

Czolbe began his literary career in 1854 with a reply to a critique of materialism by Immanuel H. Fichte, son of the famous German philosopher

Johann Fichte.[9] With one exception the article was reprinted in Czolbe's book, *Neue Darstellung des Sensualismus*, a year later. In the article Czolbe agreed with Fichte that the recent revival of materialism was not necessarily the result of the rise of natural science, for even most scientists concurred that qualities of sense were more than simply matter in motion. Sensations, agreed the scientists, depended on a purely subjective arousal of the internal state. But while Fichte was using this as a defense of a nonmaterialistic point of view, Czolbe was reluctantly agreeing that scientists, in assuming that philosophy could explain whatever appeared impossible to account for in science, were merely slavishly bowing to an age-old practice.[10] The *Neue Darstellung des Sensualismus* was meant as Czolbe's answer to such scientists.

In his preface he admitted that it might be easier to assume unknown, suprasensual categories to explain the world. He even allowed that a system based wholly on sense experience might be impossible.[11] One thing, however, was certain: he was not satisfied that the work of Feuerbach, Vogt and Moleschott had systematically and adequately represented materialism.

> For example, when they say that matter is the substance and cause of all phenomena, but do not give a satisfactorily clear concept (*anschaulichen Begriff*) either of matter or of the manner in which everything arises from it, then their materialism is little more than unintelligible talk, just as dark and incomprehensible as the suprasensual assumptions of their opponents.[12]

Czolbe's own attempt was based on two convictions. First, that one must accept only results that were *anschaulich*; i.e., intuitively clear. Over and over again he employed this Cartesian criterion. Secondly, he was convinced that whatever was intuitively clear was devoid of suprasensual elements. Czolbe announced his program in the opening lines of his book. The basic principle of sensualism was: "in all thinking to eliminate the assumption of suprasensual things (*übersinnlicher Dinge*)."[13]

Czolbe maintained that there was a choice to be made prior to any attempts to explain the phenomena of human experience, a choice which would determine the nature of the explanations given. He allowed three different kinds of explanation of unknown causes arising in science. One could declare the cause to suprasensual (dualism), one could reject the idea of a suprasensual cause altogether and seek the explanation within the sense world of cause and effect (sensualism), or one could simply name the cause without deciding whether it was *anschaulich* or *übersinnlich*.[14] Since the third alternative was not really a conclusion, but a suspension of judgment in Czolbe's eyes (requiring a relativity impossible to a naive realist), his conclusion was that a choice must be made at the outset which of the first two alternatives one would adopt.

One could opt for explanations containing suprasensual categories. Czolbe acknowledged that as far as logic was concerned this could be done without flaw. Such a decision was based on the possibility of the existence of suprasensual causes; hence no logic in the world could disprove them, the best efforts of Strauss, Bauer, Feuerbach, Vogt, and Moleschott to the contrary.[15] Why then would one choose to adopt the sensualist presupposition, which excluded the suprasensual, if one approach were of no more logical value than the other? Because, replied Czolbe, some people prefer the clear to the unclear, and Czolbe explicitly identified the intuitively clear with sense knowledge and the unclear with the suprasensual.[16] Wheareas philosophy and theology possessed partial clarity (in fact theology was clearer than philosophy because it persuaded more people than did the confused formulations of philosophy), sensualism was the consistent position for those who value clarity (*Anschaulichkeit*).

> He whose need for clarity of concepts, judgments and inferences is *unlimited* will decide for, or feel himself driven to the uniform principle of sensualism.[17]

Like his critic Fabri, Czolbe saw the choice for sensualism to be a moral one, something that Fabri appreciated and quickly pointed out. The difference, of course, was that the two chose opposite moral axioms. On the question of the existence of a supernatural soul, Czolbe denied that it had been either physiology or the principle of the exclusion of the suprasensual that had caused him to reject it. It was rather his moral obligation to the natural world order,[18] exactly the same reason given by Rudolph Wagner for the opposite position.

Sensualism had other things in its favor too. Besides being the position that demanded clarity, it also was a unified and harmonious representation of our conscious experience; i.e., it did not suffer the deficiencies of dualism. Czolbe confessed here the influence of D. F. Strauss, but more broadly he showed himself to be at one with the monistic goals of the Young Hegelians in general, most of whom he had read.[19] Sensualism to Czolbe meant realism, because sensualism did not cease its inquiry until it had found "the objective foundation" of even our inner experience. Sensualism did not permit the basis of experience to sway uncertainly in the air, but went directly after the physical cause.[20]

In working out the implications of his system, Czolbe demanded that the requirement of clarity (*Anschaulichkeit*) be met by all concepts, judgments, and inferences.

> The operation of eliminating the suprasensual from thinking consists in permitting distinctly conceivable, or obvious concepts, judgments and inferences to be constructed from sense perceptions alone, and in rejecting on principle every conclusion built on something not imaginable, every indistinct concept.[21]

But what about thinking itself? Was it not, as the idealists maintained, the clearest witness to the suprasensual? Here Czolbe hedged, for he argued that while there was a distinction between thinking and perceiving, thinking was more or less a stand-in for direct sensation, and as such it too should be *anschaulich*.

> Thinking of something is only an expedient for the immediate sensual perception of it, hence clear thinking (*das anschauliche Denken*), which stands next to perception, is also the best thinking.[22]

Czolbe was struggling here with the problem he had inherited from Feuerbach. Whereas the latter had solved the problem by minimizing the distinction between thinking and sensing, Czolbe attempted to retain the difference, but demanded that thought not create anything on its own. For some unknown reason Czolbe felt that in this he had escaped the deficiencies "of raw, or crude empiricism, crass sensualism, common mechanism, etc."[23]

To Czolbe a clear explanation was one based solely on mechanical action.[24] He tackled, for instance, the problem of the mechanical transmission of nerve stimuli. It had been argued, on the basis of Johannes Müller's notion of specific energy, that a physical explanation of sensation was impossible, since several different kinds of stimulation of the nerves (electricity, pressure, etc.) all resulted in the same sensation. What occurred, therefore, could not be a simple mechanical transmission of a physical stimulus. Vogt and Moleschott had not attempted to deal with this objection to the mechanical explanation of sensation, and Czolbe called them to account for it. His own solution was to declare that the nerves possessed a characteristic elasticity which transmitted only the motion or effect corresponding to the sensation, reflecting all other stimulation. In this way a particular frequency in the mechanical transmission could be associated with the stimulus, but the stimulus did not always have to be of the same kind.[25]

The *Neue Darstellung des Sensualismus* detailed the ridiculous extremes

Czolbe was forced to in order to come up with a mechanical explanation for virtually all natural phenomena. If he had an outstanding desire it was to be logically rigorous. The most noteworthy example, and the one that landed him in trouble with the philosopher Lotze, was his mechanical account of consciousness.

Like many of his day, Czolbe made a distinction between so-called inner and outer experience. The former referred to the mental world believed to originate within the intellect, e.g., reflection, and the latter to the thoughts and feelings awakened by stimuli from the outside world, e.g., perception. To an idealist there was in the last analysis only inner experience, for even perception required a subjective input in order to be apprehended. As a materialist Czolbe went to the other extreme. All was outer experience for him. Reflection and perception both could be explained by motions communicated to the brain from the outside world. The whole goal of Czolbe's system was to account for perceptual differences by means of quantitative differences. It should be noted, however, that "quantity" to him involved two elements of mathematics: number and form.[26]

To explain consciousness Czolbe made reference to a particular kind of motion, motion that turned back upon itself. In this way and only in this way did Czolbe feel that the unity of consciousness could be pictured and explained in a manner that was clear (*anschaulich*).[27] It was the brain's task to produce this self-returning direction to the motion within it; hence it was the structure of the brain that determined consciousness. The impact of repeated external impressions gradually created permanent changes in the molecular structure of the brain, and Czolbe used this approach to account, in a vague and unsatisfactory exposition, for the association of ideas.[28] Further, he claimed to have shown how concepts, judgments, and inferences originated "with physical necessity" from perceptions.[29]

Czolbe constructed an elaborate system to explain the mental world. Different qualities of perception were associated with varying kinds of motions, and the different intensities of these motions were correlated with the basic feelings of need (low intensity), pleasure (moderate intensity) and pain (high intensity). All mental activity possessed one quality in common called consciousness.

The will he explained by means of the basic feelings of need and pain corresponding to low and high intensity sensations. Through trial and error humans had learned to associate muscular movements with the alleviation of pain and the satisfaction of need. "The origin or beginning of an act of will is always a need or a pain to which a certain judgment is joined by association."[30] Needs and pains did not, therefore, have to be associated with the body alone. There were also intellectual needs that demanded satisfaction; e.g., the need for sociability on the one hand and egoistic requirements on the other.[31] The point was to eliminate the suprasensual notion of an immaterial will freely directing the individual.

The opposing needs of sociability and egoism corresponded to what Czolbe called the moral and egoistic will, each acting to the exclusion of the other. If social needs predominated in the individual then the egoistic needs were held in check and vice versa.

Czolbe demonstrated at this point that his allegiance to the social establishment of his day in no way would be threatened by his non-establishment world view. One would expect him to conclude, on the basis of his physical explanation of social and egoistic needs, that the individual's action was not subject to moral judgment. At one point he did in fact imply as much,[32] but by associating social needs with the "moral" will, it was evident from the outset that Czolbe would never permit the individual to escape responsibility. He went on to defend the punishment of criminals, even capital punishment, and to maintain that moral freedom consisted in the domination of social needs over egoistic ones. He urged the need for moral education, making an exception only of the insane as far as moral responsibility was concerned. All this fell somehow out of his entire system, and proved, according to Czolbe, that ethical responsibility did not depend on the assumption of an absolute free will.[33] The structure of German society was so far beyond reproach to him that he could not challenge it. It never occurred to him to question the world of traditional values.

On the basis of this position Czolbe constructed what he considered to be an immediately intuitive economic, legal, and moral model for society.[34] In it he defended the existing capitalistic division of labor, free competition, and rights of inheritance, though he buffered these principles with an appeal to equity (*Billigkeit*), which called for leniency and philanthropy for those who could not work. He rebuked those cynics who doubted the good intentions of the leaders in power, and referred to the ideal towards which the state was to strive. Here in the first book Czolbe held that no one form of government was intrinsically better equipped to achieve that ideal than any other, only that the state was a necessary part of the eternal world order. Later he changed his mind to a preference for monarchy.[35]

The question remained, would his conservatism carry over into his position on religion? Strangely enough this self-proclaimed atheist adopted

Schleiden's defense of historical Christianity. In general Czolbe felt that the course of history was proof that suprasensual systems naturally succumbed to the onslaught of the natural sciences,[36] but this did not mean that Christianity had played no constructive part whatever. The Christian religion, he said, had been responsible for the emancipation of mankind from the darkness of mythology. Czolbe went even one step further in claiming that Christianity had laid the foundation for a nonreligious natural science, politics, and morality.[37] Naturalism was in his view entirely consistent with the existing state institutions and the presence of an external church. It was in practicality "strictly conservative."[38]

As a philosophy of the real, the factual, and the natural, sensualism, Czolbe felt, was not in contradiction with traditional values; rather, it provided a better way to uphold them. One whole part of his book was dedicated to demonstrating that materialism, or sensualism as he preferred to call it, did not result from self-love, but that "the essential demands of Christian ethics are also those of sensualistic ethics."[39]

The unusual position outlined above becomes slightly more understandable when one realizes the strong naive realism of its author. Czolbe felt that he had found the "clear" explanation of the riddles of human experience, and therefore that he had found the truth. He saw but one alternative to his approach; viz., to cast aside his entire sensualistic system in favor of a dualism based on free will. Free will, however, was not only a suprasensual concept, but it involved a self-contradiction. Complete freedom of the will meant to him that one could will and not will the same thing simultaneously,[40] and Czolbe trusted too much in reason and the rationality of the universe to tolerate any solution such as that.

If Czolbe's brand of materialism proved that materialists did not always scientifically deduce political radicalism from their position, it also showed that materialists disagreed about other inferences they made from their world view. Here in the first book Czolbe's position on most things was materialistic enough. If anything, his insistence on spelling out the mechanics of organic and psychological processes went far beyond the others.

In the second part of the work Czolbe tried to produce a general mechanical explanation for physical phenomena, in particular light, heat, electricity, magnetism, cohesion, chemical affinity, etc. To explain the imponderables he made recourse to different vibratory motions of a basic substratum. To him the assumption of a special matter of heat, electricity, magnetism and light represented the tactics of suprasensualism compared to his own mechanical, intuitively clear explanation.[41] Nor was the substratum necessarily an "ether" different from ponderable matter; the ether was an idea too similar to the suprasensual soul. That there had to be some kind of substratum was certain, but Czolbe thought he could account for the different speeds of sound and light, for example, by two simultaneous wave systems in air. To do this he had to deny that Torricelli's vacuum had been perfect, and to extend the earth's atmosphere in rarefied form all the way to the sun, but since Czolbe was willing to draw unusual conclusions in order to remain consistent in his own eyes this was not an impossible move.[42]

If this kind of speculation sounded strange for a materialist, Czolbe's other major disagreement with his fellow scientists brought even more delight to the opponents of materialism. A rigorous application of the elimination of all suprasensual ideas from science meant to him that he could not accept any notion of creation whatever, be it by a divine agent or via some natural process. This meant, contrary to the nebular hypothesis, that the earth had no origin.

> If one views the elimination of the suprasensual as the basic principle of sensualism, then giving up the hypothesis of an origin of the world ... is a necessary component of the sensualistic interpretation of the world.[43]

When it was spelled out, this pronouncement meant that Czolbe was not only unwilling to accept the nebular hypothesis, but also the decline in the heat of the earth's interior, the period in the earth's history when it was without life, spontaneous generation, and the interpretation of the fossil record that supported the development of life forms from a primitive to a complex state.[44] Czolbe believed that organic matter had no more of an origin than did inorganic matter, for organic formation always presupposes plant structure.[45] Therefore one would never be able to produce organic material by means of retorts and crucibles.[46] To admit an origin to organic life on earth meant that one had to believe in spontaneous generation, by which Czolbe invariably meant abiogenesis. He used over and over again one method of reasoning to defend his conclusions: "From the unacceptability of the consequences one may infer the falsehood of the premises."[47]

Naturally he rejected the idea that species could change from one into another over time. He charged that it was impossible to make any intuitively clear concepts at all out of such an origin of species.[48] With regard to Carnot's law that heat flowed only from warmer to colder bodies and could be transformed only partially into mechanical work, and regarding Helmholtz's use of this to predict a heat death of the universe, Czolbe conveniently inverted his method of rejecting premises on the basis of unacceptable consequences: "This conclusion is to be rejected decidedly because of the very doubtful premises."[49]

The eternal stability of organic forms was logically preferable for Czolbe because it involved no mysterious creative phenomena. One always could seek for a mechanical, and therefore a conceptually clear explanation as long as creation was not involved. Although he acknowledged that his position was not capable of proof,[50] he was so convinced that the progress of science would show him correct that some twelve years after the *Neue Darstellung des Sensualismus* he anonymously put up 500 thaler as a prize for the best answer to the question: Are the facts of astronomy, geology, and biology such that they unconditionally necessitate the assumption of a temporal beginning of our solar system and in particular of the earth and its inhabitants, or do they possibly allow the assumption of their eternal existence?[51]

The response to the *Neue Darstellung des Sensualismus* was quiet. Büchner used the work in writing his *Kraft und Stoff*, citing it often in spite of the fact that he did not go along with all of it. The most severe criticism came from the philosopher Hermann Lotze of Göttingen. Czolbe was the only one of the scientific materialists Lotze bothered to discuss, since in his view only Czolbe had tried to remain consistent.[52] In the book Czolbe had said that his inspiration for the general principle of eliminating the suprasensual had come from Lotze's successful elimination of vital force.[53] Lotze, who saw the issues of vital force and the conscious soul as two very different matters, refused the honor of having inspired Czolbe.[54]

The Göttingen philosopher began by pointing out that sensualism was an epistemological position, and that Czolbe was not really a sensualist because his claims were metaphysical.[55] He showed how even this sensualistic system had been unable to eliminate all suprasensual catagories. His reply was clear: "All thinking, once it presents itself, seems to me to consist precisely in the addition of the suprasensual to the intuition (*Anschauung*)."[56] Besides criticizing Czolbe's notion of free will, Lotze hit home most severely with regard to his alleged explanation of consciousness. After all the mechanics of the brain had been spelled out, said Lotze, the question remained: Who was conscious? Motion returning to itself was not consciousness unless there was also a subject present with a capacity for it.

Was a turning wheel conscious?[57] Czolbe was sufficiently upset by Lotze's critique to attempt a reply, entitled *Entstehung des Selbstbewusstseins: Eine Antwort an Herrn Professor Lotze*, but it contained little new.

It was nine years before Czolbe published his next major work. He had two purposes in view: to overcome the Kantian rift between the objective external world and the world of thought, and to correct his position from the first two books. *Die Grenzen und der Ursprung der menschlichen Erkenntniss* did contain some basic changes in the Czolbean perspective.

Vaihinger has erred in attributing to phase two of Czolbe's thought a shift from his conviction of certainty about the elimination of the suprasensual to a recognition that other positions were also logically possible; i.e., to what Vaihinger identifies as a shift to a new, moral grounding of his program.[58] That Czolbe was well aware of the logical possibility of suprasensual systems has been abundantly documented above. What the *Grenzen* does is merely reinforce his earlier persuasion.

According to Vaihinger, that Czolbe wavered in his view of consciousness as a result of Lotze's review of the *Neue Darstellung des Sensualismus* is evident from his reply to Lotze. There he entertains the possibility that consciousness may be extended beyond organisms.[59] But once again Czolbe's flexibility had allowed him to have considered that idea already in the *Neue Darstellung des Sensualismus*.[60] The real difference between the first two works and the *Grenzen* lies in Czolbe's acknowledgement that Lotze had been correct in maintaining that there was an unbridgeable gap between a mechanical *description* of consciousness and one's *experience* of consciousness. Only in the *Grenzen* did Czolbe acknowledge this gap by raising sensations to the status of a primary element of reality, a category previously occupied only by matter. Yet sensations, in spite of being independent of matter, were not for Czolbe as ethereal as one might assume. In fact, sensations were not conceivable to him at all as nonspatial entities.[61]

To understand Czolbe one must distinguish between his reference to the basic elements of reality and, as in the title of the *Grenzen*, his reference to the limits of human knowledge. Sometimes Czolbe himself confused the two, but they must be kept separate if his message is to be consistent. The basic entities of reality in the *Grenzen*, matter and sensations, represent Czolbe's updated solution to the pre-Socratic search for a φύσις, the irreducible stuff from which all phenomena in the world of appearance derive. This, then, is a metaphysical doctrine.

Die Natur.

Zeitung zur Verbreitung naturwissenschaftlicher Kenntniß und Naturanschauung für Leser aller Stände.

Herausgegeben von

Dr. Otto Ule, in Verbindung mit Dr. Karl Müller, E. A. Roßmäßler und andern Freunden.

№ 1. Halle, G. Schwetschke'scher Verlag. 3. Januar 1852.

Zum Titelbild.

Unheimlich starrt dem Unkundigen der Schooß der Tiefe entgegen, ein gahnendes Grab, eine klaffende Hölle. Freundlich strahlt dem Kundigen aus ihrer Nacht die Gluth des Lebens, quillt der Strom einer vieltausendjährigen Geschichte.

Nicht erloschen ist die Gluth des Innern: noch bricht ihre Leidenschaft in verheerenden Flammenströmen aus den Poren der Erde.

Aber duftende Blumen verhüllen die Mächte der Tiefe, schmücken den Kampf mit dem Kranze des Friedens. Ein reiches Leben umschlingt die Elemente in ewiger Harmonie, eine rastlose Thierwelt verknüpft Land und Meer, und des Menschen Gedankenkette zieht sich von Pol zu Pol, von Ost zu West.

Aufwärts schwebt des Menschen Blick, über die Wolken hinaus, getragen auf den Wellen des Lichts, zu den Sternen in die Tiefen des Himmels. Was das Leben begann, vollendet der Geist. Er ergründet die Gesetze des Alls, die Vernunft der Welt. Er umfaßt Himmel und Erde und umschlingt sie mit geistigen Banden zur ewigen Harmonie des Friedens und der Schönheit!

Diese harmonische Welt nennt der Mensch Natur und diese Natur seine Heimath!

Die Aufgabe der Naturwissenschaft.

Von Otto Ule.

In Winterstürmen keimt die Saat des Friedens. So hat sich unter den Kämpfen der Vergangenheit die Naturwissenschaft entfaltet, und eine nie gekannte Theilnahme an ihren Schöpfungen ist rege geworden. Es ist, als habe der Sturm das Volk aus einem Traume gerüttelt, als erkenne es jetzt erst seinen unendlichen Reichthum an Entdeckungen und Erfindungen.

Da sieht man Ströme und Meere von segellosen Schiffen bedeckt und dampfende Wagenzüge auf Eisenbahnen rollen, die gleich Adern der Erde die geschäftigen Lebensströme

Fig. 8. Title page of *Die Natur*, Volume 1, Number 1. This journal, founded in 1852, was the most popular of the new periodicals dedicated to scientific materialism.

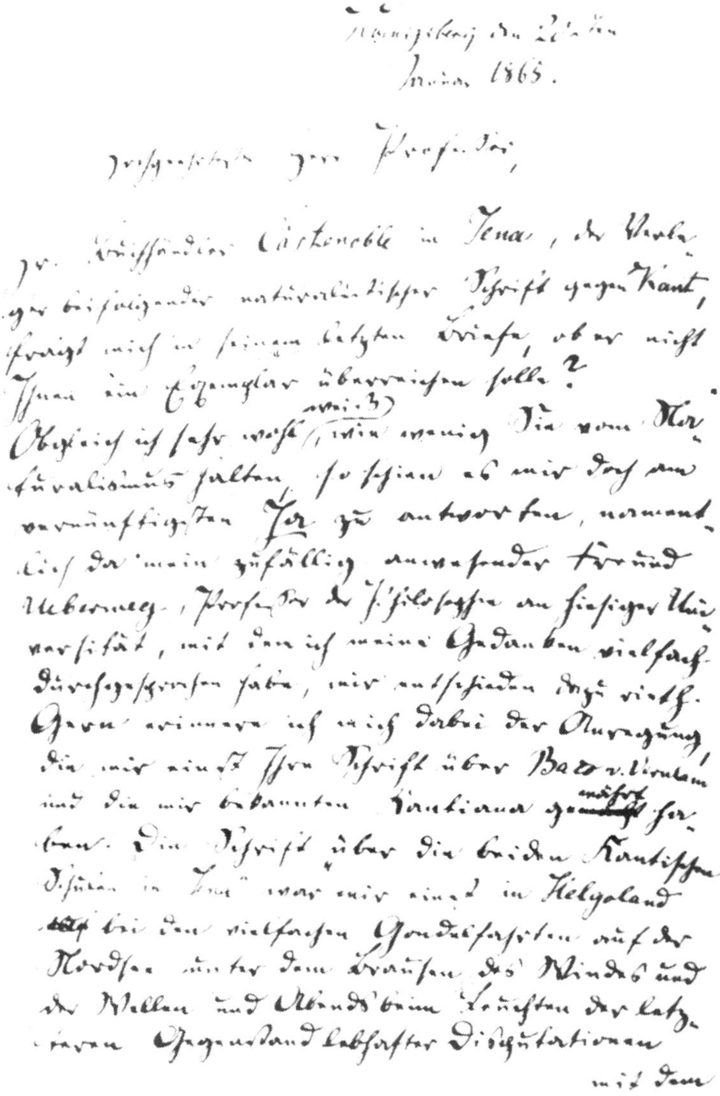

Königsberg den 20ten Januar 1865.

Hochgeehrter Herr Professor,

Dr. [illegible] in Jena, der Verleger beifolgender materialistischer Schrift gegen Kant, fragt mich in seinem letzten Briefe, ob er nicht Ihnen ein Exemplar überreichen solle? Obgleich ich sehr weiß, (wie wenig) Sie vom Materialismus halten, so schien es mir doch am vernünftigsten Ja zu antworten, namentlich da mein zufällig anwesender Freund Ueberweg, Professor der Philosophie an hiesiger Universität, mit dem ich meine Gedanken vielfach durchgesprochen habe, mir entschieden dazu rieth. Gern erinnere ich mich dabei der Anregung, die mir einst Ihre Schrift über Baco v. Verulam und die mir bekannten Kantiana gewährt haben. Die Schrift über die beiden Kantischen Schulen in Jena war mir einst in Helgoland bei den vielfachen Gondelfahrten auf der Nordsee unter dem Brausen des Windes und der Wellen und Abends beim Leuchten der letzteren Gegenstand lebhafter Disputationen

mit dem

Fig. 9. Letter of Heinrich Czolbe to Kuno Fischer, one of the founders of the Neo-Kantian movement after 1860. Within a year of each other Fischer and Moleschott had been forced to leave their posts as *Privatdozenten* at Heidelberg because of their

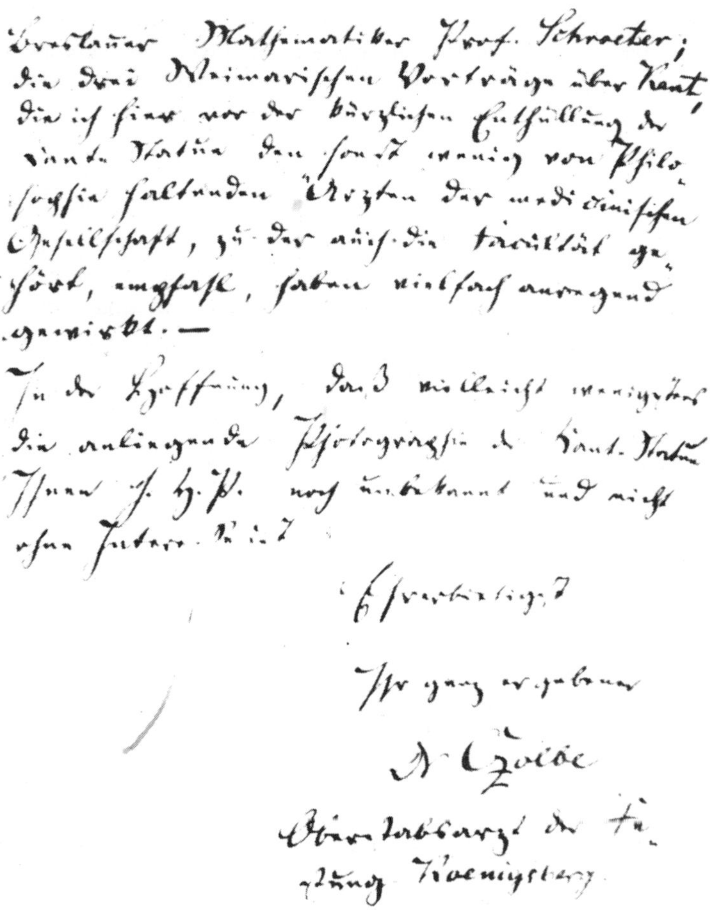
Breslauer Mathematiker Prof. Schroeter;
die drei Weimarischen Vorträge über Kant,
die ich hier vor der bürgerlichen Enthüllung der
Kant-Statue den sonst wenig von Philo-
sophie haltenden Aerzten der medicinischen
Gesellschaft, zu der auch die Facultät ge-
hört, umgefaßt, haben vielfach anregend
gewirkt. —
In der Hoffnung, daß vielleicht wenigstens
die anliegende Photographie der Kant-Statue
Ihnen h. H. P. noch unbekannt und nicht
ohne Interesse sei,
Hochachtungsvoll
Ihr ganz ergebener
H. Czolbe
Oberstabsarzt der Fe-
stung Koenigsberg.

liberal views. Czolbe asks Fischer, at Überweg's urging, to review the latest "naturalistic" critique of Kant, written by Czolbe's acquaintance Castenoble at Jena.

Finding the limits of knowledge, however, is an epistemological endeavor, and when Czolbe referred to matter as a limit of knowledge he meant to imply that the existence of matter could only be denied through misunderstanding. To matter and space as irreducible limits and conditions of human knowledge Czolbe added the world soul, a new epistemological category demanded by the inclusion of sensations as a corresponding basic entity of reality.

The *Grenzen* was also the culmination of Czolbe's acknowledgement that materialism could not escape completely from idealism. He could never simply overlook, as did Vogt, the contradiction between his belief in meaning and purpose on the one hand and mechanistic causality on the other. Already in his first work he pointed out that things stood to one another in a relation of purpose, and even there he gave an account of the link between purpose and causality.

> The concept of purpose falls within the scope of causality because all things in the world can be viewed as means, or causes from which certain purposes or effects result. . . . The relation of purpose is to some extent a higher power, or combination of the causal relation.[62]

Here in the *Grenzen* Czolbe made purpose a limit of knowledge. The existence of a purposeful world could only be denied through misunderstanding.[63]

What Czolbe was out to establish was a teleological world view that somehow was not inconsistent with a mechanical one, and he recognized that in doing so he had to oppose other materialists.[64] How he thought he accomplished this is most clearly seen in his philosophy of history. To Czolbe the eternally existing universe went hand in hand with purpose, for otherwise, e.g., in a world that came to an end, there would also be an end to purpose, or, on the larger scale, one could say that there had been no purpose.[65]

The development of culture and civilization itself was evidence to him of an all pervasive teleology. He was careful to point out that although there had been no beginning to the world or to life, there had been a beginning to history. Even in the *Neue Darstellung des Sensualismus* he had explained how the beginning of historical development was dependent on favorable accidental factors that happened to congeal at the right time, after which the development tended to preserve itself and to accelerate as lessons were learned by experience.[66] While Czolbe admitted that purpose was frequently frustrated on the individual level, and although he confessed that cultural advance could not be guaranteed, it was clear that he, like Büchner, felt that the presence of purpose (in Büchner's case progress) was a necessary assumption.

Not only did Czolbe cite purpose as a limit of knowledge, along with matter, space and sensations, but he even confessed that his monism was no longer a monism based on one elementary stuff of reality, as in materialism and idealism. His system was monistic for a completely different reason; namely, because he ordered all subjective experience by means of his eudaemonistic principle.[67] His biographer has concluded that in the *Grenzen* he forsook traditional monism because on examination it proved to lack intuitive clarity. Johnson emphasizes that nowhere in Czolbe's several works did the requirement of intuitive clarity ever lose its top priority.[68]

Nor did Czolbe ever feel he had introduced suprasensual elements into his work. He continued to hold himself apart from those who clung to suprasensual elements, even from Lotze, who appealed to causal and mechanical explanation in every area but that of the soul. Purpose for Czolbe operated according to predetermined patterns, not as the result of a free will. Likewise consciousness, although the *experience* of it was not identical to the mechanical processes that accompanied the experience, nevertheless was not free to roam beyond those processes, or to operate without them. Lotze, who was able to make room for God by doing this, earned the title 'theological mechanist' from Czolbe, precisely because he added this suprasensual element where it was not required.[69]

Czolbe had once acknowledged the logical possibility of using suprasensual categories in one's system, but here in the *Grenzen* the issue ceased being one of logical choice or moral preference and became one of true versus false morality. Now he viewed those who took the suprasensual option as *morally wrong* because they were dissatisfied with this world.[70] Once again Czolbe's naive realism played a strong role. He took for granted that he could know this world clearly (*anschaulich*) and contrasted such knowledge with the inaccessible, unknown, unclear and therefore illusory and false objects of theology and speculation. So sure of himself was he that he suggested that the New Testament's unpardonable sin against the Holy Spirit be replaced with an unpardonable sin against the world order.[71] The elimination of the suprasensual had become true morality.

The *Grenzen* had been the locus of his realization that he was no materialist in the usual sense of the word because of the philosophical difficulties of what one might call φύσις monism, or a monism that calls for one basic entity of reality. In the decade between the *Entstehung des*

Selbstbewusstseins and the *Grenzen* Czolbe concluded that Büchner's matter, the theologian's God, and the idealist's *Geist* all shared a certain mysterious power to act as the cause of all phenomena. Such explanation was for him unclear, and forced the conclusion that his system of the world would have to contain, besides matter, sensations as irreducible entities of reality, both interacting in accordance with mechanical laws.

There was a third and final phase of Czolbe's thought in which he returned to a search for a primary stuff of reality, a substance which was common to the two substances of the *Grenzen*. In an article of 1866 Czolbe prepared the ground for his final search. In it he identified the sensually clear model for all causal relations to be the parallelogram of forces, which he called the foundation of mechanics, or the "mechanical principle."[72] The immediate corollary to this was that the structure of mathematics was the ideal model of all knowledge and that all knowledge, in reducing to mechanical relationships, actually was based on the purity of mathematics.[73] He repeated his claim that sensations were extended in space, and were therefore describable mechanically. Since they did not have clear boundaries, they appeared to us to be unspatial, but such a conclusion was an error.[74] As far as mathematical entities themselves were concerned, Czolbe declared that in themselves they were neither sensual nor suprasensual, but he assured the reader that they were conceivable only because of the senses.[75] Euclidian axioms alone were permitted, for only then, in his view, could causality, mechanics and the parallelogram of forces be mutually appreciated in a conceptually clear fashion.[76]

At Friedrich Überweg's urging Czolbe tried to reformulate his ideas one final time. His last thoughts appeared posthumously in the *Grundzüge einer Extensionalen Erkenntnisstheorie.*[77] On his doctoral examination Johannes Müller had indicated to him the importance of the work of Aristotle and Spinoza. Czolbe had not understood Müller then, but now, near the end of his life, he realized why Müller had chosed that pair of philosophers. If he could combine Aristotle's emphasis on teleology with Spinozan φύσις, all the while rejecting the suprasensual, he would have recaptured the monism he had had to abandon. Czolbe saw his final task to be the construction of a bridge between these two great thinkers.[78]

Briefly, Czolbe concluded that spatially extended atoms, sensations, and forces[79] presupposed a *Receptaculum* in which all these extended entities existed. The *Receptaculum* was an abstraction from our knowledge of the world of extended things, "whence the idea of empty space lingers in us."[80]

This idea of pure extension was "an infinite substratum, or a foundation, which penetrates to every point, in which [the world] itself is located, and which had no further foundation, being in truth the ultimate one itself."[81] To the primacy of pure extension was added its temporality, without which empty space could not be thought.[82] Czolbe had come to see a distinction between temporal space and absolute nothing. The former was the pure extension that 'penetrated' matter and defined the betweenness of two pieces of matter. The latter was simply an impossibility, since one could not abstract from space.[83] Temporal space, then, was the *φύσις*, the basic stuff of all physical and psychical things.[84]

The basic entity had been found. Czolbe now had to show how the two primary entities of reality of the *Grenzen* were not in fact primary, but could be derived from temporal space. In carrying out this demonstration, however, he subtly, perhaps unwittingly, changed the manner in which he spoke of matter and sensation. Up to this point he had used sensation and consciousness interchangeably to refer to the nonmaterial primary entity of reality. Now sensation was separated from consciousness, probably because sensations, which for Czolbe were extended, spatial entities, were more directly tied to his newly found *φύσις*-temporal space. In addition, he no longer made reference to matter as such, but was now content to demonstrate that *sensations* of matter could be derived from temporal space.

By distinguishing two ways in which spatially extended sensations combined, Czolbe provided the derivation of matter and consciousness from temporal space. Sensations of matter were the sensations that combined geometrically (*nebeneinander*). Consciousness, on the other hand, was the result of mutually *penetrable* extended sensations concentrated in one place (*ineinander*), raising the intensity of the sensation involved from unconsciousness to consciousness.[85] Normal observation of things in the world involved both kinds of sensation simultaneously, and though neither could be imagined without imagining at the same time a specific content (consciousness of something, or sensation of some color or shape), still the objective nature of each could be abstracted. Czolbe felt that with such an abstraction he had reached the thing-in-itself.[86]

What shall we make of Czolbe's brand of "materialism"? One explanation, often given by his successors, was that his repeated attempts to formulate a system based on clear and distinct ideas resulted from his awareness of the philosophical superficiality of Büchner, Vogt and Moleschott. Certainly Czolbe was more concerned with philosophy than the others, but that there was more behind his complicated and rather fantastic writings than a concern for logical rigor is equally certain. What was it that drove him to the extremes he so confidently defended?

A clue to the answer lies in his use of the German word *anschaulich*. In addition to meaning "clear," or "obvious," the word also can be used to connote intuitive understanding. In Czolbe's case, perhaps clearer than elsewhere in the nineteenth century, we have a consistent attempt to spell out the intuitively held conviction that science was mechanics, and that a world view based on science was to be a mechanical world view.[87] This no doubt was the intuitive belief of many, though few, perhaps none, tried to work out the implications in such detailed fashion as Czolbe. How else can one account for his facile rejection of ideas such as heat death, which he saw to be an encroachment on territory in which only mechanical relationships prevailed?

There was still more motivating Czolbe than the idea that all parts of the system must fit together like the parts of a machine. The criterion of intuitive clarity (*Anschaulichkeit*), which a mechanical model satisfied in his view, also implied sense clarity. Czolbe demanded that he be able to see, in almost literal fashion, how things went together. Noneuclidian geometry had no chance with him, nor did other anti-intuitive notions such as an actual infinity or a point continuum.[88] Czolbe's belief that there were such things as intuitively clear starting points led him to conclusions of a highly unusual kind, conclusions that might well cause the reader to want to turn away in disgust. Yet here was a thinker who tried above all to be consistent.

Goethe too had demanded sensual clarity (*Anschaulichkeit*), and it was Helmholtz who pointed out that this requirement was a hindrance to scientific thought. Science, said Helmholtz, employed forces going beyond intuition.[89] In the twentieth century scientists have sided with Helmholtz in subordinating the reqirement of intuitive clarity to that of consistency in accounting systematically for the phenomena. No doubt much of modern physics would be classified as suprasensual by Czolbe, and even though at one time he had acknowledged the possibility that a system containing suprasensual elements could explain the world as logically as he could, his whole life betrayed that he would not have been impressed with a *merely* consistent way of saving the phenomena. He was describing the world as it really was, the thing-in-itself. Why anyone would bother his head to construct an explanation of the world based on essentially mysterious entities was beyond him. His very last printed words condemned the scientist who included in his system things which could be neither confirmed nor disproved,[90] but unlike the later verification theorists, Czolbe was not attempting thereby to eliminate metaphysics. To one who believed that the objective nature of reality could be discovered, there was simply no place for the unknowable. With his mentor Feuerbach Czolbe could say:

In what, then, does my "method" consist? In this: reducing to nature everything supernatural by means of what is human, and everything super human by means of nature, always and alone, however, on the basis of clear, historical, empirical facts and instances.[91]

PART 3

ISSUES

CHAPTER VII

OF PHILOSOPHY AND SCIENCE

The scientific materialists obstinately refused to acknowledge capricious or creative forces directing natural processes from outside nature. Büchner, for example, had an unwavering confidence in what he called the unity and regularity of natural events. No fact that would ever come to light, no advance of science, he said, would ever impair this conviction, now shared by every impartial student of nature.[1] One of his last words revealed that his major motivation had not been so much materialism, not so much monism as it had been Feuerbachian naturalism.

> It seems to me that there is but one thing [the free-thinker] cannot do without: it is the recognition of a natural world order existing by itself and tied together by the law of cause and effect, plus the knowledge of a scientific world view based on that recognition and on the results of science (*Wissenschaft*). This world view need not be materialistic or spiritualistic, realistic or idealistic, monistic or dualistic, it need be only *natural*.... The individual gaps which our scientific knowledge leaves open in the continuity of creation will be filled out more and more in time, and where this filling out is hopeless, the gaps will have to be bridged over by the inferences of reason.[2]

This appeal to naturalness and realism, with its implicit claim that only in this way was the world really understood, eventually proved to be a threat to established philosophy. Many philosophers resented the confidence with which the materialists spoke of the powers of natural science. The philosopher Heinrich Böhmer noted not only the dearth of creative philosophical ideas and the unprecedented depth to which philosophy had sunk in Germany, but he commented in addition about a movement afoot in which natural science was taking over philosophy altogether.[3] Writing in 1872, Böhmer explained that in recent time physics and chemistry had raised themselves above their sister sciences, and that although much had been accomplished by means of this emphasis, two extreme results were evident in the intellectual world: materialism and the fear of all speculation.[4]

Others as well noted the attack on philosophy. The antimaterialist Izaak Doedes summarized Büchner's message to say: "What natural science does not teach has no value as truth."[5] Not only did such an attitude have nothing to do with philosophy, said Doedes, but it had no right to call itself philosophy either. Friedrich Fabri, referring to the "Feuerbach-Virchow dogmatism," talked of the attitude according to which natural science, in its advance towards infinite perfectability, would eventually cast the last religious idea into the darkness of forgetfulness.[6] Fabri himself did not think highly of what he identified as "the comment of Virchow and many others about the 'sovereign rights' of natural science, along with the demonstration that scientific thinking really contains philosophical, theological, and legal thought within itself."[7]

The scientific materialists had indeed given philosophers cause for alarm and resentment. Moleschott, for example, dreamed how wonderful it would be if the historical sciences would join with natural science in method, adding:

> And in fact philosophers can no longer be viewed in opposition to natural scientists, because any philosophy worthy of its name laps up the best sap of the tree of knowledge, and by the same token produces only the ripest fruit of that tree.[8]

The notion that official philosophy could be replaced with natural science was part of the materialistic mentality. Moses Hess wrote to an unidentified acquaintance that experience must be merged with philosophy and philosophy with experience. "An empiricism of science (*Wissenschaft*)," he continued, "which is carried out fundamentally and consistently from our standpoint, makes philosophy superfluous."[9]

Some of the professional scholars feared that if induction was brought into philosophy as a necessary component of its foundation it would demote philosophy from its traditional position as queen of the sciences.[10] Büchner's answer to the charge that natural science had attacked philosophy was that science had nothing against philosophy, only against philosophers with their speculative darkness and lack of concern for facts.[11] Indeed it would be misleading to say that the scientific materialists were completely antiphilosophical, for they were not. They believed, with the possible exception of Karl Vogt, that philosophy had a place, even that they were philosophers themselves. But it must be philosophy as they understood it. Realism, said Büchner, far from denying philosophy a niche, would put philosophy back into the center of human knowledge, it would give rise to a "rebirth of philosophy."[12] "While philosophy has become more scientific (*naturwissenschaftlicher*), natural science has become more philosophical in its essence; namely, in the method of its research."[13]

The attack on philosophy by the materialists, particularly by Büchner as the spokesman, was directed mainly against the Neo-Kantians. Hegel and

Naturphilosophie were out of vogue by the time Büchner had established his reputation, so attacks in that direction were largely tolerated. Büchner pointed rather to the new form of school philosophy present in Germany, an epistemological scepticism identified by the motto "Back to Kant."[14]

What disturbed the materialists about Kant was the feeling of impotence he gave them. They heard the Königsberg philosopher saying that they could not know the world in itself, and that the senses were not the ultimate source of knowledge. In response they praised every anti-Kantian work that appeared,[15] and cast the works of Kant in as bad a light as they could.

For one thing, the claim of the Kantians that Kant's defense of God, the soul, and immortality was theoretically necessary and in practical life unavoidable was antithetical to materialistic canons. Because he had preserved these nonempirical notions in his system, Kant could hardly be said to be a friend of empiricism in their eyes.

Moreover the Kantian categories were openly *a priori*! Büchner accused Kant repeatedly of ignoring the facts in his defense of innate ideas. Causality, space, time, number, etc., had their origin not in the structure of the mind, but from man's experience in the world.[16] The senses, said Büchner, had been unjustifiably maligned by Kant. Rather than give in to the so-called Kantian rift between the phenomenal and the noumenal, Büchner believed that man had good reason to hold that the senses give man ample knowledge of the existing world, and that the intellect completes what is lacking to observation. He was sure of this because man "has arisen through gradual development from lower forms and through a constant reciprocal action with external nature itself, so that his entire organization stands and must stand in a necessary and lawful relationship with this nature and its diverse influences on living beings."[17]

Because the scientific materialists were so clearly naive realists, they found Kantian epistemology threatening. Kant was to them a subjective idealist.[18] If Kant and F. A. Lange were correct about the nature of the phenomenal world, then "there is no true knowledge whatsoever, no objective truth, no more science (*Wissenschaft*), and we move like dreamers or sleep walkers in a world strange and unknown to us, a world of whose true nature we will never experience a thing."[19] Nägeli was far more persuasive to them when he said:

The agreement of sense perception and our inner mental mediation with the objects causing the perception rests for a monism of the finite world in the fact that the same forces are active and the same laws obtain as in the things outside of us. Therefore the image given us by our senses cannot contradict the object, and the further modifications which our image undergoes in judgements must come ever nearer to the true essence of the object.[20]

Czolbe too rejected Kant's alleged inability to join thought and things. One of the two goals of his *Grenzen* was to show how sensualism could succeed where Hegel had failed; viz., to establish an identity of subject and object that would overcome the Kantian separation of the two, and to do it in reality as well as in thought.[21]

The position of the scientific materialists vis-à-vis Kant is seen most clearly in their treatment of the thing-in-itself. The mysterious, unknowable thing-in-itself was to them the final proof that Kant was basically a speculative philosopher. Like the theologians, he wanted above all to preserve a place for the unknown.[22] Replied Büchner: "Knowledge and not admission of ignorance is the true goal of science (*Wissenschaft*)."[23] "This so-called 'Back to Kant' is surely the saddest *testimonium paupertatis* that modern philosophy could give itself."[24]

Kant may not have been able to know the thing-in-itself, but the scientific materialists felt no such theoretical restriction. Czolbe and Moleschott stated explicitly that Kant had not ended metaphysics, and that things-in-themselves could be known.[25] Büchner contented himself with his pragmatic criterion for truth. Even if there were an unknowable category of things-in-themselves it would have little value or interest, since unknowable it could never be made the basis of any kind of science.[26] Truth, he once said, has never failed to be of service to humanity.[27]

When Emil DuBois-Reymond in 1872 delivered a lecture on the limits of scientific knowledge, Büchner took it to be yet another example of the theological attitude. Here again Büchner's anti-Kantian persuasion determined his response.

According to Cranefield it was DuBois-Reymond who of the four young reductionists Helmholtz, Brücke, Ludwig, and DuBois-Reymond remained closest to the original program of organic physics of 1847.[28] A quarter of a century later he too demonstrated that there were significant differences between the reductionists and the scientific materialists, though more than once DuBois-Reymond had been lumped together with the latter group.

What were these differences? Early in the twentieth century the physicist Paul Volkmann, looking back at developments of the previous century, declared that the excessive conclusions drawn from the natural sciences had been the fault of physiologists who had borrowed their model from physics without a sound first-hand acquaintance with the manner in which physical theory was to be understood.[29] What Volkmann had in mind was the reductionistic program of DuBois-Reymond and the others, but he might just as well have said what he did about the scientific materialists of the 1850's. In fact, it was not DuBois-Reymond and his associates who spoke the philosophical excesses Volkmann decried. That accomplishment was left to the popularizers of materialism.

While the materialists agreed with the reductionists, the converse was not necessarily the case. The reductionists argued that chemistry and physics formed the basis of physiology, and that the paradigm of physical theory, which at mid-century was molecular mechanics, should be the framework in which research on organic as well as inorganic phenomena should be carried out. The materialists had little quarrel with this. Both groups further agreed in denying vital force scientific status, and in assuming the general determinism of physical and biological phenomena.

Yet the reductionists did not actively make the claim that, as Lange put it, mathematical explanation was equivalent to philosophical explanation. Theirs was a materialism that was essentially sceptical, not one that was sure of itself. They were the "materialists" Lange must have had in mind when he spoke of the thinkers who retained materialism only as a maxim of scientific research.[30]

The scientific materialists were another lot altogether. In the first place they were not known primarily for their work in scientific research. They were the "intellects of a lower order" referred to by Merz, whose purpose was "to frame a popular philosophy - a reasoned creed which should stand in agreement with the new conceptions and be intelligible to thoughtful persons among the general public."[31] Like Volkmann, Merz separated out the first-rate scientists who handled mechanical principles with the proper respect from these other thinkers, "notably biologists," who broadened the use of terms such as mass, force, energy, cause and purpose into a wider sense than a purely mechanical one would permit.[32]

Both the scientific materialists and the reductionists shared the conviction that the task of science was to uncover the causal-mechanical relationships in nature, and both became so impressed with the success of this approach that, as Merz put it, the spiritual came to be regarded as an epiphenomenon or even a fiction.[33] The difference between the two groups was that the reductionists remained largely agnostic vis-à-vis the spiritual,[34] while the scientific materialists rarely hesitated to proclaim that the spiritual was wholly accountable in terms of the material. The message of the latter group was obviously the more sensational, and more likely to find its way to the nonscientist.

In his lecture of 1872 DuBois-Reymond specified two common errors. One was to require the aid of the supernatural in accounting for the step from the inorganic to the organic,[35] and the other was to hope that consciousness could ever be explained completely by discovering the mechanical operations of the brain.[36] Like Lotze before him, DuBois-Reymond spoke of the lack of a necessary contact between the motions of atoms in the brain and the experience of consciousness. There was, he said, a great gap between the two.[37]

His mind had been changed since the *Untersuchungen über Thierische Elektricität* of 1848, said DuBois-Reymond. There he had held that the limit of mechanical knowledge stood at the doorstep of the freedom of the will,[38] but now he recognized that there was another problem prior even to that one, the problem of consciousness.[39] He ended his lecture with a Latin word, ***ignorabimus***: we shall never know the relationship between mind and matter.[40]

DuBois-Reymond claimed that the outcry against his lecture surprised him, since he had thought that there was nothing in it that had not been dealt with already by the great philosophers, e.g., Kant.[41] But he should not have been surprised. The materialists thought it an insult to science to harp on what it could not do. Their Feuerbachian confidence in man's abilities was offended by this one who spoke of mysteries man would never know.

Büchner discussed the lecture in the second volume of his *Physiologische Bilder*.

> Like old wives the spiritualistic scatterbrains come back time and again to their proposition and blab . . . the old song . . . of the metaphysical significance of consciousness. . . , of the unfathomability of its essence and of the impossibility of its explanation through material circumstances or causes.[42]

To Büchner DuBois-Reymond was offensive because he had put his emphasis in the wrong place. It might seem impossible to explain consciousness at present, but that was not what was important. The crucial thing was that matter did produce consciousness when it reached a certain level of complexity.[43] Misrepresenting the thrust of DuBois-Reymond's remarks, Büchner declared that even the Berlin physiologist admitted this. Had he pondered on his own statements he might never have given the lecture.[44]

Unlike Czolbe, Büchner never saw the distinction Lotze and DuBois-

Reymond drew between the *description* of consciousness and the experiencing of it. On one occasion it seemed as if Büchner had understood, for he conceded that there was something fundamentally inexplicable involved. He tried to ignore the problem by repeating his denial "that nothing at all about the unreality of the relationship [between material conditions and mental processes] can be concluded from the fact that something inexplicable was present."[45] Upon closer examination one finds that this mystery refers not to the limitations DuBois-Reymond was trying to point out, but to man's ignorance of what Büchner called the fineness (*Feinheit*) of matter.[46] If man could get down to this level of matter the ultimate solution of metaphysical problems would be solved.[47] For him the issues raised by DuBois-Reymond were not fundamental difficulties, they resulted from man's ignorance.

In a surprising move Büchner chastened DuBois-Reymond for being proud of human ignorance, adding that this alone explained why DuBois-Reymond was willing to believe that something was impossible simply because it was not understandable.[48] Yet how many times had Büchner himself denied the existence of the soul for the very same reason! Lange's remarks about D. F. Strauss are relevant here: "This is precisely the standpoint of materialism, which postpones the insoluble problems and holds fast to the closed circle of the causal law, in order from here to open its polemic against religion."[49]

When the scientific materialists discussed what might today be called philosophy of science, the results could be surprising. Above all they wanted to be known as realists, but at the same time they were quick to dissociate themselves from the kind of empiricism whose concern was solely the gathering of facts. While they despised the old school of *Naturphilosophie* because it had deliberately ignored the world of reality, they thought it an equally foolish mistake to go to the opposite extreme.

Liebig's attack of Bacon and empiricism[50] was no doubt a major reason why the materialists wanted to dissociate themselves from crass empiricism. Still, their own conclusions were philosophical, and they themselves opposed refusing all connection between natural science and philosophy, as some had done.[51] Their mediating position brought with it something of a dilemma, for not only were they bound to encounter the problem of explaining how a scientific theory was generated from the facts, but with their emphasis on realism they appeared responsible for accounting for the correct theory.

It quickly becomes clear that the materialists' dissociation from crass empiricism did not mean that they denied its cardinal premise; viz., that the senses were the source of knowledge. One need only recall the subtitle to Büchner's *Kraft und Stoff: Empirisch-naturwissenschaftliche Studien.* Each of them made repeated appeals to empiricism and sensualism.[52]

While they admitted, as we shall presently see, that the scientist did not simply pile up the facts of observation, the materialists drew an over-simplified distinction between fact and theory. "Now what I want is – facts," was the English motto at the head of Büchner's *Vorwort* to the *Kraft und Stoff.* Both Büchner and Vogt claimed that a single fact could collapse systems of whole centuries. Vogt in particular referred to the "fact" that Copernicus had discovered; viz., that the earth was not at rest.[53] It was the conviction that they were dealing with facts that made them think of themselves as realists over against the illusory philosophical systems of the immediate past. For this reason too Büchner and Vogt were convinced that they had gotten rid of metaphysics.

Each of the materialists presented his own version of inductivism to explain how theories were born. Each acknowledged the primacy of facts, but each emphasized that the task of the scientist was to explain the facts. As Czolbe put it, the problem for modern empiricism was "to *explain* the facts clearly."[54]

Czolbe's explanation of the relationship between fact and theory was the least specific of any in the group. He heralded the work of Comte, Mill, and the Dutchman Opzoomer as an attempt to bring the ideas of Francis Bacon to bear on disciplines other than natural science,[55] but his concern for devising a physical mechanism to explain the production of concepts, inferences, etc., prevented him from addressing the problem of how specific theoretic notions were to be selected. On the one hand he was open to the idea that community agreement determined what theories should be accepted, yet he more than any of the others (with the possible exception of Otto Ule) came down hard on the unsatisfactory status of hypotheses.[56] An hypothesis was "a weakly or deficiently grounded inference provisionally assumed in order to relate experiences to one another, in order to explain them."[57] He went on to speak of its degree of probability and to contrast it unfavorably with the certainty of inference. Elsewhere he referred to hypotheses as uncertain inferences that were unavoidable because the premises of inferences, which were sense perceptions, were often limited. But he noted that an hypothesis could turn out to be correct.[58]

In his reply to Lotze Czolbe declared that he agreed with Newton that one must not only explain natural phenomena, but that one must find their *vera causa*.[59] His only retraction from this strongly realistic stance was an acknowledgement in one of his last works that there might be general hypotheses, e.g., atomism, which were ultimately unsatisfactory, but which science at present could not go beyond, even which science might never be able to go beyond.[60]

Karl Vogt also believed in the sharp separation between fact and theory. According to him, theory might well be altered as time passed, but the facts were inviolate.

> Every experimental science shares the same fate. Observation, when carried out properly, and fact, when it is really present, remain. Their interpretation, however, and their extension to a harmonic whole changes with the existing state of the science.[61]

Vogt took it as his goal to keep, as he put it, "clear-headed observation on one side, and reflection on the other."[62] Nevertheless he spoke of the role of the maturity of the scientist, and of the possibility that facts had been *observed* incorrectly.

> Fortunately we have progressed so far in our sciences (*Wissenschaften*) and in scientific self-confidence that everyone believes himself capable of advancing a satisfactory explanation of a fact. Everyone must accept observation (*Beobachtung*) as it is, until its fallaciousness or deficiency is shown.[63]

There was no suspicion in Vogt's mind that because the observer's self-confidence played a part, or because history seemed to indicate that facts had in the past been falsely or deficiently apprehended, that his category of the objective and neutral fact should be questioned. He *knew*, like the other materialists knew, that the facts were there. Facts were no cause for concern; it was rather the arrangement of facts into theories that could be problematical. Here uncertainty was far greater, even though it was not out of reach in principle. Notice in the following statement that Vogt equates *proper* observation with *correct* interpretation.

> Everyone will modify, correct, or leave [inferences (*Schlussfolgerungen*)] untouched according to his way of comprehending (*Auffassungsweise*), *but there may be only few who will take them up exactly as the observer sets them down.* I should not be surprised, therefore, if my own observations are used as weapons against my views; I myself will not hesitate for a moment to change them as soon as new facts have convinced me of their inadequacy.[64]

In spite of this apparently open attitude Vogt was accused of passing off his own interpretation as dogmatic truth. He did not, it was claimed by others, know the difference between hypotheses and scientific certainty.[65]

If Moleschott contributed anything unique to the discussion, it was his critique of Bacon's rules of causality. He pointed out that these rules were oversimplifications, that more frequently than not an effect had multiple causes working together. If a cause seemed sufficient to explain an effect, one must not automatically conclude that it was *the* cause. A possible cause, said Moleschott, was not necessarily a real cause. Moleschott was not challenging strict causality in any of this, he simply was pointing out that it could be a very complex matter indeed.[66]

Apart from this Moleschott concurred with the sentiment of the other materialists. He recorded his respect for empiricism, and explained that the increase in factual knowledge was the basis of the scientist's belief that he was dealing with reality.

> The natural scientist does not give in to the belief that he has created the law. He feels in his innermost being that the facts impose it upon him, and the more he analyzes them the more powerfully he feels their driving necessity.[67]

Not that every theory gave one certainty. Like the English materialist John Tyndall, Moleschott spoke of the gradual refinement of theory towards truth, though in his case the notion of truth had been inherited from Feuerbach:

> One goes from given facts to theoretical formulae which never contain the whole truth and not rarely lead to dreams; but from these dreams we will be awakened by new facts, and if these also should lead over the centuries to new dreams, the haze nevertheless becomes thinner, and one beholds the divine image of the human essence ever clearer, an essence which obeys the highest laws of natural necessity.[68]

Of the major figures Büchner went into the matter most thoroughly. He too believed solidly in facts. The sure victory of realism in philosophy was guaranteed, in his view, because the force of its proofs was due to facts.[69] Early in his career he reviewed Apelt's *Theorie der Induction* of 1854, plus a lecture by Robert Zimmermann entitled 'Philosophie und Erfahrung.' In the review it was argued that concepts were reflections of what was seen (*angeschaut*), that concepts were nothing without their objective referents, but that these referents had a content even without the concepts. Induction in Apelt's book, said Büchner, was the bridge from fact to natural law, from accidental truth to the necessary truth of reason. The power of induction lay in its ability to lead to law, the ultimate ground of explanation of natural things, from the composition of facts and observations.[70]

As Büchner proceeded he became more and more entangled in new categories. An experiment, he said, was not the mere piling up of facts along side each other. Already in experiment there was an element of reflection, hence it was possible in devising an experiment to smuggle in factors that could color the general consciousness.[71] Büchner separated out the notion of a real or true experiment (*wirkliche oder richtige Erfahrung*), in which the facts were tied together according to the laws of logic and reason. Facts, after all, were not chaotically arranged in nature, they always were subject to the laws at their base.[72]

Admittedly a real or proper experiment was not an easy thing to set up, as the history of science demonstrated. In fact, "The difficulty of performing a true experiment . . . is often far more complicated than the processing through speculation of facts once ascertained."[73]

With this contrast between proper experimentation and speculation Büchner was becoming more and more ensnared in distinctions and qualifications. In order to make clear what was meant by a proper experiment he said that through them general and extended (*allgemeine und verbreitete*) facts were derived from pure sensations. His scheme now contained raw sensations, general facts arrived at via proper experiments, and speculation or the manipulation (*Verarbeitung*) of general facts. The ordering activity, the separating of the true from the false began, according to Büchner, with experimentation, and it increased with the level of generalization.[74] The treatment of the results of experimentation should not, however, be carried out *only* with the help of induction. Deduction was often helpful and hypotheses could not be avoided.[75] In this way he felt that the debate over the contrast between philosophy and experience could be set aside. Many different kinds of methods could be used to interpret and order the facts of experience.[76]

These speculative and uncertain elements came into Büchner's system, as they did into Vogt's, "after the fact;" i.e., only after the existence of the fact had been guaranteed. To guarantee those facts Büchner quite explicitly depended upon the belief that nature was rationally apprehendable. Like Albert Einstein, who initially became interested in natural science from reading Büchner's *Kraft und Stoff*,[77] Büchner believed that there was a rational solution to nature's secrets.[78]

One wishes Büchner would say precisely how experimentation begins the process of separating the true from the false. Unfortunately he did not explain how a real experiment should be set up. He did say that there was a difference between science (*Wissenschaft*) and naked knowledge (*das blosse Wissen*), and that one should not begrudge science an element of fantasy.[79] He did say that hypotheses were to be judged according to their degree of probability.[80] And he did say:

> The naked fact . . . is in itself raw, clumsy, and, when it is not illuminated by the light of the thinking mind and grasped in its general context, mostly without value for science (*Wissenschaft*).[81]

Which side of the coin Büchner emphasized depended on his purpose. On the one hand he could insist, as he has above, that pure facts are without value to science, this to avoid the charge of crass empiricism. On the other hand he distinguished himself from the unreal musings of *Naturphilosophie* by declaring: "Whoever is always demanding explanations alone in place of the confirmation of facts or observations is no real natural scientist."[82] He pointed out that there will always be phenomena which are beyond explanation, but which nevertheless remain facts.[83] The tightrope Büchner was trying to walk between *Naturphilosophie* and crass empiricism was only possible to one who borrowed from both. From empiricism he took the notion that it was the real, objective, external world which was the goal and source of his explanations. From the tradition of *Naturphilosophie* he borrowed the strong conviction that nature was ultimately rational.[84]

The rationalistic realism of the scientific materialists, with its dependence on a strong belief in the rule of natural causality, showed itself clearly in their assumption that physics, in particular the old mechanical tradition, was the ideal model for scientific thought. Czolbe, as we have seen, was unambiguous about this. By mechanics he meant an intuitively clear set of causal relationships among entities in motion. Unlike strict materialism, he argued that there were two basic entities whose mechanical interaction explained the world, matter and sensations. There was, however, only one causal nexus according to which the motions of these entities were determined, and the model for that was, he said, the parallelogram of forces. Thus understood, Czolbe declared that mechanical relationships did not merely apply to natural science, but that they were the prototype for all knowledge.[85]

From the parallelogram of forces one could see, according to Czolbe, that the cause of a given effect was never singular, but that it was made up of several causes acting in concert, and that an effect could never be essentially different from the cause.[86] This latter condition revealed that there was a quality of certainty or tautology about mechanical relationships. Kant was wrong, he explained, to say that we never encounter the necessary in experience.[87]

Necessity was always a component of a sensually perceivable causal relationship in Czolbe's view. Hume had simply erred in his claim that causality was the result of habit, at least as far as the causal relationships in mechanics were concerned. Czolbe conceded that because of the clarity and certainty involved in mechanical causation (the parallelogram of forces), we often desire the same clarity in places where the relationship was not so intuitively clear, such as in the sprouting of a seed or the motion of an arm. In these cases, he acknowledged, we might assume causality before working out the mechanism because we have become accustomed to the sequence of events.[88]

Büchner was even more explicit than Czolbe about the equivalence between rationality and mechanics. Quoting from Wilhelm Wundt in his review of the latter's *Vorlesungen über Menschen – und Thierseele*, he wrote: "Mechanics and logic are identical," commenting that they were only different forms of the same thing, and that mechanical and logical development simply corresponded to the two modes of conception which were grounded in the nature of our knowledge.[89] In another place where he made the same claim, Büchner added: "The reason in nature is also the reason of thought," exposing his metaphysical rationalism once more for all to see.[90] Even in the *Kraft und Stoff* he spoke of "the identity of the laws of thought with the mechanical laws of external nature."[91] To the author of *Kraft und Stoff* these claims were equivalent to saying that causality ruled everywhere there was scientific thought.[92]

Büchner thought he had rid science of metaphysics. No one, he claimed, could ever expect to explain the ultimate Why, or the ultimate ground of things. That matter and force were connected inseparably was a fact, but nowhere in any of his writings had he claimed to say *how* they were connected.[93] It struck many as strange that Büchner could dogmatically assert that all activity, physical, organic and mental, corresponded to matter in motion without feeling a compunction first to find the mechanical relations he assumed could be found.

These sentiments from the last part of Büchner's life exhibit the confusion and vagueness that could arise when the scientific materialists declared mechanics to be the essence of scientific thought. It would seem, for example, that Büchner would have felt responsible, as Czolbe did, to find the How, the mechanism behind the connection between matter and force. He had instead cleverly renounced his search for mechanism by gradually shifting the meaning of "mechanical" to "causal." More and more he became convinced that the message of the inseparability of force and matter was not properly materialism, but monism.[94] Everything was at least theoretically predictable because everything was interconnected by the general laws of cause and effect.

None of the others was so bold in his claims for the mechanical paradigm as Büchner and Czolbe. Vogt spoke of "the machine of the organism," and emphasized, as did Moleschott, the mechanical process of material exchange as the basis of life.[95] But he confessed that the brain and the central nervous system were in some sense more primary sources of the vital activity of the muscles and organs, and that there was much about all this that was as yet unclear.[96] The method of solution to the question why an organism lived *was* clear, however: "Just as life is first and foremost the resultant of all the individual functions of molecules, so also must our knowledge of it proceed first from the same aspect of all individual actions."[97]

Moleschott too confined his remarks to a general discussion of the way in which physiology could benefit by borrowing from physics and chemistry. It was clear that he did not believe that his own field would be taken over by the physical sciences. He spoke of physics as providing the method for all the sciences, including metaphysics,[98] yet repeatedly he left no doubt that there was room for all the sciences. He reprimanded those who feared "that this exact science [physics] could invade the area of physiology too quickly, and become its master, as if the republic of science did not offer room for all."[99]

Even Feuerbach had had something to say about the mechanical world view. In *Das Wesen des Christentums* he had contrasted it with the religious attitude. Like Büchner he acknowledged that a mechanical structure could never explain ultimate questions such as the source of motion. There was, in his view, such a thing as a natural miracle. But that was different from religious miracle, in which God explained all inexplicable things.[100] Feuerbach, like Czolbe, demanded only that the pieces be put back together in a sensually clear manner.[101]

Strangely enough, on close examination only Büchner said explicitly and dogmatically that molecules of matter in motion explained the world. The appeal to materialism in general, and to mechanics in particular, meant that the scientific materialists trusted that the world could be dissected into its parts, and that an understanding of the world was equal to a knowledge of how these parts fit together. Indeed this is the appeal of the mechanical model, and in this broad sense the materialists were mechanists. More importantly, however, they tried to communicate to their age that in its on-going dissection of the world, science would never encounter an essentially mysterious component, neither a ghost nor a God in the machine.

While it might have been true that the majority of the scientific materialists did not maintain openly that everything in the world could be explained solely by matter in motion, this was the proposition they were all accused of propounding by their opponents on the philosophical and religious fronts. Even within natural science there arose a response to materialism. Ernst Mach, J. B. Stallo, Hans Vaihinger and others argued that mechanistic materialism did not suffice to account adequately for the unique nature of scientific thinking. The dogmatic claims of the materialists had forced the issue, and had thereby helped to call forth a new brand of reasoning about the nature of science.

In conclusion we shall examine an illustration of the scientific materialists' understanding of developments within natural science itself. The case before us concerns their teaching regarding the conservation of energy. It provides an illustration of the way in which the scientific materialists, as popularizers of science, purported to understand without hesitation the significance of developments in science, and of the way they communicated their understanding in a tone that laid claim to the authority of Germany's new natural science.

They all did not take up the conservation of energy. Vogt said little or nothing on the subject, and Moleschott confined himself to celebrating the importance of the discovery with only an occasional reference to its significance. Mayer and Darwin, said Moleschott, would be shown by history to have formed the scientific character of the nineteenth century.[102] In addition to explaining how the body, as a machine that produced heat, conformed to the first law of thermodynamics,[103] Moleschott pointed out that the interconvertibility of forces, on which principle he saw the conservation of energy to rest, was an insight that enabled men to arrange causal sequences properly. By showing which phenomena were linked to which, the conservation of energy had removed the stamp of miracle from many natural events.[104]

Czolbe had even fewer occasions to refer to energy conservation. His appeal to the idea was restricted to the declaration that physical forces too were subject to it, though his understanding of potential (*todte*) as opposed to kinetic (*lebende*) *Kraft* was not applied to psychical processes in a consistent manner.[105] Czolbe correlated sensations, which he declared were latent in space, with latent or potential force. Sensations were set free by the motions of the nerve processes of the brain, processes not equivalent to these motions, since sensations were primary entities. Czolbe's distinction between kinetic and potential force therefore did nothing but further confuse the picture. Büchner avoided this confusion, for he believed that psychical processes were derived from electrical impulses in the nerves. He inferred from this that psychical like physical life was nothing but an immense set of force transformations.[106]

It was Büchner who made conservation of energy his own. His first discussion of it came in an 1857 piece entitled 'Die Unsterblichkeit der Kraft.'[107] The "immortality" of force was not an accidental choice of words, for Büchner's understanding of the matter depended directly upon it.[108] Nature did not know standstill, Büchner explained; it was rather a never-ending cycle of motion that corresponded to diverse kinds of force. These forces were not able to be created or destroyed, but they were interconvertible one into another in such a way that their sum was a constant, and force itself was immortal. While there might easily be motion that became latent, it was never lost. A force could be transformed into another form, but the total amount of force in the world remained equal in amount.[109] Although he cited the work of Helmholtz, Grove, Faraday, and Baumgartner, it was the discussion of Friedrich Mohr he leaned upon most heavily. In describing the action of the steam engine, Mohr's (and therefore Büchner's) account was a confused exposition of the relationship between the work force of steam, friction, and the heat of combustion of coal.[110]

As time passed Büchner's message became simpler instead of more careful. He overburdened his works with examples of interchangeable forces and became more bold in his declaration that what was meant by the doctrine was that *motion* was conserved.[111] Pointing to Helmholtz, who, he said, had shown how the constant amount of force in the universe expressed itself either as the living force of moved masses (oscillations of the ether, air or molecules), or in the form of the attraction or tension between two masses, Büchner recommended that one should view force only as the motion of atoms. "Perhaps it would be better if one allowed the first expression (force) to drop completely and the word 'motion' was put in its place."[112] With this recommendation Büchner realized that he was counting on the eventual demonstration that forces as yet unexplained as motions, *Spannkräfte* in Helmholtz's terminology, would someday be explained as such. He cited the case of the elasticity of air as an example of a tensive force that had been explained successfully in terms of motion.[113]

As far as Büchner was concerned, the point was not only that there was a given quantity of force in the world that could appear under several different guises, but that this meant force, like matter, was indestructable and eternal. He began to speculate that there might be an *Urkraft* behind the various kinds of force, just as Prout, he noted, had suspected an *Urelement* into which all other elements decompose.[114]

It was perfectly clear to Büchner that force, like matter, was immortal. After all, force and matter were inseparable, and the conservation of matter had been known since Lavoisier. In reconstructing the history of the discovery of the conservation of energy Büchner's champion was Friedrich Mohr, whose paper, 'Über die Natur der Wärme,' had been rejected by Poggendorf's *Annalen* in 1837; i.e., *before* the work of Mayer and Joule.[115] There is little doubt that Büchner gleaned his understanding of conservation of energy from Mohr. For one thing, it was Mohr who helped him with the chapter on 'Unsterblichkeit der Kraft' for the later editions of *Kraft und Stoff.*[116] Through this association the two became close friends, so close that Büchner was the recipient of Mohr's papers after his death.[117] Further, Mohr's enunciation of the principle stressed, as did Büchner's, the interconvertibility of the one acting agent – force.[118]

Not that Büchner sleighted Mayer. Mayer, he said, had given the mathematical and experimental proof to energy conservation. Büchner noted that Mayer's work too had been rejected by Poggendorf, and, quoting Feuerbach, he sang the praises of the "martyrs of truth and science," including himself in their ranks.[119]

By modern standards Büchner's exposition of the conservation of energy is misleading and imprecise. He, like others of his day, did occasionally use the word energy (*Energie*). Büchner used energy as a synonym for Helmholtz's tensive force (*Spannkraft*).[120] Yet, as we have seen, he hoped to be able to explain tensive forces as motions, like other forces. In fact his hope soon became the dogmatic assertion that the tensive forces were *not* exceptions in any sense, but that there was only one activity in nature – motion.[121] It was not long before he fell into the habit of using the terms force, energy and work as if they were interchangeable.[122]

If Büchner's popular account of energy conservation is confused, it is confused only by modern standards. By the standards of his own day his exposition reflects in large measure the manner in which scientists conceived the issue, including Helmholtz, who shared many of Büchner's convictions about the mechanical basis of physical science.[123]

In Büchner's account of conservation of energy, however, there were no formulae to wade through, only the many examples of "forces" that were interconvertible. To the layman reading Büchner's exposition the claim that all forces, indeed all activity, was accountable as the motion of some primary element was just simple enough to grasp and just complex enough to seem scientific. The same could be said of other of Büchner's popular treatments of scientific issues, e.g., spectral analysis, which he took to be a confirmation of his claim in the *Kraft und Stoff* that the universe was everywhere the same.[124] His exposition of spectral analysis placed the emphasis not on Kirchhoff's law, which he never mentioned, but on the fact that the core of the burning sun was surrounded by an atmosphere which accounted for the peculiar nature of the solar spectrum.[125]

Büchner did not confine his discussion of thermodynamics to the first law, he took up the second law as well. He prefaced his treatment with a general discussion of the sun. Most of his efforts were directed toward establishing that the sun was the source of all life and activity on earth, although in the process he often reviewed the scientific information known about the sun, or discussed the religious significance that men in the past had ascribed to the sun. He explained how it was that coal could be called the "sun in the cellar," and even calculated for his readers the number of 16 HP steam engines that could be operated from the fractional amount of the sun's heat received each year by the earth.[126]

Nor was Büchner above shifting suddenly from the scientific to the moral. Light, he said, had been shown by Moleschott to have an effect on muscular activity. But light was also necessary in the moral realm. Where there was "light" there was science, education, enlightenment, truth and mental health.[127]

What about the day when the sun's light would be no more? Büchner's discussion of the second law of thermodynamics was carried out in terms of the heat death of the universe. The sun, he explained, had an enormous, but finite store of heat, much of which was radiated into cold space. Further, the heat generated on the earth through conversions from mechanical and other forces became unavailable once it was radiated into space. Hence mechanical "energy" was more and more turning into heat and being given up to the universe, gradually diminishing the amount of "force" present on earth.[128] Büchner equated the unavailability of energy here with heat that had escaped by radiation from the earth. He did see fit to refer in passing to Helmholtz's explanation, which, he explained, was based on the fact that heat could only do work when it went from higher to lower temperatures. Since by its nature heat always went from higher to lower temperatures, eventually all the temperature differences would be eliminated, thereby gradually eliminating occasions for work to be done. With the standstill that would ensue, all motion and all life would cease.[129]

The truth was that Büchner did not believe that the eventual extinction of life on earth was the last word. He criticized the physicist Gustav Hirn for overgeneralizing the significance of the increase of entropy in the universe, saying that such notions only applied to individual astronomical systems. Hirn had said that after the heat death of the universe no new formations would be possible, but Büchner declared that if that were the case, telescopes would not be uncovering the gradually evolving systems of stars which were in fact being found.[130]

For Büchner the significance of heat death was that the sun too followed the general cyclical pattern of origin and dissolution. The only difference in this application of the law of birth, life, and death was the length of time of the cycle.[131] While our solar system was doomed to die, others had not yet been born. Science confirmed the ancient myths concerning the end of the world and its rebirth. The end of the world did not refer to the entire universe – that was eternal and imperishable. Our solar system was but a wave in the great ocean of life, but even it would not die an absolute death. Büchner reviewed several cosmological speculations according to which the cold sun could become involved in the generation of a new world system.[132] He was unambiguous about it: Our world "must and will . . . celebrate its resurrection someday."[133]

This faith in the eternity of life exemplified Büchner's religious disposition. What had started out as a discussion of energy ended as a sermon on materialism. That the whole could never perish, that there was an eternal cycle in the universe he declared to be the one possible solution to the riddle of the world. The secret of existence, he wrote to a friend, was found in the figure of a circle. In his own version of religious self-denial Büchner proclaimed that the individual meant nothing in the history of the cosmos. Mankind, he explained, refused to accept this ultimately simple solution to the riddle of the world. Men had protected themselves from it by devising gradually throughout their history a complicated rational framework of reflection by which they insisted that the world must be understood. But Büchner explained that nature was not concerned to give comfort. Men must be willing to look first for truth, regardless of the consequences. He closed his treatment of the subject with the words of Feuerbach: "To faith only the holy is true, to knowledge only the true is holy."[134]

CHAPTER VIII

CONTROVERSIES IN BIOLOGY

The major scientific materialists all received their formal training in the biological sciences. Each of the four leading figures possessed a medical degree, while Vogt and Moleschott had gone on to further training in zoology and physiology respectively. It was natural, then, that in spite of their deference to physics as the model science, they would also feel at home in discussing developments in the life sciences. Three specific issues are the subject of this chapter: the controversy over the existence of a vital force, the matter of spontaneous generation, and the question of the transformation of species.

The materialists were agreed about vital force – they unanimously opposed it. The other two issues brought a diversified response from the group. Only Czolbe refused to acknowledge that Darwin had established once and for all that species change over time, though each of the others qualified his acceptance of Darwin in his own way. It was Czolbe once again who differed in his treatment of spontaneous generation, but even among those remaining, there was a marked difference in approach to that problem. More important than the individual issue, however, was the conviction that each one felt; namely, that he had remained true to the materialist's method of finding a *natural* explanation that made no appeal to mysterious categories.

According to Owsei Temkin (and Ludwig Büchner) the debate over the existence of a vital force was begun by Theodor Schwann's *Mikroskopische Untersuchung über die Übereinstimmung in der Struktur und den Wachstume der Tiere und Pflanzen* of 1839.[1] Far from being a materialist, Schwann nevertheless rejected the notion that the organism had within itself a force which directed its development according to some idea. In his view the organism was subject to the forces inherent in matter, which acted according to blind necessity.[2] Not that there was no purposefulness involved. God, said Schwann, had endowed matter with forces, which in turn, acting strictly according to the laws that governed them, produced and directed phenomena of both organic and inorganic nature. God was therefore vital force writ large.[3]

Ultimately in this scheme the question of form was explained through an appeal to a guiding Intelligence, but the immediate explanation, the one with which natural science was concerned, did not contain this element of formative power. To Schwann the appeal to a special vital force guiding the development of the embryo was as unnecessary as a special force of astronomy that gave to the planetary systems their particular order. The arrangement of the parts in both an organism and a planetary system resulted automatically from the forces inherent in matter.[4]

Other writers picked up the idea that the goals of natural science automatically demarcated ultimate from immediate kinds of questions, though few mentioned that another way of stating it was to emphasize that science had limits. Hermann Lotze's article in Wagner's *Handwörterbuch der Physiologie* of 1842, which so inspired Czolbe, reinforced the positive conclusions Schwann had drawn about the abilities of physiology to explain vital processes without reference to a vital force.

In contrast to the comparatively dispassionate views of Schwann and Lotze, the reductionist students of Johannes Müller and the scientific materialists made it their program to banish vital force from natural science. Of the latter group Vogt began the denunciations of vitalism. Early in the *Physiologische Briefe* of 1847 he noted that physiologists should be ashamed of themselves for maintaining that life was an irreducible phenomenon. That was a mere cover up for ignorance in Vogt's eyes. His ridicule of vital force revealed that, like Feuerbach, what he really opposed was mystery of any sort.

> For whenever one hit upon an inexplicable fact or mystifying phenomenon, immediately the characteristic of vital force, that inscrutable governing power of organic life, was there . . . to say: Be satisfied that organic life knows only its own laws.[5]

Vogt attacked the use of vital force to explain the production of animal heat as well. As a student of Liebig, he believed that animal heat could be accounted for without reference either to the nervous system or to hidden sources, but by means of oxidation processes.[6] The medical tradition, he said, was especially rich in mysterious categories like rheumatism, hypochondria and hysteria, "junk rooms into which we throw everything about which we have no exact knowledge."[7]

Moleschott wasted no time in joining in. In the introduction to his *Physiologie des Stoffwechsels* of 1851 he noted that scientists could sometimes recreate the circumstances under which organic matter appeared.

Vital force, he said, was a misnomer, "an unclear confusion of force with a state of affairs." It purported to account for circumstances (*Umstände*) which were produced simply by the organism's organization.[8] Fortunately this anthropomorphism, "this completely arbitrary personification of a state of affairs," no longer was used by scientists as an explanatory tool.[9]

Büchner wrote a chapter of *Kraft und Stoff* specifically against vital force. He called it the worst of those mystical and confusing concepts that weakened science and that science had overthrown.[10] If natural science permitted it a place, he continued, it would have to give up the claim that natural law was everywhere the same, and it would have to forego the belief in the unchangeability of the mechanical world order.[11] Vital force was alledged to explain what in actuality was merely the result of the properties of atoms in particular combinations.[12] Büchner went on to depict the use of vital force as an appeal to mystery, quoting Vogt, who had called it "only a paraphrase of ignorance."[13]

What were the grounds on which vital force was defended? The philosopher Julius Schaller expressed it as well as anyone. In an excellent chapter of his *Leib und Seele* entitled 'Die physikalische Auffassung des Organismus,' Schaller agreed that it was foolish to view vital force as the power directing nature. On the other hand he correctly pointed out that few scientists wanted to follow DuBois-Reymond's explanation of the very concept of force as an urge to personify the data.[14] He noted that there was a tendency among physiologists to emulate physicists, but he doubted that physiology, botany, geology and the rest of the nonphysical sciences were nothing but applied physics and chemistry.[15] If vitalists were guilty of appealing to secret forces, what about the as yet unworked out mechanical explanation of organic phenomena? Was not the assumption that such could in fact be done too much to grant? The question was, in his view, whether or not *form* could be talked about at all by the mechanists. Was it not rather that mechanists had to leave the organism in its wholeness completely out of the discussion?[16] Schaller's position was similar to that of many who retained vital force to avoid the dilemmas he had pointed out.

Much of the debate over vital force was due to a misunderstanding on the part of the participants regarding their respective explanations. If a scientist set out to explain epigenetic development, there were at least three different things he might have in mind. First, he could attempt to locate and identify the physical processes that accompanied the development. Clearly this could be done without ever leaving the realms of physics and chemistry. If, however, he emphasized that organic phenomena were somehow different from inorganic, then his task would be to show that there were unique, observable processes accompanying vital phenomena. He could if he wished point out that the laws governing these special vital functions were themselves special, but that they were nevertheless laws and therefore *natural* processes. Thirdly he might feel obligated to explain why the particular order observable in an organic form was the way it was, much as if one were to explain why the order observable in the heavens was the way it was.

The first position was that of the reductionists and the scientific materialists, but both groups went beyond the mere location and identification of physical processes. Both eliminated the distinction between the organic and inorganic, claiming that organic phenomena resulted from particular complex combinations of inorganic matter. In addition they made the assumption that the so-called unique vital processes were not fundamentally unique, but only the result of particular arrangements of the known forces and laws of inorganic matter. For them there was no qualitative difference between the development of crystal forms and the development of the embryo. The latter was just more complex than the former.

The second position was Liebig's. To him direction was required to explain the external chemical forces that act to decompose food. Only with such direction was growth and the development of a new form understandable. It was here that Liebig brought in vital force. This peculiar or vital force was not connected to an agency such as soul or mind. It was simply another force of nature, and as such it could be investigated experimentally just as all natural forces could. Furthermore, Liebig felt that vital force did not act to frustrate the laws governing forces of inorganic matter in motion, but cooperated with them. Since Liebig did not conceive of vital force as a mysterious agency directing development, it is no wonder that scholars have seen fit to qualify his vitalism as "modified" or "residual."[17] Liebig too could lay claim to naturalism, and, as one author has pointed out, his vitalism was not *logically* tied to his religious outlook.[18]

Most physiologists and organic chemists left it to the natural historians and theologians to argue over the reason why a particular organic form existed as it did. They were much more interested in the development of the form from the embryonic state, and in the preservation of the mature form in the face of external agencies which would act to alter it.

It is easy to see why there was a great deal of misunderstanding in this debate, for if one moves beyond the mere description of vital phenomena to make a pronouncement about the nature of life, as did the scientific materialists when they declared that the organic was not fundamentally different from the inorganic, and as did Liebig when he proclaimed that the two were different, then science has become metaphysics. A large portion of the debate was indeed carried out in terms of metaphysical discussion. The problem, according to Goodfield, is that the simpler inorganic systems have no counterpart for complex phenomena such as the development of the embryo, etc., and that even if the mechanisms of these processes are represented in all their complexity, physiologists "will have done nothing thereby to *explain away* the special character of the associated functions."[19] Hence a debate will continue so long as new mechanical knowledge pretends to explain this special character. What really happens, claims Goodfield, is that new knowledge often forces the crucial problem to appear under a new guise, so that with the disappearance of the old guise the problem seems to have been solved.[20]

If the debate was largely one over metaphysical assumptions, we might well inquire what influenced the metaphysical position of the participants. For the scientific materialists the motivation against the existence of vital force was clearly due to their association of vitalism with a directing agency, either God or a formative idea. These categories were extra-worldly, transcendent, and unacceptable to the Feuerbachian world view. They were unnecessary to a mechanical notion of development, for the material composition, said Vogt, determined the ideal type, not the other way around.

> The egg, once it is given, can only develop according to the way it is grounded in the structure and combination of its component parts. As soon as this material composition of the egg is changed, its ultimate development also necessarily is changed.[21]

After their general tirade against mysterious forces and excuses for ignorance, the materialists went on to reveal that their own position was far from clear. Each referred to a category of form which he assumed need not be explained because it was natural, and therefore irreducible. Vogt spoke of the organizational type (*Organizationstypus*) of species, and even of a general organizational plan recapitulated by the developing embryo.[22] Moleschott explained that once elements were organized in a particular form, they developed a capacity to persist in this form which, experience taught, could last for centuries. Yet for some reason Moleschott was able to dismiss Henle's "typical force" as a childish personification.[23]

Büchner talked of specific natural predispositions of form (*Formanlagen*) that appeared under favorable circumstances. When criticized for this he replied that the reason he could adopt such a category was because he associated it with matter and not with an external principle. To the idealist

Wilhelm in the dialogues of the *Natur und Geist* he said: "What you admire as a supernatural miracle is for me nothing more than a formal principle belonging to nature."[24] A few years later Büchner spoke of the development of cellular form differently. He confessed that one could not explain why cells which were originally similar developed differences in form. Even though this could not be done, he assured the reader, one could know with certainty that whatever the causes were, they were not external to nature.[25]

As a result of these allusions to naturalness the scientific materialists were not only as metaphysical as Liebig in their treatment of the issue, but they were very similar to him. Both made reference to irreducible categories. In the materialists' case naturalness was a cover for their metaphysically based position.

Once again it was Czolbe who was most alert to the implications of his position. For him the choice was simple: in order to be intuitively clear one had to admit openly the eternal existence of organic forms. The only other alternative was vital force,[26] and that was clearly suprasensual. Rather than take on development at all, Czolbe opted for a kind of preformationism according to which organisms were predetermined. The adult developed "from nuclei in which the form of plants and animals is represented (*vorgebildet*) in the essential parts."[27]

Vital force was opposed because, in the eyes of the materialists, it involved an agency external to nature. Were one to ask why the idea of agency was anathema to them the answer would lie to some degree in the atheistic attitude they inherited from the views of Ludwig Feuerbach. The elimination of transcendence easily could be transferred to natural science in the matter of vital force. Like Feuerbach, they were opposed to dualism in all forms, and vitalism to them represented a kind of dualism. Vital force did not provide a unified view of the whole organism, said Moleschott.

> The unity of life results much more from the deep and universal dependence which links all actions to each other, from the internal and necessarily purposeful working together of the individual parts.[28]

John Farley has pointed out that for those who did not believe in the qualitative similarity between organic and inorganic matter, abiogenesis, or the production of living organisms from inorganic matter, was an unattractive doctrine.[29] Since the scientific materialists refused to consider organic matter in a special way, it is not surprising to find that they also favored the idea that life had been produced from inorganic matter.[30]

While Moleschott was reluctant to discuss spontaneous generation, Vogt and Büchner did not hesitate to give their views. Neither made it a point to distinguish between abiogenesis and heterogenesis, but it is not difficult to extract their opinions on each matter from their works.

Vogt was highly suspicious of the word spontaneous generation (*Urzeugung*) because of its association with the tradition of *Naturphilosophie.*[31] In the *Physiologische Briefe* of 1847 he argued at length against those who claimed that infusorians could be produced from organic matter. One would have to prove, said Vogt, first that these organisms could *not* appear if the seeds (*Keime*) from which one thought they developed were completely destroyed, and secondly, that their ability to reproduce was sufficiently great to produce thousands of individuals in a few hours or days.[32] He then discussed experiments that had been made which in his view disproved the spontaneous generation of infusorians and intestinal worms. Explaining that parasites can and do survive outside the host, he described the different methods by which they could be transplanted from host to host.[33] His conclusion was clear:

> If we look over the previously cited facts we see that even the history of intestinal worms offers no basis for the assumption of a spontaneous generation; rather, that all the facts, even the most complicated, can be explained by maternal procreation, by a birth from eggs, and by the metamorphosis and migration of the worms.[34]

His position had not changed when he published his translation of Chambers' *Vestiges* in 1851. In this work he inserted notes to correct the favorable attitude Chambers displayed towards the spontaneous generation of infusorians and worms.[35] But a year later, in the *Bilder aus dem Thierleben*, something had changed his attitude. Now he was unsure and confused about the experiments he had previously thought decisive. Schwann, he said, had once convinced him that infusorians were produced not because already existing organic material "took on definite forms again," but because the organic matter acted "as a substratum for the nourishment of seeds . . . which find a favorable ground for their development and increase in fluids."[36]

He described the experiment that once had persuaded him. By heating the organic matter, water, and air all seeds were destroyed. As long as the flask was kept sealed, no infusorians were produced. Air was then introduced into the flasks. In one beaker the air was passed through sulphuric acid, caustic potash or a glowing tube, all means of killing any seeds in the new air. No infusorians ever appeared in these flasks either, but in the flask into which fresh air had gone they abounded. Q.E.D.? No, said Vogt, because one assumes in this that the composition of the air is not altered by passing it through such reagents. If spontaneous generation requires pure air, water and organic matter, then the air must not be changed in any way, and there was a question about the effect of these powerful chemicals on the incoming air.[37]

Further, he questioned the views of Ehrenberg, "the most active opponent of spontaneous generation."[38] Quibbling over Ehrenberg's use of the term eggs instead of seeds, Vogt declared that infusorians did not produce eggs as Ehrenberg implied. Next he attacked Ehrenberg's belief that infusorians reproduced by division, not realizing that in so doing he was demonstrating that Ehrenberg had not meant "eggs" to be understood as a reproductive term. What Vogt was trying to establish here is unclear, for in casting doubt on division as the reproductive means of these organisms he contradicted what he himself would maintain later in his book. Division, he said here, probably did not exist in the animal kingdom as a means of reproduction. "The phenomena explained by division do not belong there, but belong much more to the process of blending (*Verschmelzung*)."[39] Until the matter was cleared up, he concluded, the experiments regarding the spontaneous generation of infusorians remained indecisive.[40]

In spite of his openness here in 1852, he later returned to his former doubts, though perhaps without the old dogmatism he once displayed. The second edition of his translation of the *Vestiges* in 1858 contained the same corrective notes as did the first edition. Further, his *Vorlesungen über den Menschen* of 1863 emphasized that it would take a great deal to convince him that there was actually a spontaneous generation of these lower organisms.

> In spite of all the contrary claims, the spontaneous generation of organic beings from primitive matter has remained up to now beyond the scope of observation and experiment. As much as I should like to accept a demonstration of such a spontaneous generation, . . . I must acknowledge on the other side that only the most complete factual evidence can lead me to its assumption.[41]

Vogt was much more certain about the existence of an abiogenesis. His acceptance of the production of life from inorganic matter was, strangely enough, tied closely to his aversion to all theories of descension. Because descension reminded him of *Naturphilosophie*, he was predisposed against it, favoring such traditional positions as the fixity of species and geological catastrophism.[42] As we have seen above, Vogt's catastrophism made no reference to an agency of any kind, neither a Creator nor the impersonal Reason of *Naturphilosophie*, to explain the progressive quality of the fossil record.[43] But if the catastrophes spread death to older worlds in order to make room for new, more beautiful creations,[44] how were these new creations possible? The answer lies in Vogt's unique version of the production of life from inorganic matter.

Close examination of the *Bilder aus dem Thierleben* of 1852 shows that there were three different levels to his treatment of the subject of spontaneous generation. First there was the philosophical question – had life been produced from the congealing of inorganic elements? Next he considered whether chemistry could produce organic matter artificially. Finally there was the problem of the production of organic form.

His answer to the philosophical question was clearly affirmative, and his understanding of abiogenesis was all the more remarkable because of the kind of life that was produced.

From the philosophical standpoint it cannot be denied that the possibility of such a spontaneous generation indeed is not only given, but also that this spontaneous generation must have taken place repeatedly, at many times, with the same elements of which our earth and its inhabitants presently consist. . . . There must have been an epoch when the oak originated; i.e., when the same chemical elements which form oak wood, its bark, leaves and roots came together in that organic form which we recognize as oak. The same conclusion goes for animals. The entire creation surrounding us now did not exist in an earlier epoch of the earth's history – it must have once stepped into life (*sie muss ins Leben getreten sein*) – there must have been present a point in time (*Zeitpunkt*) when the elements forming the bodies of animals came together in this form.[45]

One must be careful not to read too much into these words. Although he did refer to a point in time in which the oak tree first existed, he did not say that the first life on earth was as complex an organism as a tree. That question is simply not discussed.

Pointing to the artificial production of urea (*Harnstoff*), Vogt boldly proclaimed that daily new chemical products were being artificially prepared that were identical to various compounds produced in animals and plants.[46] The artificial production of organic *forms*, however, was a different matter. Up to now, Vogt confessed, organic forms had been the uncontested products of nature, but he did not rule out the possibility that man might someday produce at least muscle fiber.[47]

In summary, Vogt did not doubt that an abiogenesis had once occurred, but with regard to the spontaneous generation of infusorians and intestinal worms his position changed from dogmatic opposition to uncertainty to respectable doubt.

Büchner took up where Vogt left off. His first remarks on the subject leave no doubt that he considered the production of life a fact of the earth's history, and that he was open to the possibility of the spontaneous generation of infusorians and worms. In the *Kraft und Stoff* the ninth chapter was entitled "Urzeugung," and Büchner wasted no time in stating his position.

There was a time when the earth as a glowing ball of fire was not only incapable of producing living beings, but must have been frankly hostile to the existence of plant or animal organisms. . . . With the appearance of water, and as soon as the temperature permitted it at all, organic life developed.[48]

The relationship between the external conditions of the earth's development and the origin, spread and growth of organic beings still was observable in the present, said Büchner, citing the cases of intestinal worms and infusorians as examples.[49] A few pages later he qualified this bold assertion. In spite of recent scientific research that cast doubt on the production of organic beings from the accidental coming together of inorganic elements and natural forces, it was not improbable, he said, that such production was still possible for the lowest of organisms. *Omne vivum ex ovo* might be true for all higher organisms, but this did not mean that their *first* production, i.e., their *Ur-zeugung*, also followed such a law.[50]

In the *Kraft und Stoff* Büchner confined his remarks to the production of life from favourable external conditions without distinguishing between abiogenesis and heterogenesis.[51] When he spoke of the subjective certainty of "the spontaneous origin of organic beings,"[52] he did so because of his confidence that life had a beginning on earth. The organic matter allegedly necessary for the production of infusorians and intestinal worms was not singled out as special in any sense. It was one of the external conditions required.

In his next book Büchner did not change his position, if anything he became bolder. The idealist Wilhelm in the dialogues of the *Natur und Geist* argued that natural science itself had shown that organic beings could not develop "arbitrarily here or there from the matter and forces of dead nature," but "that in order to be able to arise, everything organic or living requires a seed from a parent organism already extant."[53] To this the materialist August replied feebly that Wilhelm was right, but that the issue had not yet been decided. He added that there were phenomena supporting spontaneous generation, and he then went out on a limb to say that he personally believed in it in spite of the lack of scientific proof.[54]

Because he did not distinguish between abiogenesis and heterogenesis Büchner was forced to act as if doubts cast on the latter also discredited the former. When Wilhelm the idealist spoke as if he had established that a creative act had taken place some time during the earth's history, August was compelled, in his defense of the *natural* origin of life on earth, to argue that conditions in the present were no longer favorable for the law that governs the production of living things to express itself.[55]

In the *Physiologische Bilder* of 1861 Büchner continued to argue that scientists had not yet found out what conditions were necessary for life to appear. He specifically explained that Virchow's axiom that all cells come from cells did not present a philosophical difficulty to the origin of organic matter from the general mechanical properties of inorganic matter.[56]

Spontaneous generation was still a problem for him in his book on Darwin. He credited the Englishman for having restricted the issue of spontaneous generation to the lowest forms of life.[57] Even the cell, said Büchner, was probably too highly organized to be a product of spontaneous generation. The life forms most likely produced were not cells, but forms "from clumps of an animated and almost wholly unformed phlegm."[58]

The third edition of Büchner's *Aus Natur and Wissenschaft* contained a review of George Pennetier's *L'Origine de la vie* of 1868. Pennetier argued that the air contained no invisible seeds that could give rise to infusorians, that primitive life forms appeared in solutions containing no sign of living organisms, that the origin of these organisms depended on the nature and amount of the putrefactive material, but not on the air, etc. Büchner did not simply accept Pennetier's conclusions, but he did comment favorably on them.

> But in any case we learn from works like Pennetier's that the cry of triumph which the opponents of spontaneous generation have begun to utter everywhere as a spin-off of the Pasteur approach has been *premature*, and that some drops of perspiration will still have to run down the foreheads of scholars and researchers before the important question is decided definitively.[59]

Throughout his life Büchner was confident that spontaneous generation would someday be proven true. The alternative was in his eyes to accept creation or to go along with Czolbe's refusal to acknowledge a beginning to life on earth. Büchner knew in his heart that life on earth had had a beginning, and he knew that the origin of life had not required special forces. Twice he said that he believed in spontaneous generation in spite of the lack of scientific evidence.[60]

> One cherishes the well-founded hope that in time we will succeed in producing protoplasm, that primal organic substance from which all things are developed, directly

or indirectly out of the elements. The much debated question of spontaneous generation will then be solved definitively, and the age-old ghost of a vital force, which puts only a word in the place of an explanation and which unfortunately still haunts some muddled heads, will be buried forever.[61]

By the time Darwin's *Origin of Species* had come along in 1859 the temper of German materialism had already been established. It was not difficult, however, for the Germans to assimilate Darwin into their program without having to change that program significantly.

Darwin did not create the sensation in Germany he had in England. The message communicated by the scientific materialists in Germany was already perceived, at least by their opponents, as a sensationalized degradation of man. Once Germans had been told that man's mind could be compared to urine, it came as no shock that man was now supposedly related to apes. In addition, some of the materialists, notably Büchner, had argued in favor of a transmutation of species, and Vogt's translation of the anonymous *Vestiges* had undergone two editions before Darwin's work appeared. The materialists did not accept Darwin's theory without criticizing it. By the time they had explained it to the German public, it had received their imprint in the form of a qualification of the scope and significance of natural selection. Besides, Germany's intellectual tradition of criticism had accustomed German society to radical ideas as far back as the eighteenth century.

The earliest of the scientific materialists, Karl Vogt, was the most confident in his pre-1859 statements about species. Up to the eve of the appearance of the *Origin* he asserted that the transformation of one species into another was impossible.

Temkin has argued that Vogt and Virchow were two Germans who realized before the *Origin* that a theory of transmutation would fit better than any other into a mechanistic biology.[62] It must be agreed that Vogt in several places acknowledged the theoretical possibility of what he called the theory of succession (*Successionstheorie*) as opposed to his own view, the theory of revolution.[63] The former, he said, was the view of the anonymous author of the *Vestiges*, also referred to elsewhere as "the view which has come forth from *Naturphilosophie*," and the "the old Lamarckian theory."[64] But Temkin communicates a false sense of development in Vogt's thought by not realizing that the *Altes und Neues aus Thier- und Menschenleben* of 1859 was merely a compilation of three different works Vogt had issued separately earlier. Temkin points to Vogt's comment that the theory of transformation of species one into another "would offer much greater inner credibility, if the hitherto known facts did not thwart and hinder,"[65] which Temkin says was made on the eve of the *Origin's* appearance in 1859. The statement was originally made in the *Bilder aus dem Thierleben* of 1852. Elsewhere in this book, and in the *Zoologische Briefe* of one year earlier, Vogt took his stand *against* the transmutation of species.[66]

Throughout his translation of Chambers' work Vogt was careful to oppose the English author's support of descension. He denied that Chambers' view on the recapitulation of the embryo was correct, he ridiculed the old Lamarckian theory, and he challenged Chambers' ground for denying the maxim *natura per saltum nihil agit.*[67] Even when he entertained the possibility of descension, he always turned it down.[68]

Why then had Vogt translated the *Vestiges*? Chambers had attracted Vogt because he had opposed special creation by divine agency to explain the history of life on earth. But, in addition to his ideas of descension Vogt did not like Chambers' claim that a catastrophist had to be a theist, that only a theory of descension appealed to natural law,[69] and that while the Creator was not involved directly in the operation of nature, He must have started things going.[70] Contrary to Temkin's claims, on the eve of the *Origin's* appearance Vogt was still very much against the succession theory, as can be seen from his note to the second edition of his translation of the *Vestiges*, which came out in 1858. In it he corrected the Englishman once more by noting "that no species has gone over into another."[71]

Vogt's definition of species did not change fundamentally either before or after Darwin. It was given in terms of common characteristics, and *not* in terms of the fertility of offspring. He emphasized that all of zoology rested on the concept of species, "an ideal being," which he defined as follows: "To one and the same species (*Art*) belong all individuals which descend from similar (*gleichen*) parents and which, by themselves or through their descendents, become *again similar* to the progenitors (*den Stammeltern wieder ähnlich werden*)."[72] In 1851 he was quite willing to acknowledge that many mistakes had been made in the past, and that the practical application of the definition in individual cases was often very difficult. Characteristics, for example, could alter with external conditions, but he held that observation had shown that these changes were never sufficient to alter the species.[73] In the second edition of the *Lehrbuch der Geologie* in 1854 Vogt confessed that his definition resulted in an unambiguous decision only when one observed the developmental phases of the organism involved, a process totally impossible where fossils were concerned, and often impossible for zoologists dealing with living things.[74]

Vogt, we recall, had on one occasion been interpreted as a defender of transmutation because he viewed species as an abstraction.[75] When, therefore, it came time for him to consider reversing his stance on the fixity of species, his hesitation to think of species as something given readily to the senses made it an easier step to take.

By the time of the *Vorlesungen über den Menschen*, written three years after the German translation of the *Origin*, Vogt had changed his mind. Now he acknowledged that species could and did change over time, and that natural selection was a mechanism for determining the direction the change would take.[76] But never did he become an out and out Darwinist. He emphasized, for example, that there could be retrogression as well as progress in charge,[77] and insisted on recasting Darwin's discussion of species in terms of race. Accidental abnormalities might arise, and might by propagation and transmission give rise to a variety. This variety could in turn, by the continued existence of its distinguished characters, become a race and be propagated as such.[78] Races and species were for Vogt terms representing different avenues of approach to the same phenomenon.

> One assumes races when the common origin is known or is believed known, one employs species when it is lost in the night of time.[79]

Races and species were actually identical.[80]

Vogt approved and adopted the idea of "raceless animals" that had been defined by Hermann Nathusius, the student of Johannes Müller. Heterogeneous races of man could produce raceless masses, people with no fixed characteristics. This was brought about either by the transportation of natural races to foreign parts with different conditions, or by the intermixture of races.[81] New mongrel races arose from inbreeding within the raceless masses, but Vogt emphasized that it would take a long time to match the constancy of the characteristics of the original race or species. In this way he felt he could explain extinction, the renewal of creation in different epochs, the fixity of types over long intervening periods, and the development of more perfect types.[82]

Vogt's recasting of the discussion of the production of new species in terms of race was more confusing than it was helpful. Sometimes he spoke of the intermixture of races as much more important than the effect of climate, etc., in producing raceless masses. At other times he treated them with equal respect, but never was he as precise as one wished him to be. The reason for Vogt's insistence on race as a category is clear; viz., the polygenist limb on which he had perched since his feud with Rudolph Wagner. Throughout the *Vorlesungen* he argued for polygenism to the delight of his English translator James Hunt.[83]

Vogt spoke highly of Darwin in the preface to the French translation of the latter's *Variations of Animals and Plants in the Domestic State,*[84] but his qualified acceptance of *Darwinism* became more open when he observed what it had become in Haeckel's monistic and religious interpretation. Later he chose to make explicit his own position.

> Do not believe that I have abandoned the flag which I have known for so long a time; I am more a Darwinist than ever, I fully admit the theory of descension, transformation, struggle for existence, natural selection, and I have not renounced at all the *fundamental* bases on which the doctrine is supported. My heresies are not related to these bases, but to the excessive exaggerations, to the ill founded applications, to the adventurous conclusions and the illogical deductions which have too often been wished upon us as irrefutable norms.[85]

A great deal of the materialists' attraction to Darwin was not as much the uniquenesss and power of Darwin's ideas as the omission of reference to a personal Creator in Darwin's work. After Darwin the idea of creation, so bothersome to nonreligious minds, was provided with a respectable alternative. Vogt drew comfort from the fact that the English were catching up to the Germans as far as materialism was concerned. He applauded and sided with the "younger school of English scientists" led by Thomas Huxley.[86]

Moleschott's training as a physiologist had emphasized organic chemistry. Not unexpectedly, he did not concern himself with biological evolution until after Darwin's book, and even then his most extended treatment of the subject came in a popular lecture. The first editions of *Der Kreislauf des Lebens* dealt with species in Feuerbach's sense of the word, as a moral term referring to all of humanity. Only in the later, expanded editions was the issue of the transformation of species a concern.[87]

Perhaps the most revealing aspect of the German materialists' assimilation of Darwin is the manner in which Darwin was inserted into the history of biology. Of the three major figures Moleschott was the most appreciative of *Naturphilosophie*, most likely because of his early sympathies for Hegel and German idealism. First Lamarck, then Schelling received adulation in his commemorative speech on Darwin, Lamarck for having defined life in terms of physical forces, and Schelling for having made the concept of development fruitful for all creation.[88] He complained that Schelling's service was no longer appreciated enough, although he acknowledged that Oken and the *Naturphilosophen* had forced scientists to become hostile to speculation and to become lost in fact gathering.[89]

Before getting down to the details of Darwin's theory of natural selection, which he represented fairly and quite accurately,[90] Moleschott compared Darwin and Lamarck in basic outlook. It was a matter, he said, of choosing between a plan of creation and a *natura naturans*, in which nothing was created or directed by an act of will. Both Darwin and Lamarck took the latter alternative, yet there were differences between the two. As they reasoned back from the present diversity of existing organic forms, both thinkers, said Moleschott, had arrived at a common parent for plants and animals. Here, however, the two parted company, Darwin opting for a creative act of will to account for this primal progenitor, and Lamarck explaining it as the result of physical and chemical forces.[91]

Clearly Lamarck was being presented in a superior light, but Moleschott tempered his implied criticism of Darwin by suggesting that it was uncertain that the famous English naturalist had simply wanted to rescue the idea of creation. It might have been, said Moleschott, that his foresight as a natural scientist held him back from accepting spontaneous generation until doubts about it were cleared up.[92] Though Darwin was guilty of a lack of consistency in any case, the Dutch materialist added that his contribution was so powerful, "that he deserves no less admiration because he presupposes as his starting point an act of will in place of natural necessity."[93]

The significance of natural selection, in Moleschott's eyes, was that it had completed the work that Lamarck had begun, and that was not only to demonstrate that species changed over time, but to point up "the uniformly progressive control over the genesis (*Werden*) of organisms exercised by natural forces all qualitatively alike."[94] There was no doubt but that Darwin had better worked out the laws involved, thereby proving the natural necessity of the present arrangement of organic life, but when it came to significance, Moleschott referred to "the Lamarckian and Darwinian view of life."[95] Darwin and his colleagues were not the first ones to examine the workings of nature, and whether Darwin liked it or not, he had shown free research to be one of the most powerful weapons against blind faith.[96]

It is true that Moleschott thought Darwin's work to be highly significant. In fact, he listed it along with Mayer's contributions to the conservation of energy as the two great achievements of the nineteenth century. Yet when one inquires what it was about Darwin that he revered, it turns out to be less the idea of natural selection in itself than the general naturalistic attitude Darwin had helped to advance.

If the German materialists saw in Darwin something less than did the English, then Czolbe was an extreme example. Granted, he was the most nonconforming one of the lot, but here was a materialist who did not accept Darwin's theory. Given his views on the eternity of organic forms, this should come as no suprise. Czolbe's references to Darwin were indeed few; only in one place was he dealt with explicitly. Czolbe called into question the logic behind what he thought to be Darwin's assumption that the existence of variations was sufficient grounds to infer that the number of original species was minimal. Comparing the notion of species with the ideal motions of the heavens, Czolbe declared that such an inference was equivalent to a denial of Kepler's laws on account of the existence of perturbations.[97]

He listed other objections as well, the major ones of which revealed that he knew he was on the defensive. His position, after all, was the direct opposite of Haeckel's, for he did not agree with the idea that development and progress were not possible if there had been no beginning to life.[98] Since his discussion of Darwin came hidden in a highly philosophical work, it did little to alter the commonly accepted association of Darwinism with materialism.

The author who did the most to reinforce this association was Ludwig Büchner. He himself had already broached the subject of transmutation of species in the popular *Kraft und Stoff*. Although he had dissociated himself from *Naturphilosophie* in the preface, in the text he argued in such a fashion that it became clear that he did not see the issue as Vogt had. For Vogt the transmutation of species (*à la* Chambers) was associated with the school of Schelling. If that were not bad enough in itself, transmutation in Chambers had turned out to lead to the Deity.

Büchner was more sympathetic to philosophical thought than was Vogt, and his criticism of *Naturphilosophie*, though similar in result, was cast in Feuerbachian terms. The *Naturphilosophen* presupposed, said Büchner, that nature was the realization of a creative intelligence, and such an approach shared all the deficiencies of speculative philosophy. Once he had sworn his allegiance to facts as the basis of his own ideas, Büchner brashly presumed that he was qualified to judge when one view was more factual than another. This amounted to his own brand of presupposition, although, of course, he did not recognize it as such.

Vogt had objected to Chambers' deism, and since this deism was linked to a theory of transmutation, he was inclined to side with the "facts," which to him spoke against descension. Büchner, however, in associating the transmutation hypothesis with *Naturphilosophie* did not view the latter deistically. The "creative intelligence" of this speculative school was not God, but the pantheistic *Schöpfergedanke*,[99] the ultimate creative Reason that lay at the basis of reality. What Büchner disliked was not that conservative religious implications could be derived from *Naturphilosophie*, but that its starting point was given *a priori*.[100]

The result was that while Büchner admitted that no species change had yet been observed,[101] he argued at length that the facts pointed to a law of gradual transition.[102] Paleontology, comparative anatomy, and embryological investigations all provided evidence for this law according to Büchner, who cleverly quoted Vogt as an opponent of the transmutation theory to the effect that animals in primitive times might have produced young differing in many respects from their parents.[103] After citing several examples of changes in species, including Müller's discovery that snails were produced from holothurians (which "is likely to raise beyond all doubt, even in our own time, the possibility of an enduring development of one animal species from another"),[104] Büchner concluded that the law of transformation must be acknowledged, "according to which the development present is not completely gradual, like the old school of *Naturphilosophie* wanted it, but a development which always has to be more discontinuous, and this already in the embryonic state."[105] He even mused that according to this law an ape, depending on existing conditions, might have given birth to a man.[106]

In all this Büchner felt that no supernatural or essentially mysterious forces were at work. It was hardly shocking after this to hear him defend the essential similarity between man and brute, or to adopt Feuerbach's explanation of the idea of God.[107] These were spectacular points intended to raise the ire of conservative opponents.

Büchner dealt with the matter of teleology in nature in the context of the problem of freedom. There was a sense in which man was free, he argued, but in every province other than the mental life of humanity there was no freedom at all, and if no freedom, then no intrinsic purpose. To allow purpose would amount to the rule of eternal reason and to a denial of the laws of nature. Büchner developed here the same kind of perspective that Darwin depended on: "If the deer has long legs for running, he has preserved them not to be able to run fast; rather, he runs fast because he has long legs."[108] Furthermore, Büchner had already explained that changes in external conditions over long periods of time were the primary cause for the appearance of new attributes.[109] He concluded:

> A group of structures in nature which to us look purposeful are nothing but the result of the effect of external natural relations and the conditions of life on natural beings which are in the process of forming or have already been formed.[110]

From the *Kraft und Stoff* it was not obvious that Büchner's ideas on transformation contradicted those of *Naturphilosophie*. In the *Natur und Geist*, which immediately followed in 1857, Büchner challenged himself to spell out the differences between his view of natural development and that of the older school. The difference was, according to August the materialist, that the *Naturphilosophen* believed that external influences affected the already formed individuals, while such a notion clearly went against the facts. The proper view was that over a long period of time, with its accompanying changes of conditions, there were effects unobservable in our short perspective that altered the organic seeds (*Keime*). These seeds, or nuclei, contained in them a potential for higher and higher levels of development. The level they achieved was dependent on how far external circumstances allowed them to progress. Hence two different species (*Gattungen*) could arise from identical seeds.[111] This theory, which first came out two years before Darwin's *Origin*, Büchner never gave up; in fact, it became the focus of Büchner's dissatisfaction with Darwin.

In 1860, the year Darwin's book appeared in German translation, Büchner wrote a critique of Agassiz's position on materialism in which he opposed Agassiz's attitude towards the fixity of species and towards the effect of external conditions on the development of species. One could see from Agassiz's theological solution to the problem, observed Büchner, that to oppose transmutation of species landed one in unscientific territory. Büchner, therefore, issued a call for more research on this question before the scientific community decided totally against the possibility of species change.[112] When the *Origin* came out Büchner wrote a review of it in which he declared it to be the research he had called for.[113]

Büchner was not saying that Darwin had solved the problem completely. Darwin relied too much on natural selection in Büchner's eyes, and while this earned him praise for getting rid of teleological factors, it went too far in overlooking external conditions that themselves might bring about variations in species.[114] Then Büchner added something quite curious. It could be, he acknowledged, that we are just too used to the old perspective to realize that the solution has been found by Darwin, but he would go along with the German translator Bronn, who argued that Darwin had overvalued natural selection and undervalued the role of external conditions in accounting for species change.[115] There was one more thing he had against Darwin, the same thing that had bothered Moleschott. The English naturalist had opted for a miracle to explain the origin of life, but, since the debate over spontaneous generation was not yet over, the issue would no doubt be settled by the use of natural forces alone.[116]

Why had Büchner qualified his acceptance of Darwin as he did? The main reason was his curious theory of how species change over time. The key there was the role of external circumstances, precisely what he accused Darwin of undervaluing. A closer look at this creation of Büchner's makes this clear.

Büchner took up his theory again in the *Sechs Vorlesungen über die Darwin'sche Theorie* and in *Der Fortschritt in Natur und Geschichte.*[117] His main concern in the latter tract was to show that modern science lent no support to the comfortless view that history was not progressing. Such a teaching was fortunately false, said Büchner, and must be so. To show why, Büchner reconstructed the theory first developed in the *Natur und Geist*, this time borrowing the model of a tree from both Darwin and the "German Darwin," Ernst Haeckel.[118] It was the growth of the whole tree that Büchner associated with progress, not the growth of any particular branch.

According to Büchner's model, the seeds of life corresponded to the root or stem of the tree, while the *development* of the seeds to differing points of maturity corresponded to the development of the branches. Each new branch symbolized an extension of the life of the seed. Each individual path would therefore retrace previously existing paths before becoming unique by extending into a new offshoot. In an idea he might have taken over from Vogt, Büchner explained that each species, i.e., each path from the stem to the end of a branch, had a cyclical existence just as did all forms of matter. Each developed to the highest point it was capable of reaching "according to its nature."[119]

The implication was that extinct species, according to their nature, had died young, while species existing in the present had been allowed to develop longer. The important point was that each path was independent of all others, in spite of the fact that up to a certain level it was identical to some of the others. Realizing this, one can understand what Büchner meant when he said that man as a species was old, and that one should not look for his roots in another existing species. This was Büchner's way of emphasizing that man had a common ancestor with other species, but that his roots did not lie in other existing species.[120] Büchner's model also correlated the length of a path with the levels of biological classification. If one traced back the path defining a particular species to the stem, it became identical with other paths before it, thereby indicating the family, order, class, etc.

Both in his lectures on Darwin and in *Der Fortschritt in Natur und Geschichte* Büchner began his explanation of the model by dissenting from the transmutation doctrine of *Naturphilosophie*. What he wanted to accomplish was to find a way to permit the transmutation of species without having one species transform directly into another from the one before it. His solution was different, he maintained, because each species developed *independently* from the original stem.[121]

Further, his model was unlike *Naturphilosophie* because it was based on growth (*Heranwuchs*), which he said was a natural, not a supernatural process.[122] Growth was obviously something that was regulated by the external circumstances in which it occurred. It was not at all that natural selection was inconsistent with his scheme, for Büchner could have used that to explain how the new paths extended or grew beyond the older ones. But if he had given in to Darwin's mechanism completely, he would have had to drop the explanation he himself had used before Darwin – external conditions. Büchner gleefully quoted Darwin's admission to Moritz Wagner in Munich that he had undervalued the role of external conditions.[123]

Of course Büchner was far more interested in guaranteeing progress than Darwin was. In the subtitle to his book on Darwin appeared the phrase, "the relation of this theory to the doctrine of progress." His optimism and confidence in science to set the standards for the future required that mankind be assured of advancement. His model, as he saw it, accomplished this. While he acknowledged that growth was only certain in the overall picture, he nevertheless had simply neglected to mention how individual regresses could be accounted for in the tree analogy. Each species represented a point in the continual growth of the original seeds of life. The potential of each species, or the degree of maturity attained by the seeds, existed from eternity, and external circumstances could at best regulate when that latent potential would make its appearance. That the line of progress was not smooth he readily admitted, but overall progress had to come about.[124]

Natural selection, if taken as the sole mechanism for the production of species, did not guarantee progress at all. What Büchner did, then, was entirely predictable. He incorporated natural selection into his own model, while retaining the elements he needed to guarantee progress; viz., the explanation of the development of new species as a *growth* process regulated by external circumstances. This came out clearly in his judgement that Darwin had overvalued the role of external circumstances.

In addition to avoiding any threat Darwin's natural selection might have posed for his predisposition towards progress, Büchner showed his bias again in his evaluation of Lamarck, whom he once called the real founder of descension theory,[125] and in his explicit acceptance and emphasis on the inheritance of acquired characteristics. To a thinker, or a group of thinkers, who possess a strong commitment to the inevitability of progress, belief in the inheritance of acquired characters is certainly not uncommon.

In his lectures on Darwin Büchner compared Darwin and Lamarck. The virtues of the latter were not only that he was the first to show that species were not fixed, but in giving various external causes for this.[126] Lamarck accepted a law of progressive development, postulated spontaneous generation, and was not afraid to apply his idea to man, even to the point of holding that the root of the human race was a human-like species of ape.[127] But Lamarck also fell short of Darwin as well, said Büchner, for Darwin had not depended on the self-activity of the organism as had Lamarck, but had emphasized more than he the significance of external conditions.[128] Although Darwin was admittedly the more empirical of the two, he nevertheless came at the end of a long list of forerunners, which began with Lamarck and ended with Büchner himself.[129] Darwin's explanation of the giraffe's long neck by natural selection was admittedly much better than the proverbial Lamarckian stretching,[130] but, said Büchner, the two agreed at several crucial points: they both denied the idea of species, they both rejected catastrophes, and they both put great weight on inheritance.[131]

It was this view of Darwin's work as an important support of the power of inheritance that most clearly marks Büchner. In his exposition of Darwin's theory, Büchner implied, though he nowhere said so explicitly, that Darwin believed in the inheritance of mental properties such as drives and talents.[132] Later in a work entitled *Die Macht der Vererbung*, Büchner made his own convictions explicit. Darwin, he said, had made the process of inheritance famous.[133] Büchner's goal in detailing the different kinds of inheritance and their significance was to show the influence of heredity on the physical and mental progress of mankind.[134]

The great significance of heredity was only recognized after Darwin, he said, when the idea was applied to man. In addition to physical or bodily heredity, there was also mental and moral heredity, important factors in distinguishing man from other beings. Habits, tendencies, drives, talents, instincts, feelings, character, moral sense, intellect, all were examples of inheritable characteristics.[135] While it was true that the laws of heredity were not yet worked out, Büchner did not doubt the significance of inheritance as far as progress was concerned.

In fact, Büchner later admitted that Darwin alone was not sufficient to solve the "problem of progress." The explanation of progress was only complete when one brought in the force of progressive inheritance, which he equated with the inheritance of acquired characters.[136] Although he rejected the various speculative attempts to account for it, including Darwin's pangenesis, he simply declared that characters acquired during life were communicated to the seed. Once again he had found "facts" that as yet had escaped adequate explanation, but which were nevertheless certain.

Büchner conceded that such talents as the acquired ability of a pianist could not be passed on, but he did give examples of what acquired characters could be transmitted. Among these were immunity against infectious disease, a tendency to drunkenness, and other qualities.[137] Without the inheritance of acquired characters one was left with a Darwinism reduced to the natural selection of chance variations, and that would never do.[138] In an article late in his life vindicating Lamarck as the real champion, Büchner said it quite simply.

> It is a great weakness and inconsistency in Darwin that individual or random change – or the formation of varieties, which the whole course of nature strives to obliterate through cross-breeding or through the death or unfit individuals – should be the forerunner of new species.[139]

In so far as the German materialists brought Darwinism to Germany, and they did play a considerable role, they imparted to it a particular German stamp. All of them, except Czolbe, went along with the times in declaring Darwin's work a great achievement, but as time passed at least Moleschott and Büchner played down its uniqueness in light of Darwin's predecessors. Vogt and Büchner did not hesitate to apply Darwin's ideas to man and to proclaim Darwin's theory as proof of the qualitative similarity between man and animal. But most important of all, Darwinism exposed the optimism and idealism of the scientific materialists.

Vogt, Moleschott, and Büchner may have agreed that Darwin had helped to eliminate teleology from science, but they certainly did not mean by this that the world contained no purpose. They all believed in the progress of history. None of them assumed that the world and the development of life was the result of pure chance. What they opposed was a teleology resulting from external agency. Once again they appealed to the natural and the factual as cover-ups for their lack of mechanisms to explain development. What they were fighting for was an *attitude*.

Moleschott readily acknowledged that he was not so blind as to deny that there was purpose in nature,[140] but he felt that the way in which purpose should be understood ought to be from within nature's own structure.

> Harmony of the universe is inherent to the beginning of things; and if we could embrace them in a single glance, we should see that first causes correspond with final causes; and teleology and causality would merely be the two faces of the same medal.[141]

Büchner's ideas were similar, yet even more explicit. From what he himself said, it was evident that he must have taken to heart the position of his elder brother Georg, as spoken in the latter's trial lectures at Zürich in October of 1836: "Everything that exists does so for its own sake. To seek the law of this existence is the goal of a viewpoint opposite the theological, one which I call the *philosophical*."[142] Georg was contrasting here his philosophical view with the English tradition of the argument from design, and he emphasized the necessity of omitting external purposes.

Ludwig felt that he had eliminated teleology where it counted – in the short run. But that he believed in a general purpose was clear from his convictions regarding progress. On one occasion he associated purposefulness with the survival of the fittest. Even Empedocles knew long ago, he said, that "it lies in the nature of that which is purposeful to preserve itself, while that which lacks purpose must perish." This, he continued, was nevertheless first generally accepted only after Darwin.[143]

Büchner appealed to the process of growth, a natural process he thought, to support his claim that no external agencies need be involved in the explanation of living nature. Purpose was built in and predetermined, yet it was still purpose. The world, Büchner said, was "not something constructed, but something growing;" it and everything in it "has not been created, but has evolved, has not been made, but has developed."[144]

External agency, the thing he most wanted to eliminate, was an arbitrary interruption of natural processes. Just as there was no vital force directing the growth of the individual organism, there was no guiding force directing the growth of the universe. Nature was no artifact, but the resultant sum of billions of years of development. The course of this development was self-produced; it could not have turned out in any other fashion than it had in reality.[145] The protestations against *Naturphilosophie* had a very hollow ring about them, for Büchner's estimation of Darwin, determined as it was by his belief in the progress of history, was born from his unconscious sympathy for the best of German Romanticism.

CHAPTER IX

MATERIALISM AND SOCIETY

In this chapter we shall examine the way in which the scientific materialists used science, more properly how they used their conception of science, to draw conclusions about the social and political realm. Before attempting this, however, a word about the significance of such an enterprise.

Even if we are successful in locating points at which the materialists appeared to reason directly from science to society, we cannot on this account conclude simply that their ideas about science were the basis of their social convictions and actions. It is not immediately obvious which influenced which. Might it not be equally possible that the social and political turmoil of the years following 1830 was more influential in shaping the materialists' conception of science than science was in shaping their conception of society? Or could one not argue that the materialists were first and foremost anticlerical figures, and that their fight against religion quite naturally involved them in an opposition to the throne joined to the altar? In this case their appeal to science, indeed their very understanding of the nature of science, might be a symptom of a deeper antireligious, or at least antiauthoritarian temperament.

If the German materialists were as profoundly affected by Feuerbach's religious world view as has been asserted, then the last alternative, that of viewing the scientific materialists primarily as anticlerical and anti-authoritarian figures, would seem the most appropriate way to understand them. Feuerbach's convictions in social and political matters, and the manner in which he arrived at them, were tied directly to his opposition to the religious ideas of his day. The scientific materialists, heavily influenced by Feuerbach as they were, adopted a social view much like Feuerbach's. Hence it is natural to suggest that they did so for reasons similar to Feuerbach's; viz., their opposition to the German religious establishment of the 1850's.

All the materialists were humanists and, in their sense of the word, naturalists. That is, they all explicitly adhered to traditional moral values, and they all felt that to uphold these values one need have no reference to a transcendent deity. In this they were followers of Feuerbach, whose emphasis on love and the divinity of man had persuaded them that traditional religion was not only hypocritical, but heretical.

The social and political views of the scientific materialists, then, were not determined by their understanding of science. If anything, their use of science to defend their political opinions was determined by their already formed social and political convictions, and these in turn resulted from their Feuerbachian perspective on philosophy and religion. When a given materialist's social and political views differed from Feuerbach's, or from those of another scientific materialist, his use of science to provide the framework for solution to social and political problems reflected that difference. In other words, there was no one "scientific" understanding of society among the German materialists of the nineteenth century. Rather, science was used to support different conclusions depending on the scientific materialist involved.

The diversity of opinion on the significance of science for society is visible not only among the different materialists, but it is also evident from the changes of position over time in the views of the individuals. Czolbe, for example, in carrying his program of the elimination of the suprasensual from science over into society, came up with an unabashed, "intuitively clear" defense of bourgeois capitalism. But as time passed he changed his mind about the form of government that should preside over this capitalism. Again, Vogt was certain in 1850 that science indicated that society should be anarchistic, only later to drop this idea in favor of a more traditional liberalism. Each one had his own views on the matter. Their agreement came not on the lesson science held for the structuring of society, but in the opposition to traditional theism. Beyond that, both in their interpretation of the nature of science and the significance of science for society, each went his own route.

Feuerbach had remained notoriously apathetic in the political arena in spite of the fact that he had called for politics to replace religion and had condemned the separation of theory and praxis.[1] Marx's flattery did not penetrate his aloofness. In 1848 he left Frankfurt in disgust over the pitiable spectacle of politics at work. He himself was not unaware of this inconsistency; he even tried to justify it on one occasion.[2]

Feuerbach had initiated the call to action in the 1840's. Action seemed to fall naturally out of his system of thought. As one scholar has summarized it: "Humanity was divine; humanism meant worshipping man because man was God. And worship meant action."[3] For Feuerbach the belief that there was a reality outside or above the ordinary world of our senses, the belief of theology and idealistic philosophy, meant not only that one landed in a logical dualism, but also in moral-political subjugation.[4]

Religion, as the traditional guardian of morality, based its commandments on a false belief, an artificial authoritarianism. When the real situation had been exposed, religion, so thought Feuerbach, would wither away and die, and politics and technology would replace it.[5] The basic drive of modern man was for freedom from artificial and unnecessary subjugation in both religion and politics.[6] The realm of truth was no longer theology or God, but nature,[7] and nature was republican.[8]

To hear him speak one might have thought Feuerbach a thorough revolutionary. Socialism and communism, he said, were children of Christianity.[9] But they were illegitimate children, for they strove to set up an absolute equality in a world that did not know such a thing. In 1849 Feuerbach acknowledged that he was a defender of a constitutional monarchy in practice, but, he said, he viewed the development of a republic as an ultimate goal.[10] He politely dissented from Ferdinand Lassalle's bitter estimation of the German progressive party,[11] and in the end Feuerbach, who has been said to have inspired revolutionary socialism,[12] earned a place in Marxist thought as a materialist in philosophy, but still an idealist where society was concerned.[13]

If anyone was close to Feuerbach's moral position it was Moleschott. Even their periods of extreme radicalism coincided. It was Moleschott's *Die Lehre der Nahrungsmittel* that gave rise to Feuerbach's 'Die Naturwissenschaft und die Revolution' in 1850, and both men soon quieted down from these extreme reactions to the failure of the Revolution.

In *Der Kreislauf des Lebens* natural science was called the Prometheus that gave man dominion over the elements.[14] Here as elsewhere Moleschott displayed how he used science to support the social position he had come to. He did so by comparing the life of society to the view of organic life that he had described in his major work. Borrowing from Liebig, he explained how in an organism each part had a natural right to the free application of its work force, how no one part hindered or restricted another, and how the application of these ideas to society meant that unequal taxation was condemnable in the state.[15] Moleschott made his point in language similar to that of his mentor Feuerbach:

> A free and just distribution of force and matter, that is the goal which all recent movements have pursued more or less dimly, a distribution of matter which makes labor possible and through labor an *existence worthy of human beings*.[16]

Although he denied that communism was the natural political result,[17] Moleschott did say that the future belonged to socialism, and that this was guaranteed, as was the impossibility of communism, by the unassailable fact that force followed matter in the material exchange (*Stoffwechsel*) of society.[18] Later he explained what he meant. Whoever demanded a certain amount of work was morally obligated to provide the nutriment to support this work. This was the kernel of every socialistic rule of life.[19]

Moleschott came to believe that technological improvements, which he credited to the advancement of science, had a humanizing effect on man, that man's capacity for evil was being eroded by advances in knowledge. "In this way," he wrote, "it is a fact that every application of science develops the moral force of man."[20] Probably because of this naive faith in Feuerbachian man, he opposed the use of force to achieve social equality. The main task of social theory (whose method was the same as that of natural science) was to examine the factors affecting life and to better social conditions as the result of the gradual, uninterrupted development in progress.[21] Moleschott fits the general pattern defined by the scientific materialists. Like the rest of the scientific materialists, Moleschott started out more radical politically than is indicated by the more tempered position he later defended. This change of attitude is even more visible in Karl Vogt and Ludwig Büchner, the two materialists most active on the social and political scene. Moleschott was the one who appealed to Feuerbach most explicitly, but the hostility against clericalism and traditional religion was a motivation to the others as well.

Karl Vogt was the one scientific materialist who was almost totally uninterested in philosophy, let alone theology. Prior to Frankfurt, where he met Feuerbach personally, he knew his works through his friends Herwegh and Herzen, both enthusiasts of Feuerbach's ideas. Of all the materialists, however, he alone never specifically appealed to Feuerbach to support his ideas. Nevertheless Vogt's opposition to authority was, like Feuerbach's, based as much on his aversion to unfounded religious authority as it was to the authority of the state. The two went hand in hand. His ridicule was directed at their combination, represented in the Prussian King's understanding of the Christian state.

As Feuerbach did in 1850, Vogt pictured natural science as the antithesis of the religious and political status quo, writing that each advance in science, because of the materialism on which natural science was based, contributed to the erosion and inevitable downfall of the Christian state.[22] Each new law, every new truth uncovered in research helped to tear down the propositions of catechism and bourgeois law. Many scientists themselves, said Vogt, were deluded in thinking that science was restricted to the limits that Humboldt had drawn for it in the first volume of his *Kosmos*, limits beyond which faith reigned. Our age, he said, had torn down the wall between the material and the immaterial.

In this it is like the railroads. One has to build them to take care of the material interests of the people, and in so doing one cannot prevent ideas from being smuggled into the land along with goods and persons.[23]

Misteli sums up his analysis of the significance of Vogt's statements here in the preface to the *Ocean und Mittelmeer*, written on the eve of 1848, by claiming that Vogt clearly and simply proclaimed "to do science means to make revolution (*Naturwissenschaft treiben heisst Revolution machen.*)"[24] Further, Misteli prefers to understand Vogt's political radicalism along the lines suggested at the outset of this chapter; viz., as an extension of his *antireligious* materialism.

The purpose of this revolution, however, coincides rather well with the conclusion of the materialistic Bible: unrestrained war against church and religion! This proposition need only be broadened slightly to read: war against church *and state*. . . . Hence [let there be] war to the church in order to knock from its high pedestal the proud state [lurking] behind it. That is how the new solution goes.[25]

One of the issues that helped to establish Vogt's reputation as a delegate in Frankfurt concerned the future status of the church. According to Misteli, by his second day in that city Vogt was well known as an atheist.[26] But it was his speech of August 22, 1848, with its thoroughly sarcastic and polemical overtones, that totally antagonized both Center and Right. Here perhaps more than in any other single instance Vogt demonstrated how opposition to established religion was easily extended to the established political structure.

The matter centered on the separation of church and state, truly a radical step in Germany. Vogt stated that he favored the separation of the two institutions, but only under the condition that everything known as the church disappeared from the earth and withdrew into its home in heaven where it would remain inaccessible to life on earth. Continuing, he spelled out his opinion of the church.

Every church, without exception, is as such a restriction of the free development of the human spirit, and because I desire a free development of the human spirit in all directions and without limits, I want no restraint on this freedom, and therefore I want no church.[27]

After going on to deny that true morality could be based on a coercive institution like the church, Vogt declared that genuine morality was born only out of true freedom. There were too many examples of those who said that they believed in the freedom of the church from the state, but who in the end had only a limited concept of freedom. In *his* explanation of the separation of church and state Vogt displayed once again his ability to move from a religious issue to a political one.

Gentlemen, there is but one means of making this freedom of the church, this separation of church from state harmless. Sanction wholly unconditional freedom, give completely unconditional development to democracy in all directions and in all consequences; then you will need not fear the separation of church from state, then you need not fear those who grub about in the name of God and religion.[28]

Complete freedom, freedom to be an atheist without fear of legal punishment, that is what he went on to demand. Turning to the faction on the Right, he exhorted them not to rely on the past. Addressing the Center, he urged that even the standards of the present not be the basis of Germany's future. Then, in the midst of applause from the Left and the gallery, he proclaimed that he sided with the future, whose belief in unlimited freedom would prevail.[29]

That Vogt's oratory was understood as the same old message Feuerbach had initiated at the beginning of the decade became clear from a cartoon that appeared soon after the speech was given. The cartoon depicted two ferocious-looking characters, followers of Vogt, bearing placards that summed up Vogt's message in the slogans "The beyond is no more," and "Heaven only on earth,"[30] precisely the sentiments Feuerbach was most often understood to have preached. The cartoon was thoroughly antagonistic to Vogt's revolutionary ideas, however, for in the background two elderly onlookers asked, "Who will bring it?" One of the old onlookers thought he saw in Vogt's followers the old Parisian Goddess of Reason, once again demonstrating how easily the move from a radical position on religion to revolutionary political ideas could be made.

Throughout the duration of the Parliament Vogt continued to appeal to science in support of his political stance. The catastrophist language he had used in his inaugural lecture at Giessen continued. In fact, others of the Left also exploited the language of natural science to argue their positions.[31] In each case Vogt's appeal to science was directed towards what Mistelli calls his political idealism. Minimally it was an attempt to enlist natural science in support of the Left.

Once on the floor of the National Assembly Vogt took the step he had hesitated to take in his Giessen lecture of 1847. He explicitly declared that revolution as a means of *political* change was supported and justified by the fact that progress in nature was not gradual, but violent. His words to the Assembly left no doubt.

> There is in nature, and therefore also in political relations in my opinion, no other development than through revolution. There is no progressive development of stable conditions, there is no peaceful, lawful development in nature. A given state of affairs, produced by revolution, remains and persists for a considerable time until it perishes from the marasmus of its own being, and then this state must be purged and transformed through a new revolution: that is the law of nature, and that is also the law of political progress.[32]

More and more he had exploited science to defend his position. Here in his enthusiasm he equated the two. After the Revolution had failed, when his enthusiasm had been dampened, he no longer held that what was true in nature was therefore true "also in political relations." The heat of the excitement had led him into this extreme position.

At Frankfurt Vogt did not connect himself with the most radical members of the Assembly. With the split of the Left over the issue of Polish freedom, Vogt, along with Robert Blum and others, began meeting in the *Deutscher Hof* to distinguish themselves from Ruge, Ludwig Simon and others farther to the left. One of the most frequent spokesmen at Frankfurt,[33] Vogt urged the formation of a national army, and supported such liberal measures as the abolition of capital punishment and cancellation of the ranks of the nobility.

That he was serious about his anti-authoritarianism is evident from his position against constitutional government. Even so-called democratic authority made him suspicious. A constitutional regime was no more than a continuation of bureaucracy, he said on the floor of the Parliament. "The claim that the rule of the majority is the principle of the constitutional system is an official lie. Ruling through minorities: that is the whole secret of doctrinaire constitutionalism."[34] A state should perish, he said, if it ever denied equal rights to all citizens.[35]

His preference for anarchism, or the elimination of all state control, was no doubt picked up from his association with Bakunin and Herzen in Paris. It reached its zenith immediately following the failure of the Revolution. Having brought science into the picture in defense of his stance at Frankfurt, he did so once more when the failure of the Revolution was certain. He even began to speak of such cross disciplines as "social physiology" and "physiological statistics" in support of his beliefs.[36]

The most outstanding example of Vogt's linking together of political anarchism with natural science came in his *Untersuchungen über die*

Thierstaaten, begun in Switzerland in December of 1849. For twelve pages of the preface Vogt lashed out in vehement disgust against the Gothaer, the party of the right-center at Frankfurt, which he believed to be the cause of the demise of the Revolution and the victory of the reactionary forces.[37] In comparison to the confusion at Frankfurt, the singleness of purpose according to which animal states were organized embarrassed him. Ants, bees, and other insect colonies were not fooled; they knew where the basis of social organization lay as well as Vogt.

Material well being is the bond which holds together all animal states; the satisfaction of all the needs of the individual by the whole (*Gesamtheit*) is the single stipulation of its peaceful continuation.[38]

The quality of the material conditions was also of crucial importance, for there was in Vogt's mind a direct relationship between diet and government. Writing at the very time Moleschott was compiling *Die Lehre der Nahrungsmittel*, the two came up with identical messages, only Vogt took pains to make it explicit. From the fact that one could alter the composition of secretions by changing diet, Vogt argued that the same could be done with the secretion of the brain – thought. Noting that the study of nutrition represented a new field in science, he called for scientific investigation of the ties between diet and political and social ideas. He believed that the result would be the banning of political hatred and persecution, since an aristocratic or revolutionary disposition would no longer be a party matter, but a phenomenon that could be controlled by diet.[39] The same thing was true of religion, "for since belief is only a characteristic of the body's atoms, a change in beliefs depends only on the manner and kind of the replacement of the atoms of the body."[40]

The clergy in particular was not spared ridicule. In a section similar to Orwell's *Animal Farm*, Vogt depicted the clergy as cockroaches, lovers of darkness who consumed what others worked for. Their natural enemy was the social democratic ant, which, when repressed beyond limit, would unite to fight the cockroaches until all of them had been exterminated. Vogt's description of the attack of the social democratic army against the cockroaches was brimming with the vengeful language of sadistic violence.[41]

Next Vogt turned his attention to the social organization found in nature, and in doing so he did not shy away from explicit political imagery. Constitutional monarchy, for example, was represented by the bee state. The cruelty exhibited among the bees showed that in a constitutional monarchy even the citizens might be sacrificed to keep the leaders in power. Vogt noted, however, that even here no worker ever went hungry, and he called it to the attention of those who favored constitutional monarchy in Germany.[42]

The point he wanted to make as an anarchist was that no governmental organization would do. It is important to realize that in spite of his description of the social democratic annihilation of the clergy, Vogt was not a socialist. Peaceful socialism, as taught by Saint-Simon, Fourier and Weitling was partially realized in nature, Vogt said, by the loosely organized water organisms he called the *Blasenträger*. Even here, however, the optimal situation was not realized, for the individual had very little freedom.[43]

With his deterministic materialism clearly showing, Vogt announced that there were but two alternatives open to those who wanted to continue fighting for freedom. One could devote oneself to natural science, and continue the erosion of the Christian state from within, or one could rededicate oneself to forceful means. What was needed was a combination of the two:

> Only when the young generation marches to battle with the sword in one hand and the codex of animal capacities (*Thierfassungen*) in the other will new times be possible. . . . The revolutionary zoologists and the revolutionary generals are still going into the schools and training academies – one will have to wait until their education is completed.[44]

Confidence in the erosive effect science had on authoritarianism spurred him on. The lesson would be learned from the "codex of animal capacities," he said. But what was this lesson? If Vogt continued to believe, as he had at Frankfurt, that conditions in nature were normative for society, one might expect him to describe some animal state that was organized according to standards he favored, and to argue on that basis that human society should take a lesson from nature. Indeed, if Vogt's scientific investigation of nature were the source of his ideas, and not a tool used to support a position already held, this tactic would not be surprising. But such a move was never very convincing, since nature presented such a variety that virtually any position could be found exemplified somewhere if only the investigator looked far enough.

The lesson Vogt had in mind was two-sided. First, as he has already indicated, he wanted to underline that the basis of social structure was material conditions. This was a direct inference from his materialistic outlook, and was equally evident in all the different forms of social structure in the animal world. Every member of the society must have his material needs provided for or the arrangement simply could not last.

Secondly, he wanted to point out that the animal kingdom was *different* from human social organization. "Gentlemen," he had said even at Frankfurt, "there is a difference between human society and animal society."[45] Animals, by which he meant mostly insects, displayed a tendency opposite that which man exhibited, for with them the better the society, the more the individual was subordinated to the whole. There *was* a scale of perfection, some animal societies giving more place to the individual than others. But all animal states, be they socialist, republican, aristocratic or monarchial, restricted the individual.[46]

Again at the end of the work Vogt repeated what he had said at Frankfurt: man did not have to live like animals. The lesson was that man could deliberately control the structure of society because he could control its material base.[47] The supreme value for humanity was not the same as the highest good in an animal state. Among the animals top priority went to a smoothly working structure, but in human society freedom from state control should be prized above all else. Although at the time the Marxist program was still relatively young and unknown, it was clear from his denunciation of those whom he called false prophets that Vogt's convictions, so close really to *laissez faire* ideology, could never mesh with those of Marx. His words were not directed at Marx, but eventually they would apply to him.

> They are false prophets who think they can bring about the salvation of humanity through laws, systems, arrangements of state. They are false prophets who really believe they can make the people happy by changing the regime and by systematic benefaction from above! The progress of humanity to better things lies only in anarchy, and the goal of [humanity's] striving can only be anarchy.[48]

Vogt was not among those who believed in scientific socialism or even in the existence of laws governing society. The salvation of humanity would not come about because of such restriction, but because of what Vogt conceived the ultimate example of freedom – anarchy. The most he would admit was that anarchy might appear to be chaos to the short-sighted, but to the far-seeing eye it was the image of the harmony of the spheres.

> Come then, sweet world-delivering anarchy, for which the depressed spirit of ruled and ruler longs, you, the only savior from this state of stupefaction.[49]

Marx was correct to recognize readily that the scientific materialists were not sympathetic to his brand of materialism. There were indeed several differences between scientific and dialectical materialism. First, it was the dialectical materialists who sought for "scientific" socialism, while the scientific materialists, apart from a few exceptions, shied away from seeking *in nature* verification of their ideas. "A man can read in the book of nature whatever he pleases, just as in the Bible," said Vogt later in his life.[50] Darwin's ideas might have been an indication of the dialectics of nature for Engels, but as far as the elderly Vogt was concerned, "Darwinism is neither socialistic nor aristocratic, neither republican nor monarchical."[51]

Vogt had learned his lesson with catastrophism. He did not again make the mistake of using a scientific theory to "prove" a political position. In the religious sphere Vogt was concerned to fight the traditional, exalted view of man by linking him to the beast, but on the political and social level his use of nature was different. In the latter case man was to learn from nature that he could rise above the animals. *How* he was to be different from them Vogt thought was obvious to anyone who knew the truth.

Secondly, the scientific materialists were not the complete revolutionaries the Marxists were. They might have agreed with Marx and Engels that theism and the monarchy had to be replaced, but they stopped short of eliminating private property, or condemning the bourgeoisie. As early as the *Thierstaaten* Vogt took a jibe at Marx and Engels for being able to find proletarians everywhere.[52] At best, thought Vogt, communism might work for factory workers, but never for farmers.[53]

Although somewhat sympathetic to socialist ideas, and although rabidly antireligious, the scientific materialists to a man opposed the use of force to gain political advantage. Vogt's vacillation is typical. When Prussia ignored the Frankfurt National Assembly by agreeing without consultation to the Malmö Truce in 1848, a delicate situation arose at Frankfurt. On September 5 the ratification of the Truce was refused by the National Assembly, then eleven days later this decision was reversed. The Left, including Vogt, naturally had fought against submitting to Prussia, and the change in the decision was taken as a defeat for them. On September 17 and 18 riots broke out in Frankfurt, and 2400 Austrian and Prussian soldiers arrived to quell it. They were too late to save the lives of Prince Lichnowsky and the Prussian general Auerswald, both brutally murdered by the mob. Predictably, the Left was blamed for the tragedy by the rest of the Assembly. Vogt was commissioned by both the *Deutscher Hof* and the *Donnersberg* to defend the Left against such charges, which he did in a pamphlet entitled *Der 18. September in Frankfurt am Main.*

The most significant thing about this work is Vogt's insistence that the Left had never condoned the use of force. Citing Ludwig Simon, Robert Blum and others, he described the efforts of the Left to contact the

Archduke concerning the riots before the disaster had occurred. The murders, he said, were senseless mob action; the Left bore no responsibility.[54]

Misteli is most likely correct to conclude that Vogt's position prior to September 18 was "decided revolution, but without force."[55] After the riots, according to Misteli, Vogt reconsidered his view. By October 23 his language had become less cautious: "Anarchy and its horrors are unavoidable evils on the way back to freedom," he declared.[56] Still, he distinguished between the anarchy "from above" and that "from beneath." Only the latter was his goal.[57]

Later, at the height of his anger, Vogt briefly gave in to the use of force. Saint-Simon, Fourier, Weitling and other peaceful socialists overlooked, he maintained, the fact that opposition to reasonable progress always came from those who would rather die than be convinced by reason. "And do you think," he asked sarcastically, "in [your] comfortable delusion that your reforms . . . can be carried out without battered heads?"[58] It was a position he could and would later retreat from without embarrassment. He had not said that the reformers *had* to use force, only that heads would have to roll somewhere. But his criticism of peaceful socialism left little doubt as to his meaning at the time.

Finally, Vogt's run-in with Marx crystallizes for us the difference between scientific materialism and the revolutionary dialectical materialism. None of the other scientific materialists was ever as radical politically as Vogt had been around 1850. The differences between Vogt and Marx, therefore, should bring to light the contrast in the kinds of materialism both represented.

The mutual dislike between Vogt and Marx showed itself early, and soon grew into mutual hatred.[59] Tactical and ideological differences emerged quickly. Vogt stood for generally liberal tactics, including the education of workers in science, literature and aesthetics before urging them to demand their share of progress.[60] According to his son Wilhelm, it was Vogt's work of 1849, *Die Aufgabe der Opposition unserer Zeit*, with its assessment of liberalism and its proposals for liberalism's future course, that originally set Marx against him.[61] Needless to say, the naive anarchism Vogt held around 1850 must have seemed ultimately bourgeois to Marx, and as the decade progressed, Vogt's reversal of his Frankfurt opposition to the Prussian crown in favor of a *kleindeutsch* solution to German unification further aggravated Marx's antagonism.[62]

The story of the formal clash between Vogt and Marx is too involved to receive here the full attention it deserves. The account that follows provides the most important relevant material.

In the spring of 1859 Vogt began soliciting the aid of German liberals in support of a program aimed at keeping Prussia from joining with Austria in the inevitable war between France and Austria. After Napoleon III's New Years Day greeting to the Austrian ambassador,[63] France's course was clear: Napoleon would support Italy in the struggle for independence against Austria. Vogt was ostensibly motivated by his sympathy for Italian independence and by his hatred for Catholic Austria. Austria's despotism, he said in the statement of his program, was eternal, while French despotism was temporary.[64]

Vogt's idea was to found in Geneva a new journal, *Die Neue Schweiz*, and to pay prominent Germans to write articles supporting the anti-Austrian position.[65] He called for the crushing of Austria and the exclusion of the extra-German provinces from the *Bund* in order to achieve unification. These sentiments were further spelled out in his book, *Studien zur gegenwärtigen Lage Europas.*[66]

In the *Studien* Vogt referred to the matter and force in the life of the state, and he appealed to the lesson of physics that force must be organized in order to accomplish anything.[67] Austria, Germany's most dangerous enemy, the "cancer of the continental situation," the "obstacle of every liberal and uniform development of Germany,"[68] must be defeated. By locating the problem with Austria's Catholicism, Vogt once again displayed his overwhelming antireligious motivation, and therefore his profound disagreement with Marx concerning the source of society's ills.

Marx first learned of Vogt's program when Vogt tried to enlist the services of Ferdinand Freiligrath, not realizing that the latter was part of the London group.[69] A few weeks later, at a public meeting on the Italian question in London, Karl Blind happened to mention to Marx that Vogt was being reimbursed by the French to carry out his program of paying writers to urge Prussia against Austria, thereby keeping France free from worries about Prussian intervention.[70] Blind assured Marx that he could prove his allegations, and soon both men, without intention, became part of an attempt to indict Vogt publicly as a traitor to Germany.

When a week later a short article incriminating Vogt appeared in a newly founded paper, *Das Volk,* Marx allegedly was surprised. He had been approached about contributing to the paper at the same meeting he had spoken to Blind, but he had declined because he did not know the founder, one Dr. Biscamp,[71] and because he was busy with other things. From this point on the intrigue becomes extremely complicated, but the upshot of the matter was that Vogt got hold of the article, and in turn delivered a public warning to his Swiss followers not to believe the lies of the fugitives in London led by Marx.[72]

Two more articles in *Das Volk* and a flyleaf entitled "Zur Warnung" appeared in June and July of 1859, all directed against Vogt. The flyleaf, allegedly authored by one of the German fugitives living in London, was picked up by the conservative, pro-Austrian *Allgemeine Zeitung* and reprinted on June 22 as "Karl Vogt und die deutsche Emigration in London."[73] The same charges were levelled against Vogt, but this time in a respectable German periodical. Forced to react by the widespread publicity, Vogt filed suit for libel against the *Allgemeine Zeitung*, convinced that the charge of treason it had printed was too difficult to prove.[74]

In this Vogt was correct. In spite of the fact that he probably never hoped to win his suit in a south German court, Vogt no doubt knew that there would be no incriminating evidence forthcoming. Pressured by Marx for the evidence he once claimed to have, Blind backed down, and eventually asked Marx not to involve him publicly. The *Allgemeine Zeitung* in turn pressed Marx for the incriminating documents he told them he possessed, but these turned out to provide merely circumstantial evidence. The defense for the *Allgemeine Zeitung* repeatedly dragged forth Vogt's radicalism in an attempt to establish his betrayal of the Fatherland, including the claim that his support of the *kleindeutsch* plan of unification was itself sufficient to show his treasonous attitude.[75]

For his part Vogt repeatedly pointed out that no documents were forthcoming. He delighted in forcing the editors of the *Allgemeine Zeitung* to admit that they were working with Marx in London. In fact, this was the reason why Vogt decided to file his suit in Germany and not in England.[76] Further, Vogt was successful in forcing the editors to admit that there were no documents proving that he had been paid by France to work in her interests.[77]

The court's decision was predictable. It ruled that the charges were much more severe than simple injury or libel. Treason, the judge argued, did not come under the jurisdiction of the Augsburg district court. The matter would have to be settled in the court of Schwaben and Neuburg. Vogt's suit was refused, and he was instructed to pay his own costs.[78]

The trial took place in October of 1859, and Vogt's account of it, with his own unproven charges against Marx and "die Schwefelbande,"[79] appeared in December. Marx, reported Vogt, had forged the original notice attributed to Blind. When the Berlin *National Zeitung* repeated Vogt's charge in its review of Vogt's account, entitled *Mein Prozess gegen die Allgemeine Zeitung*, Marx filed suit against both it and the *Daily Telegraph*, which reproduced the *National Zeitung* claim in England. The Prussian court turned down Marx's case, and he thereupon resolved to print his version of the story as *Herr Vogt*, which came out in December of 1860.[80]

Marx took Vogt's charges in *Mein Prozess* personally. Early on Engels had cautioned him not to let the Vogt affair cause him to fall behind in his work, but in his return letter Marx told Engels that Vogt was trying to make him look like an insignificant "vile bourgeois villain," and if successful it would be a "*grand coup*" for vulgar democracy. "It must, therefore, be answered in turn with a *grand coup*."[81]

This was also the justification given in the preface to the *Herr Vogt*. Vogt was very influential in democratic circles, said Marx. As a scholar, natural scientist, and political figure he was well known, just as his defense lawyer at the trial had claimed. Marx saw it to be especially necessary to unmask Vogt in Germany, where there was a fierce battle in progress over unification. He wrote to Freiligrath in late February, 1860 that the fight with Vogt was decisive "for the historic vindication of the Party and for its later place in Germany."[82]

Marx's concern to expose Vogt and all those who felt as he did, including even his earlier collaborators Ruge and Herwegh, has been taken by official Marxism as evidence of Marx's farsightedness in separating off petty bourgeois ideology from revolutionary socialism.[83] In spite of the high praise later given *Herr Vogt* by Engels and Lassalle, it became clear as 1860 passed that they and others did not share Marx's enthusiasm for the task. Funding the work was a problem,[84] but even more to the point was Engels' comment to Jenny Marx in mid-August that the great effort Marx was expending was not worth it, since public interest in the matter had faded.[85] Engels was right, for ten years after it was published, it was impossible to find *Herr Vogt*.[86]

While much of the enormous detail in *Herr Vogt* might be, in Richard Reichard's words, "a waste of intellect,"[87] this surely cannot be said about the work in general. The considerable amount of biographical and historical material concerning the early phases of the German communist movement, which Marx collected to correct Vogt's memory of those years, make it, along with Marx's *Enthüllungen über den Kommunisten Prozess zu Köln*, an invaluable reference work.

Still, it must have been frustrating to Marx to have to resort to the same tactics used by the *Allgemeine Zeitung* in its fight with Vogt. Due to the lack of solid incriminating evidence, Marx too was forced to deal with Vogt on circumstantial grounds. He thoroughly analyzed Vogt's book, *Studien der gegenwärtigen Lage Europas*, denouncing it at every turn, and demonstrating how consistent Vogt's position was with the charges that had been leveled against him.[88] Turning the tables, he challenged Vogt to produce just one piece of evidence that he, Marx, had ever engaged in blackmailing prominent Germans who had once shown sympathy for him.[89] In the last analysis, however, all he succeeded to do was to arouse the liberal press in favor of Vogt and against himself.[90]

The ten years from 1860 to 1870 brought significant changes in both Germany and in France. Defeated at the hands of a new, unified German Empire, the government under Louis Napoleon collapsed. In the financial records of Napoleon III's rule, printed soon after the defeat of France, Vogt was included on the civil list of expenditures as the recipient of 40,000 francs.[91] The payment had been made in August of 1859; i.e., after the charges against Vogt had been made. Still, Marx in *Herr Vogt* had recorded Vogt's trips to Paris in August of that year, though he did not know that it was then that money changed hands.[92]

It did not take Marx and Engels long to publicize the above revelation. In a short article entitled "Abermals Herr Vogt," Engels made public the confirmation of Marx's earlier charges, and included a reference to Vogt's present anti-Prussian attitude.[93] Again reproaches from liberals poured in to Liebknecht, editor of the *Volkstaat*, where the matter had also been made public on April 15, 1871. Vogt's disclaimer, sent to Liebknecht in Leipzig by the *Schweizer Handels-Kourier*, admitted only that it was possible that his name had been misused, implying that the entry "Vogt" might refer to someone else with that name.[94]

Vogt was the most active politically of all of the scientific materialists. Although his name was tied to that of Marx, Vogt himself was hardly a revolutionary in the same sense. Even at the height of his anger in the early 1850's he was working for anarchism, or the destruction of the state. But if what Vogt preached resembled anything, it was the Manchester school of *laissez faire* capitalism. His radicalism was therefore deceptive. It took only a few years for it to adapt itself to the more respectable cloak of German liberalism.

Ludwig Büchner's views on society represent yet another brand of German liberalism. In the 1860's Büchner became involved with Ferdinand Lassalle in the latter's attempt to rise to the leadership of the workers' movement in

Germany. Büchner had founded a workers' union in Darmstadt in February of 1863, which soon grew to 400 members. From Büchner's later account, the period from 1862–1863 was one of general excitement in Europe. There was revolution in Greece and an uprising against Russia in Poland. There was a civil war in America, while closer home a war between France and Prussia seemed entirely possible. Within Germany, the elected representatives were struggling for power against the Crown. All this had less interest for the workers' clubs, however, than the fight between the followers of Schulze-Delitzsch and those of Lassalle.

Wilhelm Schulze-Delitzsch, active on the political scene from 1845 on, had proposed the formation of workers' unions based on the self-help model of the English. Lassalle, on the other hand, thought it foolish to stand on such idealistic ground, and proposed first that the workers form a political party of their own, distinct from the progressive liberals who took forever to get anything done. Secondly, this party was to work for broader voting rights. By means of the vote, the party would be able to guarantee state intervention on the workers' behalf. The state would supply the initial capital required to construct a system of workers' associations. Later these associations could, on their own, demand their rightful share in the profits of their labor.[95]

Lassalle got in touch with Büchner and tried to persuade him, as head of the Darmstadt union and popular figure in German society, to come out in support of his program. The letter was well timed, for Büchner had been asked to evaluate Lassalle's program for the district central committee of the workers' unions. He was to give his report in six days on April 19, 1863. Lassalle praised Büchner's literary work, and noted that he had every reason to hope for the support of those who were having such a great effect on the position of the workers from the side of natural science.[96] He had written to Büchner, he explained, to bring him to his side, and to set him straight about some questions Büchner had raised elsewhere in print concerning the program of his brochure, which had come out in March.[97]

Büchner's address to the workers' assembly in Rödelheim, reprinted as *Herr Lassalle und die Arbeiter*, showed that Lassalle had not swayed Büchner to his side yet. He criticized Lassalle for flippantly exploiting the results of science for his own ends (!), and in general sided with Schulze-Delitzsch's humanistic goals against Lassalle's pragmatic demands. Not only did he question the accuracy of Lassalle's statistics, but he pointed to problems with Lassalle's state supported factories: Where would the capitalists come from who would direct the venture, seeing that they would not be working for themselves? How is the state to decide on the kind of factories to set up? Are the lazy, the dumb, and the incapable to have the same shares? Further, said Büchner, this program would increase the federal bureaucracy even beyond the excesses of the present.

Büchner listed fifteen different reasons why Lassalle's proposal was inadequate,[98] but the major error he designated was Lassalle's answer to the question, "Who is the state?" Lassalle declared that while 1/2% of the population earned 1000 thaler per year, 72 1/4% made under 100 thaler, and 16 3/4% brought in from 100-200 thaler per year. 89% of the people were hard-pressed.[99] When he proclaimed that his program was meant to help the state, it was clear whom he had in mind.

Büchner explained why this way of putting things was deceiving. Country people made a great deal by hand, and no one ever reported all of his income to such surveys. In addition, Lassalle had misused the term "worker," according to Büchner. Everyone was a worker except for a few fat fellows who had inherited their wealth. Büchner was speaking to a workers' club, and he told them that if Lassalle was successful, it would not be *they* who ran the country, but the great rural population, which Lassalle had cleverly included under the rubric "worker."[100]

A meeting was set up in May for Lassalle personally to defend his views. In one last attempt to swing Büchner to his side, Lassalle paid him a personal visit at Darmstadt shortly before the Frankfurt convention. But to no avail, as Büchner stubbornly resisted Lassalle's intimidating personality. Büchner was in the end closer to the progressives whom Lassalle hated so intently.[101]

Were one to believe Büchner's description of the Frankfurt meeting, it had been a sad day for Lassalle. He spoke too long (5 hours!), he stuttered, he was boring, he was so slanderous that Büchner, as the presiding officer, had to interrupt him twice.[102] Yet somehow, when all was said and done, Lassalle had swung the majority of the clubs to his side, even the Darmstadt one![103]

Lassalle's death soon thereafter cut short any possible future cooperation between the two, but Büchner never changed his mind about Lassalle's program. From an end-of-the-century perspective, he looked back over the developments in the interim, and although he declared that Lassalle had been the founder of the whole social democratic movement, curiously omitting reference to Marx and Engels, he argued that achieving a broader franchise had not in fact worked, just as he had warned it would not.[104]

Büchner's opposition to Lassalle and his preference for Schulze-Delitzsch's idealistic program reveal that Büchner too had settled down after 1848 into a humanistic, liberal position. In 1863 Büchner took a stand against democracy for fear of rule by the ignorant, and defended the liberal party because it opposed both feudal aristocracy and the military state.[105] He felt that Lassalle was trying to accomplish with one stroke what might take centuries to achieve. What was needed was more discussion. At the conclusion of his Rödelheim speech he proclaimed that education was the key: "Only unity makes one strong, only education makes one free."[106]

Later it would become clear that Büchner was dead set against the use of force to accomplish one's goals. Lassalle had argued that agitation would be necessary to achieve universal suffrage, and this worried Büchner. By the time of his *Darwinismus und Sozialismus* he opposed force in principle. Anything achieved by force could not last long, "since governing civilization through lack of civilization is an absurdity and is possible only temporarily."[107] It is no wonder, then, that his brother praised him with words that might have applied to Moleschott, but never to Vogt. He was "the ideological materialist in whom Christian love of humanity and obstinate acceptance of the barest facts flowed together so wonderfully."[108]

Büchner had his own plans for society. In *Der Mensch und seine Stellung in der Natur* of 1872 he reiterated his opposition to Lassalle, this time clearly coming out in favor of a modified form of capitalism. The capitalist, he argued, had a right to more of the profits than did those who worked for him. "The so-called capitalistic manner of production is but a necessary and unavoidable result of our given social relations."[109]

While on the one hand he thought it foolish of the working class to call for the elimination of capital, he denounced even more the Manchester school. Such bourgeois attitudes ignore the social conditions that could lead to a revolution by force, being content simply to revile all revolutionaries. Büchner had far more sympathy for social injustices than this. He felt that such ignorance on the part of the middle class was because the bourgeoisie, whose existence in Germany he had denied a few years before, resented its roots in the lower strata of society.[110]

Like Vogt before him, Büchner did not feel that what was good for animals was necessarily good for man. His conclusion regarding the significance of Darwinism for society was uniquely German, totally unlike the notorious lessons soon to be drawn by the Americans. It was true that man's *origin* tied him to the animal world, but from now on man's struggle would be intellectual, not physical.

The farther [man] removes himself from the point of his animal origin and relationship and allows himself to assume the place of the power of nature, which power has been ruling him unconditionally, the more he becomes *man* in the geunine sense of the word, and the more he nears the goals which we must view as the *future of man and of the human race.*[111]

As he went on to speak of man's natural destiny in terms of the species he sounded more and more like Feuerbach himself. This destiny, he explained could never be realized as long as man felt himself an individual being guided by mere egoistic motives. "Man is a gregarious or social being and can achieve his destiny, and with it his happiness, only in association with those of his own kind, that is, within human society itself."[112]

Büchner was optimistic. He believed that the impulse towards progress would prevail even under adverse circumstances, and that the end of history was the paradise of the ancients. The only difference from the myths of old was that Büchner's paradise was real, and it occurred at the end of history, not at the beginning.[113]

These were general considerations. When it came to practical matters he was not lax in providing some ground rules for the ordering of society. Both in *Der Mensch und seine Stellung in der Natur* and late in his life in a booklet entitled *Darwinismus und Sozialismus* he set forth three proposals to rectify social injustice. In his eyes these proposals possessed the capacity to preserve individual incentive, something which social democracy and communism failed to do.

If Büchner had his way, ground rent would be eliminated, and the land would be brought back into the possession of the community. In addition, rights of inheritance would be limited (in the later work he said eliminated). Finally, the state would become an insurance organization (*Versicherungsgesellschaft*) against sickness, accident, old age, and death.[114]

The social injustice within the capitalistic society of his day, especially within the Germany of the last three decades of the century,[115] motivated him to suggest these correctives. He noted that the adage, "Whoever does no work does not eat" did not cover the cases of those who did no work, but ate well, or of those who did work and did not eat enough. Some of his criticism of capitalism became quite caustic indeed.[116] Among other problems he thought that the position of women in Germany society was inexcusable.[117]

At the same time he clearly defended many capitalistic principles. The natural inequalities among men were in fact "in the nature of man and of things themselves."[118] To overlook or to deny them was to him equivalent to a denial of freedom. The division of labor was not only operative in society, but it was a general principle of all progress.[119] He opposed state regulation of work, for that would require a bureaucracy ten times as complicated as the one at present, and it would end in tyranny. He could not understand those who demanded equal profit sharing by workers, since this would eliminate the incentive of the entrepreneur. Should the day laborer, he asked, who had no other task than to place one stone on another, receive the same share as the capitalist who provided the required means?[120] As for the lazy member of society who refused to take advantage of opportunity, he deserved his fate. "He perishes not from his circumstances or from the injustice of society, but because of himself."[121]

Büchner emphasized that man had a right to be proud of modern technology and his mastery over nature. Natural science had been the most significant of all the sciences influencing modern society, witness the innumerable discoveries of the last century, the aids to technology, and the tendency towards experimentation over against speculation.[122]

In light of this it is no wonder that Büchner judged the SPD and Marxism harshly. He would have nothing to do with communism and the cancellation of private property, not only because it ruined individual freedom, but surprisingly enough because he considered it utopian.[123] One *could* imagine a communistic society, he admitted, where work was voluntary and property was held in common. Indeed, every society had socialistic elements within it, and some communistic societies in old Peru, Africa, New Zealand and elsewhere were able to operate successfully.[124] At one point Büchner pointed out that communism, viewed generally as the greatest enemy to culture, was basically a Christian idea.[125] Communism and socialism, explained Büchner, were not the evil monsters they were made out to be by the rich. He openly acknowledged that the rise of social democracy had made people aware of the problems in society. Yet he was clearly against the party's narrow conception of the worker, and its potential for violent totalitarianism.[126]

In addition, Büchner could not accept the party leaders' reply when they were asked to describe the future state. They said that they could not now predict how things would later be structured, but that the task at present was clearly to tear down the class state. To Büchner an inability to prescribe laws for the future state was too great a weakness. It was not unfair of him, he said, to demand to know what the future organization would be.[127] This kind of criticism earned him the complete disgust of Engels.[128]

At the outset of his approach to the solution of the social problem Büchner was motivated by Feuerbachian convictions. Man had no one to thank for his existence; he must seek the purpose of existence in himself, in his own welfare, and in the welfare of his race.[129] Büchner had no difficulty placing man above all else, including nature. "The real task of humanism . . . ," he once proclaimed, "rests in the war against the struggle for existence, or in the *replacement of the power of nature by the power of reason*."[30] There was an enormous difference between natural law, which operated unconsciously, and the conscious activity of man.[131]

The progress of man's development was ever before him, a progress he measured by the distance man could remove himself from theological and metaphysical ideas. The world view he defended involved in necessary fashion a thorough reform of the half-theological, half-metaphysical tenets which hitherto had formed the basis for the ideas of state, society, education, religion and morality.[132] The one thing that motivated him more than anything else was his hatred of the encroachments of religion where they did not belong. In his lecture on the concept of God, he began to sound like a preacher. Human reason stood on its own feet, he proclaimed. It was the judge over itself and truth.

> All truth lies therefore solely in us ourselves and in our free thought, which is incompatible with every kind of faith in authority, and on which one must not dare to wish to place any limitations.[133]

He went on to say that history showed that theism led to monarchism and priests. Only atheism led to reason, progress, and the recognition of genuine humanity. "No longer as before, Head down! but Head up!"[134]

This incredible confidence in the power and ability of reason to arrive at truth, supported by his naive realistic understanding of the power of science, was motivated in the beginning by the Feuerbachian concept "*Homo homini deus est*." It made Büchner blind to his own biases, and it allowed him to consider himself objective.

This is evident nowhere more clearly than in his thoughts on morality. On first reading, Büchner sounds as if he is defending the relativity of morals; in fact, he declared that this was his position. He confessed that *some* moral law was necessary for a society to exist, but on numerous occasions he pointed out that moral codes differed among different societies, thereby proving that Kant had erred in taking morality to be something innate. Like all intellectual capacities of mankind, morality was learned from experience.

Since the existence of some moral code was "a necessary result of the nature of things themselves,"[135] and since man's reason *could* arrive at the truth, it is not surprising to hear Büchner begin to speak of "the real or essential morality."[136] This kind of morality required education, and it became clear that Büchner did not at all rate western values on a par with those of the other societies he had discussed. The highest moral code was the humanism he had described. Moral codes not as advanced as this, including much in nineteenth century society, could be bettered not through preaching and threatening, since history had shown those tactics were unworkable. But morality could be advanced by making the mental life of mankind better. This could be done by means of moral training and an education carried out under the teachings of humanism.[137]

How does science fit into all of this in Büchner's mind? As with Vogt at Frankfurt, Büchner's references to nature and natural science are not always applied to society in the same way. Sometimes he appealed to natural conditions as models for humanity; at other times he used nature as a lesson for man in the way that Vogt once did. At one place he positioned man and human reason above nature; in another he argued that man could know nature as it really was because he too was part of nature. Once again like Vogt, Büchner tried to fight theological views in which man was raised above the animals, while at the same time he proclaimed that man was indeed more than animal.

Of the two attitudes, Büchner emphasized the one that placed man above nature, and the power of reason above the power of nature. He believed that the humanistic world view directly implied that in human society the means with which men are equipped to struggle for existence should be distributed equally, and that natural science pointed the way to this conclusion. What followed logically for Büchner in nineteenth century Germany might well elude modern patterns of thought.

> Here natural science once again shows us the right way. For if, as was demonstrated, the essential task of humanism, or of the further development of humanity in opposition to the brutal state of nature, rests in the war against the cruel struggle for existence, or in the replacement of the power of nature by the power of reason, *then it is clear* that this goal must be attained above all through our seeking to bring about the greatest possible equalization of the means and conditions under which and with which every individual has to fight in his struggle for existence or in his competition for his standard of living.[138]

We must not allow Büchner's simple declaration that natural science was behind his social position to blind us to the moralistic motivation hidden in these conclusions. He was well aware that to replace the power of nature with the power of reason man would have to transform what was formerly an individual matter into a concern of the species. "A *common* effort, ruled by reason and justice, a *social* struggle for the conditions of life must replace the brutal war of nature," he announced.[139] Man in community was not *nature's* concern, but humanity's.

Social man had the right to restrict individual man by equalizing the means with which he entered social competition. There should be nothing inherited from predecessors, no private ownership of what rightfully belonged only to the community, and all should be equally insured against sickness, death, etc. Beyond these things the individual was free to operate for himself. Büchner revealed more than he realized when he said that his position was theoretical, and that he was not defending any particular party.[140] His thoughts on society remained safely ensconced within a humanitarian tradition that permitted him to live a respectable life in the very society he so often criticized.

CONCLUDING REMARKS

In summarizing the significance of the scientific materialists it is important to note that 'materialism' is hardly an appropriate label for the message they preached. Like everyone else, they were men driven by ideals. They were hardly determinists who believed that nothing could be done about the course of events.

What they said was one thing, but, their denial of the freedom of the will to the contrary, what they did revealed that they were men of action, not words. J. T. Merz has aptly noted that man cannot give up his belief in the existence of a supreme and unalterable moral standard. "It seems contrary to human nature," he continues, "to rest content in the region of practice with a fluctuating and merely temporary rule."[1]

The scientific materialists found their moral standard in the new morality of Ludwig Feuerbach. They were attracted to Feuerbach because he had called for facts. As men who identified with the natural sciences they appreciated that emphasis. They were also attracted to Feuerbach's new religion. Not only did it expose the falsehoods on which traditional religion was based, but it provided a way of preserving the ideals and moral values they had never intended to throw away.

The overwhelming trademark of the scientific materialists, as far as the historian is concerned, is not their materialism, but their atheism, more properly their humanistic religion. As unique defenders and propagators of Feuerbach's message, they brought to that message the tools of their trade. As far as *they* were concerned, their message had come from the world of science. In reality it had been born within the context of the 'religious' disposition they adopted.

Nowhere is this identifying feature of the scientific materialists more evident than in the treatment they have received recently in East German Marxist literature. A re-evaluation has been called for to correct the harsh judgment the vulgar materialists earned from Marx.

In a collection of articles on nineteenth century German scientists, the concern of the editor is to make visible "that progressive spiritual heritage of German natural science." This tradition, it is said, has long ago been discarded by the ideologues of the imperialistic bourgeoisie everywhere but in the DDR.[2] Büchner is credited in this work not only for his criticism of capitalism, but for his materialistic opposition to the systems of idealistic philosophy.[3] He is treated in the same manner Feuerbach is handled by Marxists – a materialist in philosophy, but an idealist in social thought. Büchner never understood the revolutionary significance of the working class.[4]

In another recent word on scientific materialism from East Germany, Dieter Wittich emphasizes over and over again that one of the great services rendered by Vogt, Moleschott, and Büchner was their popularization of atheism. His goal in reissuing the major works of these men is, he says, to re-evaluate the theoretical relationship between vulgar materialism and dialectical materialism.[5] Although the vulgar materialists committed the unpardonable sin of remaining metaphysical materialists at a time when dialectical materialism had become not only possible, but reality,[6] the scientific materialists' attack on theism was an important contribution to the development of history.[7]

If the materialists are remembered primarily for their polemical use of science against religion, we should not overlook that this attack is not without its effect, a positive effect, even today. Speaking of atheistic humanism, Julian Casserley likens it to the prophetic element in Christianity. The scientific materialist was quick to denounce crude and inadequate systems of theology, and to demonstrate "a passionate insistence on the duty and capacity of man to make for himself a life-determining decision about matters of ultimate faith."[8]

NOTES

INTRODUCTION

[1] Charles Breunig, *The Age of Revolution and Reaction, 1789–1850* (New York, 1970), p. 257.

[2] *ID*, p. 140. This characterization was not restricted to Germany. Dostoyevsky's novel, *The Possessed*, written between 1870 and 1872, contains a passage in which the works of this trio, illuminated by church candles, were exhibited on three pedestals in the lodging of an eccentric sub-lieutenant. Fyodor Dostoyevsky, *The Possessed*, trans. Constance Garnett (New York, 1936), p. 353.

[3] This claim will be defended below. For the present, general works that analyze the influence of Feuerbach on the scientific materialists are: H. Böhmer, *Geschichte der Entwicklung der naturwissenschaftlichen Weltanschauung in Deutschland* (Gotha, 1872), pp. 122, 161; Otto Zöckler, *Die Geschichte der Beziehungen zwischen Theologie und Naturwissenschaft*, 2 vols. (Gütersloh, 1879), II, p. 401; Sidney Billing, *Scientific Materialism and Ultimate Conceptions* (London, 1879), p. 313; Heinrich Treitschke, *History of Germany in the Nineteenth Century*, trans. E. and C. Paul, 7 vols. (London, 1919), V, p. 597; Merz, *A History of European Thought in the Nineteenth Century*, 4 vols. 1896–1912. Reprinted (New York: Dover, 1965), IV, pp. 427, 692–93; Friedrich Jodl, *Ludwig Feuerbach*, 2d ed. (Stuttgart, 1921), pp. 21–22; Adolph Kohut, *Ludwig Feuerbach: Sein Leben und seine Werke* (Leipzig, 1909), p. 436; J. M. Robertson, *A History of Free Thought in the Nineteenth Century*, 2 vols. (New York, 1930), II, p. 196; Fr. de Rougemont in Wilhelm Vogt, *La Vie d'un homme: Carl Vogt* (Paris, 1896), pp. 157–58 (For W. Vogt's dissenting opinion cf. p. 100); Emanual Hirsch, *Geschichte der neuern evangelischen Theologie im Zusammenhang mit den allgemeinen Bewegungen des europäischen Denkens*, 5 vols. (Gütersloh, 1954), V, p. 583; S. Rawidowicz, *Ludwig Feuerbachs Philosophie* (Berlin, 1931), pp. 148, 334. The Russians saw Vogt, Moleschott, and Büchner as the "students of Feuerbach" (Rawidowicz, *op. cit.*, p. 111, n. 2).

[4] Early critiques of Hegel include two by Friedrich Bachmann, *Über Hegel's System und die Notwendigkeit einer nochmaligen Umgestaltung der Philosophie* (1833), and *Anti-Hegel* (1835), plus one by Franz Dorguth, *Kritik des Idealismus und Materialèn zur Grundlegung eines apodiktischen Real-Rationalismus* (1837). Other critiques appearing later in the forties can be found in Rawidowicz, *Ludwig Feuerbach's Philosophie*, pp. 79–81.

[5] Eugene Kamenka, *The Philosophy of Ludwig Feuerbach* (London, 1970), p. 71.

[6] *Ibid.*, p. 36.

[7] After the translation in Sidney Hook, *From Hegel to Marx* (Ann Arbor, 1966), p. 222.

[8] Kamenka, *op. cit.*, p. 53.

[9] 'Der religiöse Ursprung des deutschen Materialismus,' *Werke*, X, p. 155.

[10] *Werke*, II, p. 217.

[11] *Ibid.*, pp. 218–219.

[12] *Ibid.*, p. 219.

[13] Hegel himself, of course, was not guilty of conferring a separate, alienated or

transcendent existence on his abstractions in the same way this was done within historical Christianity. Hegel's God, for example, was hardly the personal God of Luther. Feuerbach found fault with Hegel less because of an error of commission than because Hegel had neglected to place the experience of sensations on equal footing with the experience of consciousness.

[14] *NS*, p. 1.

[15] A. Clerke, 'Alexander von Humboldt,' *Encyclopedia Britannica*, 11th ed. Vol. XIII (Cambridge, 1910), p. 874b.

[16] William Langer, *Political and Social Upheaval, 1832–1852* (New York, 1969), p. 535.

[17] See Carlo Paoloni, *Justus von Liebig. Eine Bibliographie sämtlicher Veröffentlichungen* (Heidelberg, 1968), pp. 106–13.

[18] On the dissipation of *Naturphilosophie* see the following: Büchner, *SV*, pp. 29–31; Treitschke, *History*, V, p. 587; F. Rosenberger, *Die Geschichte der Physik*, 3 vols. in 2 (Braunschweig, 1882–1890), III, p. 9; Merz, *History of Thought*, III, p. 177; and J. Ben-David, *Scientist's Role in Society*, (Englewood Cliffs, 1971), p. 117.

[19] On the nature of the sharp disjunction made between natural science and philosophy by scientists in this period see M. Schleiden, *Über den Materialismus der neueren deutschen Materialismus, sein Wesen und seine Geschichte* (Leipzig, 1863), pp. 45–46; Hermann Helmholtz, *Popular Lectures on Scientific Subjects*, trans. E. Atkinson (New York, 1873), p. 8; J. Moleshott, *UW*, pp. 3–4; H. Böhmer, *Naturwissenschaftiliche Weltanschauung*, p. 143; and Rosenberger, *op. cit.*, p. 327.

[20] Three of these writers of popular science were closer to Feuerbach in age than were the scientific materialists. Emil Rossmässler (1806–1867) wrote popular science to earn money when he was forced into early retirement from teaching by the state. His popular writings included *Systematische Übersicht des Thierreichs* (1846ff) and *Der Mensch im Spiegel der Natur* (1850ff). Bernard Cotta (1808–1879), Professor of Geology at the *Bergakademie* in Freiberg, which he had attended as a youth, wrote *Geognostische Wanderungen* in 1836–1838, a text on geology in 1839, *Geologische Briefe aus den Alpen* in 1850, and a series, *Briefe über Humboldt's Kosmos* from 1850 to 1860. Cotta, though completely unsympathetic to religion, was no materialist. He opposed Vogt, for example, on behalf of the *Allgemeine Zeitung* in 1859. Hermann Burmeister, (1807–1892) was professor of zoology at Halle from 1837 until his permanent departure for Argentina in 1861. His *Geschichte der Schöpfung* of 1843 went through seven editions by 1867. Later he wrote *Geologische Bilder* (1851, 1853) and accounts of his travels in South America. Vogt's *Im Gebirge und auf den Gletschern* (1843) and his textbook on geology belong to this same literary genre.

[21] While Feuerbach did declare that the philosophy of the future was based on natural science, his own program was expressed in the terminology of his profession; i.e., as a philosopher and not as a scientist. Because of this, and because his message was viewed as theological and philosophical heresy, Feuerbach never was able to overcome the natural suspicion with which most people regarded him. He did not succeed in becoming the leader of a movement as did the scientific materialists, whose attraction had come by way of popular science. Natural science, after all, had long been used in support of religion; Humboldt's *Kosmos*, for example, was hardly seen as an antireligious work.

[22] For example Johann Voigt's *Magazin für den neuesten Zustand der Naturkunde*, vol. 1, 1797; *Hamburgisches Magazin, oder gesammelte Schriften zum Unterricht und Vergnügen aus der Naturforschung und den angenehmen Wissenschaften überhaupt*, vol. 1, 1746; *Neue gesellschaftliche Erzählungen für die Liebhaber der Naturlehre, der Haushaltungswissenschaft, der Arzneikunst und der Sitten*, vol. 1, 1758; *et al.*

[23] Cf. Herbert Krause, *Die Gegenwart, 1848–1856. Eine Untersuchung über den*

deutschen Liberalismus (Saalfeld, 1936). On science in *Die Gegenwart*, cf. pp. 66–70. Interestingly, chemistry, not physics, occupied the position of highest rank among the sciences covered (pp. 66–67). For the liberal, humanistic goals and the moderate political stance of this journal, see pp. 12–15, 70, 80, 89–90. Krause himself, as a loyal German of the 1930's, disapproved of the journal's pro-Semitic position (p. 80).

[24] Known as Radenhausen's *Isis*, Lange called it an "excellent naturalistic system." F. A. Lange, *A History of Materialism*, trans. E. C. Thomas, 3d ed., 3 vols. in 1 (London, 1925), III, p. 32.

[25] Carl Diesch in his *Bibliographie der Germanistischen Zeitschriften* (Leipzig, 1927) does not list any of these journals of popular science. Nor does he mention any other popular scientific periodicals. It would seem that this type of journal was a product of the 1850's, but we shall have to await the completion of Joachim Kirchner's edition of *Bibliographie der Zeitschriften des deutschen Sprachgebietes*, 4 vols. (Stuttgart, 1969ff) to be more certain.

[26] 'Zum Titelbild,' *Die Natur*, I (1852), p. 1. A glance at the contents of *Die Natur* is instructive. In addition to major essays on subjects from 'Die Atome' to 'Die Physik im Krieg' and 'Die Ehre im Spiegel des Naturgesetzes' (all from volume three), the journal provided a review section of works on science and, in the supplement, coverage of scientific developments outside Germany.

[27] Moses Hess, *Briefwechsel, 1825–1881,* ed. Edmund Silberner unter Mitwirkung von W. Blumenberg ('s Gravenhage, 1959), p. 316.

[28] *Ibid.*, p. 342. This in spite of the fact that *Das Jahrhundert* was banned in Prussia. Cf. Ule's complaint to Hess, p. 354.

[29] *Ibid.*, p. 322.

[30] *NH*, p. 194. This work, though published in 1885, was actually written much earlier. See below, pp. 114f.

[31] Karl Fortlage, 'Materialismus und Spiritualismus,' *Blätter für literarische Unterhaltung*, no. 30 (1856), p. 541. Fortlage wrote a series of articles on materialism in order to expose its superficiality; hence he was no friend of the movement he described.

CHAPTER I

[1] The name Feuerbach is well known to historians of theology and to scholars interested in the development of Marx's thought. There are several standard biographical works which may be consulted, for example Wilhelm Bolin's Biographische Einleitung to *Ausgewählte Briefe von und an Ludwig Feuerbach*, vol. XII of *Werke*, pp. 1–211; Wilhelm Bolin, *Ludwig Feuerbach, sein Wirken und seine Zeitgenossen* (Stuttgart, 1891); Adolph Kohut, *Ludwig Feuerbach, sein Leben und seine Werke* (Leipzig, 1909); Friedrich Jodl, *Ludwig Feuerbach*, 2d ed. (Stuttgart, 1921); Karl Grün, *Ludwig Feuerbachs philosophische Charakterentwicklung, Sein Briefwechsel und Nachlass*, 2 vols. (Leipzig, 1874).

[2] *Werke*, XII, p. 215.

[3] *Ibid.*, pp. 216–17.

[4] Franz Schnabel, *Die protestantischen Kirchen in Deutschland. Deutsche Geschichte im neunzehnten Jahrhundert*, vol. 8 (Basel: Herder Taschenbuch, 1965), pp. 121–22.

[5] In an autobiographical sketch of 1846, written for the *Jahrbücher für speculative Philosophie*, Feuerbach relates that he had gone to Heidelberg primarily to hear Daub. Reprinted in Bolin, *Ludwig Feuerbach*, p. 12.

[6] Cf. Laura Guggenbuhl, 'Karl Wilhelm Feuerbach,' *Dictionary of Scientific Biography*, ed. C. Gillispie, vol. IV (New York, 1971), pp. 601–02.

[7] *Werke*, XII, pp. 12–15.

[8] Kohut, *op. cit.*, pp. 35–39.
[9] *Werke*, II, p. 363.
[10] *Ibid.*, p. 363.
[11] Cited from a manuscript by Bolin in *Werke*, XII, p. 17, n.
[12] *Ibid.*, p. 17.
[13] The 'Fragmente' are found in *Werke*, II, pp. 358–91, and 'Zweifel,' the heading given 1827–1828 in this collection, on pp. 362–64.
[14] *Werke*, II, p. 363.
[15] Quoted from the to me unavailable *Kleine Schriften* (Frankfurt, 1966) by Z. Hanfi, trans. *The Fiery Brook: Selected Writings of Ludwig Feuerbach* (New York, 1972), p. 8.
[16] *Ibid.*
[17] Kohut, *op. cit.*, p. 63. Kohut discusses the contents of the Erlangen lectures on Descartes, Spinoza, logic, metaphysics, and history of philosophy, pp. 63–68.
[18] Schnabel, *op. cit.*, p. 306, calls Erlangen, Rostock and Dorpat the "strongholds of orthodoxy." The significance of Erlangen's conservatism will become evident below.
[19] Eugene Kamenka, *The Philosophy of Ludwig Feuerbach*, p. 24.
[20] Cited in *Werke*, XII, p. 25.
[21] Friedrich Jodl, 'Vorwort,' *Sämtliche Werke* by Ludwig Feuerbach, 2d ed., III, p. viii.
[22] *Ibid.*, p. ix; Feuerbach, *Werke*, III, p. 33.
[23] *Geschichte der neueren Philosophie. Darstellung, Entwicklung und Kritik der Leibniz'schen Philosophie* (Ansbach, 1837); *Pierre Bayle, nach seinen für die Geschichte der Philosophie und Menschheit interessantesten Momenten dargestellt und gewürdigt* (Ansbach, 1838).
[24] Several of these are listed in Rawidowicz, *Ludwig Feuerbachs Philosophie*, pp. 79–81.
[25] F. Bachmann, *Über Hegels System und die Notwendigkeit einer nochmaligen Umgestaltung der Philosophie* (Leipzig, 1833), p. 145.
[26] *Ibid.*, p. 146.
[27] *Ibid.*, pp. 146–47.
[28] *Ibid.*, p. 147. To better understand Hegel's treatment of experience as a necessary beginning of philosophy, see Hanfi's Introduction to *The Fiery Brook*, pp. 10–11; and F. Schelling, 'Introduction to a First Sketch of a System of Natural Philosophy,' trans. Tom Davidson. *Journal of Speculative Philosophy*, I (1867), p. 197.
[29] F. Bachmann, *Anti-Hegel* (Jena, 1835), p. 14. Cf. above, n. 14. Feuerbach later used the same example Bachmann did to establish this point in his own critique of Hegel; viz., the relationship between a bud and a blossom. Cf. *Werke*, II, pp. 160–61, and Bachmann, *Anti-Hegel*, pp. 12–18.
[30] *Werke*, II, pp. 22–23.
[31] *Ibid.*, p. 23.
[32] *Ibid.*, p. 61.
[33] *Ibid.*, p. 39.
[34] *Ibid.*, pp. 56f, 80.
[35] To Hegelians Feuerbach's review must have been a sign of increasing loyalty, for their surprise at Feuerbach's defection later was equal to Feuerbach's zeal in his critique of Bachmann. See, for example, Rosenkranz's amazement as recorded by Löwith in his Einleitung to *Sämtliche Werke*, *Werke*, I, p. xii.
[36] *Werke*, IV, p. 143.
[37] *Ibid.*, p. 141. Emphasis mine.
[38] *Ibid.*, p. 142. Emphasis mine. Cf. also pp. 143–145.
[39] Kamenka, *op. cit.*, p. 71. My emphasis.

[40] *Werke*, II, p. 300.
[41] In the edition of the work on Leibniz published in the *Sämtliche Werke* of 1847, Feuerbach added a footnote to clarify the change of perspective he had by then acquired. Here Feuerbach, after taking a position against *a priori* conceptions and in favor of the empirical origin of ideas, opposed Leibniz's position that the generalized idea came from the *Verstand* and not from the object itself. Leibniz had denied that generality could come from the senses by induction, for induction had no certainty until all the cases had been checked. Hence the *Verstand* for Leibniz performed the task of generalizing the data of the senses. But Feuerbach pointed out that the senses also are capable of generalizing: "Do the senses show me only leaves, not trees as well; only trees, not also forests; only stones, not as well cliffs; only cliffs, not also mountain chains? Is then perception 'blind,' the senses dull and silly?" *Werke*, IV, p. 194. Cf. also pp. 188–193.
[42] Cf. Kamenka, *op. cit.*, p. 170, n. 4.
[43] Of Bachmann's influence we hear nothing. It is true that Feuerbach derided Bachmann all his life, but this was for Bachmann's lack of understanding of Hegel's significance. Cf. Löwith, *op. cit.*, p. xii.
[44] This is not to say that he did not criticize empiricism or materialism after 1839 and the critique of Hegel. See, for example, his derision of the truths of revelation as "the worst empiricism of all" in *Das Wesen des Christentums* (*Werke*, VI, p. 256. Cf. also p. 63). Also in *Werke*, II, p. 210 (1841): "Were it only a matter of the impressions of the object, as spiritless materialism and empiricism hold, then animals could, even must already be physicists."
[45] *Werke*, II, p. 132.
[46] *Ibid.*, p. 133.
[47] *Ibid.*, pp. 133–34.
[48] *Ibid.*, p. 136.
[49] *Ibid.*, pp. 140–141.
[50] *Ibid.*, pp. 143–44.
[51] The Berlin *Jahrbücher* is called the organ of Hegelian thought by T. Ziegler, 'Zur Biographie von D. F. Strauss,' *Deutsche Revue*, XXX, Jahrgang Vol. II (1905), p. 199, n. On the founding of the *Hallesche Jahrbücher*, cf. W. Brazill, *The Young Hegelians* (New Haven, 1970), pp. 73–78.
[52] In Feuerbach, *Werke*, XIII, p. 180.
[53] Kamenka, *op. cit.*, pp. 69–80; Hanfi, Introduction to *The Fiery Brook*, pp. 7–26; Brazill, *op cit.*, pp. 143–144; Sidney Hook, *From Hegel to Marx* (Ann Arbor, 1966), pp. 226–233. Brazill is not as concerned with this issue as the others, though he is overly concerned to see Feuerbach as a proponent of the Young Hegelian revisionism that separated religion and philosophy.
[54] 'Zur Kritik der Hegelschen Philosophie,' *Werke*, II, p. 187. Earlier in the same work he had said: "Hegel begins with the concept of being, or with abstract being. Why should I not be able to begin with being itself; i.e., with real being?" (p. 165).
[55] *Ibid.*, p. 187.
[56] *Ibid.*, p. 185.
[57] *Ibid.*, pp. 181, 183.
[58] From the preface to the second edition of *Das Wesen des Christentums*. Cf. *Sämtliche Werke*, VII (1849), pp. 9–10.
[59] *Ibid.*, p. 10. Brazill, in his treatment of Feuerbach as a Young Hegelian, plays down the break with Hegel by pointing to the Hegelian spirit of *Das Wesen des Christentums*. He sees it in several places–in the importance of man's self-alienation for historical

development, in the comprehension of the divine in human terms, and in the fact that "spirit had its life within human history, not outside it." The enemy was dualism. Brazill, *op, cit.*, pp. 149–150.

[60] *Sämtliche Werke*, VII (1849), p. 11.

[61] *Werke*, VI, pp. 43, 56–57, 139, 264.

[62] *Sämtliche Werke*, VII (1849), p. 12.

[63] Friedrich Engels, *Werke*, XXI, p. 272.

[64] *Werke*, XII, pp. 80–81.

[65] Cf. *Werke*, XII, pp. 86 ff. Rau reports that Feuerbach did not expect the *Wesen* to reach the public as it did, but that once that had happened he wrote no more for scholars, but for the universal man. Albrecht Rau, *Ludwig Feuerbachs Philosophie, die Naturforschung und die philosophische Kritik der Gegenwart* (Leipzig, 1882), p. 2.

[66] Other works around this time include short pieces entitled *Über den Anfang der Philosophie* (1841), *Die Notwendigkeit einer Reform der Philosophie* (1842), and *Vorläufige Thesen zur Reform der Philosophie* (1842). In 1846 appeared the first two of ten volumes of collected works, and within eight years the *Das Wesen des Christentums* came out in three editions. By the end of the 1840's there was scarcely anybody in Germany who had not heard of him.

[67] *Werke*, II, p. 309. Cf. also Rawidowicz, *op. cit.*, p. 148.

[68] *Werke*, II, p. 232.

[69] *Ibid.*, pp. 322–23.

[70] Georg Herwegh, *Georg Herweghs Briefwechsel mit seiner Braut*, ed. Marcel Herwegh, 3 parts (Stuttgart, 1906), p. 252. For Feuerbach's reply see *Werke*, XIII, p. 110. The new journal never got going. Cf. p. 110, n.

[71] *Werke*, XIII, pp. 24–25, n. 4. The feeling was mutual. Cf. Herwegh's 'Seinem Ludwig Feuerbach,' in Georg Herwegh, *Werke*, ed. H. Tardei (Berlin, 1909), III, p. 155.

[72] *Werke*, XIII, p. 46.

[73] Kohut, *op. cit.*, p. 146. Cf. the 1837 letter to an acquaintance in Erlangen in *Werke*, XII, p. 309: "For a long time now I have sensed a lack, a great lack in being so behind in the natural sciences. . . . The philosopher must have nature for his friend."

[74] *Werke*, XIII, p. 35. Feuerbach added that he had hoped to write on the philosophical, ethical, and pedagogic significance of the natural sciences the next summer, but no such work ever came forth. Although Feuerbach's enthusiasm for science remained great throughout his life, it must be said that his lack of genuine understanding of science meant that his references to science were at best naive and superficial.

[75] *Ibid.*, pp. 44–45.

[76] *Ibid.*, p. 151.

[77] Quoted from the Frankfurt newspaper *Didaskalia* of April, 1848 in *Werke*, XII, p. 114. William Brazill, *op. cit.*, p. 152 is simply in error when he makes Feuerbach an elected delegate.

[78] *Werke*, XII, p. 115.

[79] In a letter to the publisher Otto Wigand, cited in *Werke*, XII, p. 121.

[80] *Ibid.*, p. 118.

[81] For the influence of Feuerbach's lectures on Keller, cf. Heinrich Hermelink, *Liberalismus und Konservatismus, 1835–1870. Der Christentum in der Menschheitsgeschichte von der Französischen Revolution bis zur Gegenwart*, Vol. 2 (Stuttgart, 1953), p. 454, and Hirsch, *Geschichte der neueren Theologie*, p. 610.

[82] For a treatment of 'Die Naturwissenschaft und die Revolution' see below, p. 91–92, and especially p. 234, n. 60.

In 1867 a teacher in Stuttgart asked him in a letter for the difference between his atheism and the materialism of Vogt, Moleschott and Büchner. It was, he replied, the difference between human history and natural history. Chemistry and physiology knew nothing of the soul or God, only human history had introduced such notions. While, therefore, both he and the scientific materialists regarded man as a natural being, his concern was "the entities of thought and fantasy which originate in man, and which serve in man's thought and tradition as real entities." *Werke*, XIII, p. 339.

[83] Kamenka, *op. cit.*, pp. vii–viii.

[84] *Ibid.*, p. 91. On Feuerbach's lack of organization and the aphoristic style of his work, see also pp. 90, 95, 97, and Sidney Hook, *op. cit.*, p. 226.

[85] See Vogt's letter to Deubler in Konrad Deubler, *Tagebücher, Biographie und Briefwechsel (1848–1884) des oberösterreichschen Bauern-philosophen*, ed. Arnold Dodel-Port, 2 vols. (Leipzig, 1886), II, p. 79. Vogt had already established correspondence with Deubler independently the previous year.

CHAPTER II

[1] The old Latin form was: "Ubi tres medici, duo sunt athei," and its modern reformulation can be found in J. B. Meyer, *Zum Streit über Leib und Seele* (Hamburg, 1856), p. 16; and A. Wagner, *Naturwissenschaft und Bibel* (Stuttgart, 1855), p. 5. Wagner, incidentally, goes on to defend natural science against such charges.

[2] Two more works published later, one by Schleiden and another by the philosopher Lange, will be discussed here briefly because of their relevance to the issues of *early* German materialism as opposed to the literature that poured forth on Darwinism after 1859. Several of the works referred to above were unavailable in the United States, forcing dependence on reviews. Only works personally examined by the author are listed in the bibliography.

[3] In the same year the less well known *Neue Darstellung des Sensualismus* by Heinrich Czolbe was published, while the other famous work of scientific materialism of the century, Moleschott's *Der Kreislauf des Lebens*, had been written three years earlier.

[4] Karl Fortlage, 'Materialismus und Spiritualismus.' *Blätter für literarische Unterhaltung*, Nr. 30 (1856), pp. 541–548; Nr. 49 (1856), pp. 889–899; Nr. 19 (1857), pp. 337–347; Nr. 12 (1858), pp. 205–215.

[5] Karl Fischer, *Die Unwahrheit des Sensualismus und Materialismus* (Erlangen, 1853), pp. xii–xv. Fischer attacked Vogt for inconsistently arguing that mind was a secretion of the brain (therefore an unconscious and determined phenomena), but also that man was master of voluntary muscular activity.

[6] *Ibid.*, pp. 51–52.

[7] "The scientific literature whose appearance is due to sensualism and materialism is very plentiful, and it is generally agreed that Feuerbach ... has become the most significant representative of sensualism." (*Ibid.*, p. vii) Others also identified Feuerbach as the original inspiration for scientific materialism; for example, Adolph Cornill, *Materialismus und Idealismus in ihren gegenwärtigen Entwicklungskrisen beleuchtet* (Heidelberg, 1858), pp. 42ff. J. B. Meyer also saw Feuerbach as "the philosophical founder of contemporary materialism." J. B. Meyer, *Zum Streit über Leib und Seele*, p. 15. By materialism Meyer meant Vogt, Moleschott, Büchner and Czolbe. Cf. Lecture two, pp. 25–48. Meyer's attack on them centered around atomism. (pp. 28–32).

[8] Fischer, *op. cit.*, p. 8.

[9] *Ibid.*, pp. iii, 4, 5–6n, 44–47. This reasoning helps explain the move from philosophical materialism to what later became known as Darwinism in the minds of nonscientists. It is certainly true Darwinism did fall under the rubric of materialism for many. Feuerbach, incidentally, *did* claim that man and animal were qualitatively similar. Cf. *Werke*, II, pp. 349–50.

[10] *Ibid.*, pp. ix–x, n. Fischer apparently had not read Feuerbach's essay 'Wider den Dualismus von Leib und Seele, Fleisch und Geist,' in which Feuerbach challenged the value of specifying the difference between the living and nonliving by means of a principle separate from the body. Cf. *Werke*, II, pp. 336–38.

[11] Quoted from the *Philosophie der Zukunft* in Fischer, *op. cit.*, pp. ix–x, n.

[12] *Ibid.*, pp. 45–47. Moleschott had accounted for sin on the basis of the unnatural, not the will.

[13] *Ibid.*, pp. 48–50.

[14] *Ibid.*, pp. xvii, iv.

[15] *Ibid.*, p. 18.

[16] *Ibid.*

[17] Peter Gay, *The Dilemma of Democratic Socialism* (New York, 1952), pp. 141–42.

[18] Julius Frauenstädt, *Der Materialismus. Eine Erwiderung auf Dr. L. Büchner's* 'Kraft und Stoff' (Leipzig, 1856), pp. vi–xiv.

[19] *Ibid.*, p. 43.

[20] *Ibid.*, pp. 39–40.

[21] *Ibid.*, pp. 44–51.

[22] *Ibid.*, p. 79. Cf. also pp. 66f.

[23] *Ibid.*, pp. 108–10, 139f.

[24] *Ibid.*, pp. 192–201.

[25] For a fuller account of Wagner's ideas, see below, pp. 73ff.

[26] J. Frauenstädt, *Die Naturwissenschaft in ihrem Einfluss auf Poesie, Religion, Moral, und Philosophie* (Leipzig, 1855), pp. 7–10.

[27] *Ibid.*, pp. 55–75. There were similar conclusions regarding the ultimate compatibility between science and morality, and science and philosophy in later chapters. Basically Frauenstädt's proof was indirect: if these disciplines were incompatible, then science either should have led to immorality or taken over philosophy.

[28] Lange, *History of Materialism*, II, p. 249.

[29] *Ibid.*, II, pp. 306–07, 332. Another philosopher who shared Lange's defeatist attitude was Johann Erdmann. Cf. his *A History of Philosophy*, trans. W. S. Hough, 3 vols. (London, 1880), III, p. 131. The first German edition was in 1870. Erdmann held that materialism had begun with Feuerbach's *Philosophie der Zukunft* (p. 130), thus joining Treitschke, Merz and others in naming Feuerbach as the father of German materialism. Lange, incidentally, reserved this title for Moleschott. Lange, *op. cit.*, III, p. 26.

[30] Cf., for example, his 'Kant and F. A. Lange,' in *NW*, II, pp. 331–42, and also his treatment of the new journal founded by Haeckel and others in the same work, pp. 7ff.

[31] Julius Schaller, *Leib und Seele. Zur Aufklärung über Köhlerglaube und Wissenschaft* (Weimar, 1858), p. 243. First edition 1855. For his sympathetic, yet critical discussion of atomism, see pp. 240–45.

[32] *Ibid.*, pp. 161–66. In this Schaller was fully aware of the hypothetical or conditional nature of his assumptions. Nor was Schaller alone in his similarity to Whitehead in nineteenth-century Germany. Paul Kuntz has detailed the strong affinity between Whitehead and Hermann Lotze of Göttingen. Cf. Paul G. Kuntz, Introduction to *Lotze's System of Philosophy* by George Santayana (Bloomington, 1971), pp. 25–26, 68ff. This is a re-issue of Santayana's dissertation, which was originally completed in 1889.

[33] *Ibid.*, pp. 34–43, 65–71.

[34] *Ibid.*, p. 96.

[35] *KW*, 4th ed., p. li.

[36] Cf. M. Heinze, 'Jakob Frohschammer,' *Allgemeine Deutsche Biographie*, **49**, 1904. Reprinted (Berlin, 1971), pp. 172–77.

[37] *KW*, 4th ed., pp. lxii–lxiii.
[38] Jakob Frohschammer, *Menschenseele und Physiologie. Eine Streitschrift gegen Professor Karl Vogt* (Munich, 1855), pp. 124–28. Frohschammer thought it extremely presumptuous that the materialists tried to explain thought by means of brain processes when they themselves admitted that scientific knowledge of the brain was still *terra incognita*. (p. 137)
[39] *Ibid.*, pp. 80–81, 155f. Cf. also pp. 117–19.
[40] *Ibid.*, p. 161. Vogt had ridiculed Frohschammer for saying that astronomy would have overstepped its boundaries if it claimed that the sun had *not* stood still in Joshua's day. Frohschammer, noting this criticism, demanded to know what difference there was between Vogt's contention that the sun had not stood still, and his argument that an immaterial soul existed. (p. 48)
[41] *Ibid.*, pp. 119–21.
[42] Adolph Kohut, *Justus von Liebig. Sein Leben und Wirken*, 2d ed., (Giessen, 1905), pp. 340–41.
[43] Moleschott, *KL*, 3rd ed., p. v.
[44] More on this encounter is contained below, pp. 73ff.
[45] This in a short note to the *Allgemeine Zeitung*, in whose pages Wagner had opposed Vogt prior to the Göttingen lecture. Vogt of course delighted in Wagner's silence. It was the best reply, he said, for one who had nothing with which to counter. *KW*, p. xxxiii.
[46] Andreas Wagner, *Naturwissenschaft und Bibel, im Gegensatze zu dem Köhlerglauben des Herrn Carl Vogt, als des wiedererstandenen und aus dem Französischen ins Deutsche übersetzten Bory* (Stuttgart, 1855), pp. 9–16. The reference in the title is to the French naturalist Bory de St. Vincent (1780–1846). Cf. pp. 19, 54. For Vogt's account of Wagner's review of the *Bilder*, see *KW*, pp. xxxviii–xxxix.
[47] *Ibid.*, p. 7. Cf. also pp. 6–8.
[48] This is discussed below, pp. 177–178.
[49] Wagner, *op. cit.*, pp. 24–25, 32. Cf. also pp. 26–27, 31–33.
[50] *KW*, 4th ed., pp. xli, xliii.
[51] *Ibid.*, p. xlviii. The question of the scientific materialists' position on the fixity of species is discussed below, pp. 175ff.
[52] A. N. Böhner, *Naturforschung und Culturleben* (Hannover, 1859).
[53] *Ibid.*, p. 130. Cf. also pp. viii, 128, 304.
[54] *Ibid.*, pp. 134–35. He even linked materialists with St. Simon's and Fourier's 'communism.' (pp. 138–39)
[55] *Ibid.*, pp. 141–57. One must not forget that in the nineteenth century there was an unprecedented expansion of missionary activity and evangelism. Western Christianity was spreading over the world as never before, so much so that the missions historian Latourette called this the greatest century since the first. See Alec Vidler, *The Church in an Age of Revolution* (Baltimore: Penguin Books, 1965), pp. 246–256.
[56] Hermann Klencke, *Die Naturwissenschaft der letzten fünfzig Jahre und ihr Einfluss auf das Menschenleben. In Briefen an Gebildete aller Stände* (Leipzig, 1854), pp. 15–16.
[57] *Ibid.*, pp. 20–21. Elsewhere: "The working classes are kept from immorality by natural science." (p. 89)
[58] *Ibid.*, pp. 17–18, 278, 294–98.
[59] *Ibid.*, pp. 1–3, 71–72, 76, 87f. Klencke thought that the philosophical basis of natural science lay in its ability to express empirical results in mathematical terms. (p. 73) He had a high regard for the senses as a source of knowledge, and felt that this empirical emphasis of modern science made it seem as if there were hundreds of years between, for example, the old and the new physiology. (pp. 74–75)

[60] *Ibid,*, pp. 17–19. Creation, for example, was a miracle, but the date on which the earth had been created was not part of the miraculous side of the question.
[61] *Ibid.*, pp. 95ff, 120, 323.
[62] Karl Fortlage, 'Materialismus und Spiritualismus,' *Blätter für literarische Unterhaltung*, (1856), Nr. 30, pp. 546–47. Reichenbach's work was entitled *Köhlerglaube und Afterweisheit* (Vienna, 1855).
[63] This is the argument of Otto Woysch, *Der Materialismus und die christliche Weltanschauung* (Berlin, 1857). Of similar tone is D. A. Hansen, *Der moderne Materialismus und die evangelische Volksschule* (Oldenburg in Holstein, 1857). Fortlage's reply was to turn the argument around and blame scientific materialism on pietism because its archaic anti-intellectualism had forced the church to retreat instead of advance. German idealism, according to Fortlage, was the best hope of religion in modern times. Fortlage, *op. cit.*, Nr. 12, pp. 213–214.
[64] W. Menzel, *Die Naturkunde im christlichen Geiste aufgefasst* (Stuttgart, 1856). This work is covered by Fortlage along with another of a similar nature, *Kosmologie. Der Mensch im Verhältnisse zur Natur und zum Christentum* (Wien, 1856), by J. Benedikt. Cf. Fortlage, *op. cit.*, Nr. 19, p. 334. Other comparable tracts were the anonymous *Kurze populäre Widerlegung der neuern materialistischen und angeblichen Sterblichkeit des menschlichen Geistes* (Berlin, 1857), covered in Fortlage, *op. cit.*, Nr. 12, p. 212, and C. F. Göschel, *Der Mensch nach Leib, Seele und Geist diesseits und jenseits* (Leipzig, 1856), in Fortlage, *op. cit.*, Nr. 19, p. 346.
[65] F. Michelis, *Der kirchliche Standpunkt in der Naturforschung* (Münster, 1855), pp. 16, 85. Michelis had in mind particularly Schleiden's *Studien: Populäre Vorträge* (Leipzig, 1855), 2nd ed. 1857. What no doubt pleased him about this work was Schleiden's favorable comparison of Catholicism with Protestantism in the years immediately following Copernicus. Cf., for example, Schleiden, *op. cit.*, pp. 277–78. In addition, Schleiden was not irreligious; he defended the existence of a pure form of Christianity. (p. 380)
[66] Michelis, *op. cit.*, p. 3. Hermann Klencke also dated public enthusiam for science from Humboldt's efforts, but he went back to the public lectures on which the *Kosmos* was based. Klencke spoke of the social pressure on educated people to learn how the steam engine, the gas light, etc., worked. Klencke, *op. cit.*, pp. 4–5, 26–27, 78.
[67] Michelis, *op. cit.*, pp. 30–31.
[68] *Ibid.*, p. 79. No Catholic felt redeemed because he believed in the system of Copernicus, Michelis argued, yet he could be a good Christian without it. (p. 80)
[69] See, for example, Schleiden's statements about the role of superstition and revealed truth in Christianity. Schleiden, *op. cit.*, n. 16, p. 322; n. 5, p. 377; and n. 8 pp. 380–82. Büchner pointed to Schleiden's antimaterialistic emphasis from the other camp. Specifically, he complained of Schleiden's 1856 contribution to *Westermann's Monatshefte* in which Schleiden had come out against materialism. Cf. Büchner, *NW*, II, pp. 69–77. Later in 1863 Schleiden wrote a thoroughly Neo-Kantian critique of materialism, *Über den Materialismus der neueren deutschen Naturwissenschaft, sein Wesen und seine Geschichte* (Leipzig, 1863). For the heart of his Kantian approach, see pp. 20–31,
[70] Heinrich Czolbe, *ES*, p. 40, n. Cf. also Hermann Dörpinghaus, *Darwins Theorie und der deutsche Vulgärmaterialismus im Urteil deutscher katholischer Zeitschriften zwischen 1854 und 1914* (Freiburg, 1969), pp. 267, 277.
[71] Heinrich Hermelink, *Der Christentum in der Menschheitsgeschichte*, II: *Liberalismus und Konservatismus*, pp. 401–02, 413, n. 1.
[72] Dr. Mantis, *Goethe im Fegfeur. Eine materialistisch-poetische Gehirnsekretion* (Stuttgart, 1856). Even Czolbe got into the act (p. 26).
[73] F. Fabri, *Briefe gegen den Materialismus* (Stuttgart, 1855), pp. 127–28. The second

edition in 1864 contained two additional chapters, one against Darwin's theory and the other against Lyell's geology.
[74] *Ibid.*, p. 25. For an example of the supreme confidence the materialists had of their ultimate victory, see the preface to the fourth edition of Büchner's *Kraft und Stoff.*
[75] *Ibid.*, p. 26.
[76] *Ibid.*, p. 10.
[77] *Ibid.*, , p. 8. Vogt's discipleship was documented from his *Bilder aus dem Thierleben* of 1852, Moleschott's from his *Kreislauf des Lebens* of the same year, while Büchner was identified as an immediate follower of Moleschott (pp. 11–19).
[78] *Ibid.*, p. 13.
[79] *Ibid.*, pp. 28–31. This facile association of egoism with scientific materialism infuriated the materialists, especially Büchner, some of whose ideas were quite socialistic.
[80] *Ibid.*, pp. 24, 47.
[81] *Ibid.*, p. 23.
[82] *Ibid.*, pp. 67ff.
[83] *Ibid.*, p. 7.
[84] *Ibid.*, p. 171.
[85] *Ibid.*, pp. 194–97.
[86] *Ibid.*, p. 205. Reason can only reinforce our experience, but if the experience is not present, reason can only deny it without foundation.
[87] *Ibid.*, p. 209.
[88] *Ibid.*, pp. 215–16. Emphasis mine.
[89] See, for example, his handling of Vogt's position on the nonexistence of an original pair, *Ibid.*, pp. 199–200.
[90] In the preface to the fourth edition of the *KS*, (see fifth edition, pp. lviii–lix). Cf. also Fortlage, *op. cit.*, Nr. 30, p. 545. Yet this little publication went through two editions in one year, and its major content was printed in two numbers of the *Allgemeine Zeitung*. Cf. *Der Humor*, p. 52. The second edition of *Der Humor* contains a reply to Büchner's preface to the fourth edition of *KS* on pp. 67ff.
[91] *Der Humor*, p. 70.
[92] *Ibid.*, pp. 36–37.
[93] Und in der Hütte wie auf Throne
Muss dieser wahre Glaube wohnen:
Wenn nicht - so flieht die Religion
Mit dem Gewissen bald davon
Und nur in wüster Anarchie
Verblutet Volk und Monarchie.

Ibid., pp. 9–10. Samples of other poems in the work are 'Dr. Büchner's Erfahrungswissenschaft ohne Erfahrung,' and 'Der Logos und der Stoff.'
[94] *Ibid.*, p. 38.
[95] *Ibid.*, pp. 81–90. His hero was Oersted, of whom he said, "Yours is the spirit of science." (p. 91)
[96] Fortlage, *op. cit.*, Nr. 49, p. 899. Drossbach's work was entitled *Das Wesen der Naturdinge und die Naturgesetze der individuellen Unsterblichkeit* (Olmutz, 1855).
[97] Anonymous, *Helionde, oder Abendteuer auf der Sonne* (Leipzig, 1855). Cf. Fortlage, *op. cit.*, Nr. 49, p. 898.
[98] This was Moritz Müller's *Betrachtungen und Gedanken über verschiedene wichtige Gegenstände vom religiös-politischen Standpunkte* (Karlsruhe, 1857). Müller's work went through three editions within four and a half months. Cf. Fortlage, *op. cit.*, Nr. 12, p. 212. Somewhat similar was the anonymous *Briefe aus X über den Schlüssel zum Weltall* (Berlin, 1856). This author also rejected belief in miracles and claimed to have

left out the Creator in his explanation of the universe. Fortlage showed that he had smuggled God and miracle in by way of such principles as his "guiding thought." Fortlage, *op. cit.*, Nr. 49, p. 899.

[99] Wilhelm Schulz-Bodmer, *Der Froschmauskrieg zwischen den Pedanten des Glaubens und Unglaubens* (Leipzig, 1856).

[100] Quoted in Fortlage, *op. cit.*, Nr. 30, p. 546.

[101] Mathilde Reichardt, *Wissenschaft und Sittenlehre. Briefe an Jakob Moleschott* (Gotha, 1856). p. 47.

CHAPTER III

[1] Vogt has left us, as has Moleschott, an uncompleted autobiography written at the end of his career. In Vogt's case the work covers roughly only the first third of his life; i.e., up to 1844. Fortunately the remainder is dealt with in his son Wilhelm's biography, published not long after the autobiography appeared. See Karl Vogt, *Aus meinem Leben. Erinnerungen und Rückblicke* (Stuttgart, 1896), and Wilhelm Vogt, *La Vie d'un homme. Carl Vogt* (Paris, 1896). Both of these works, especially the latter, must be read with a sensitive and critical eye in order to screen out the obviously biased intent of the authors.

[2] *AL*, p. 16.

[3] *Ibid.*, p. 6.

[4] *AL*, p. 30. Vogt comments: "That this inclination (to independence) also carried over to us children was understandable." (p. 30).

[5] Hermann Misteli, *Carl Vogt. Seine Entwicklung vom angehenden naturwissenschaftlichen Materialisten zum idealen Politiker der Pauluskirche, 1817–1849* (Zurich, 1938), p. 8; Karl Vogt, *AL*, p. 13. That Feuerbach also knew him personally is clear from a letter to Feuerbach from one Konrad Haag. Cf. Feuerbach's *Werke*, XIII, p. 265.

[6] Willi Sternfeld, 'German Students and their Professors,' pp. 128–61 in *In Tyrannos*, ed. Hans Rehfisch (London, 1944), p. 134. Follen eventually had to resign due to his outspoken stand on the emancipation of the slaves in the 1830's.

[7] Quoted by Mistelli, *op. cit.*, p. 9.

[8] *AL*, p. 20.

[9] *AL*, p. 62. Cf. pp. 26ff for a description of life in Giessen.

[10] *Ibid.*, p. 43.

[11] *Ibid.*, p. 70.

[12] *Ibid.*, pp. 97, 102.

[13] *Ibid.*, p. 98. Cf. also pp. 98ff.

[14] *AL*, p. 19. On the historical differences between north and south Germany see W. H. Bruford, *Germany in the Eighteenth Century: The Social Background of the Literary Revival* (Cambridge, 1935), pp. 144–148; 162–174.

[15] *AL*, p. 119. One of the Büchner brothers, Alexander, attended Giessen over a decade later and gave a similar report of student life there: "Here I enjoyed . . . the golden freedom of idleness . . . and still I earned an academic degree." Alexander Büchner, *Das 'tolle' Jahr: Vor, Während, und Nach 1848* (Giessen, 1900), p 110.

[16] *AL*, p. 121. Cf. also p. 23.

[17] *AL*, pp. 58, 114. For a description of Liebig's laboratory, see pp. 124–26. Also see A Büchner, 'Vorwort,' in Ludwig Büchner, *Im Dienst der Wahrheit* (Giessen, 1900), pp. xi–xiii. Alexander Büchner spoke of Liebig as the one bright spot in the university's life. Liebig's students, called "poison makers" by the rest of the student body, were forced to live in isolated quarters since the students feared explosions, one of which had cost Liebig himself an eye. (p. xii)

[18] *AL*, pp. 126–27.
[19] *Ibid.*, p. 115.
[20] *Ibid.*, p. 119.
[21] Karl's father had also been pursued by police on his trip to Switzerland. After he had left Giessen authorities there received news that the elder Vogt possessed certain forbidden manuscripts which he intended to publish in his new homeland. For the story of Vogt's father's adventure see *AL*, pp. 140–41, and for Karl's account of his own intrigues, see pp. 141–57.
[22] Vogt noted in particular the predominance among the Germans in Bern of graduates of south Germany's famous Tübinger Stift. *AL*, pp. 160, 176. Moleschott at Zürich also noticed the resentment created by the mass importation of Germans to fill Swiss university positions. Jacob Moleschott, *FMF*, p. 280.
[23] Later in his *Ocean und Mittelmeer* Vogt characterized zoologists, anatomists, physiologists, and botanists as medical men who did not want to practice because of their love for natural science. *OM*, I, p. 131.
[24] For example Misteli, *op. cit.*, pp. 35–36.
[25] *Études sur les Glaciers* (Neuchâtel, 1840). The German edition, *Untersuchungen über die Gletscher*, came out in 1841.
[26] His dissertation, 'Zur Anatomie der Amphibien,' was published as a pamphlet in Bern in 1838. Vogt, primarily motivated by a lecture he had given at a meeting of Swiss naturalists in Bern, but also further stimulated by some observations made on first arriving in Neuchâtel, published separately an article he had written in the fourth volume of *Neue Denkschriften der allgemeinen schweizerischen Gesellschaft für die gesammten Naturwissenschaften*, which expanded on his dissertation, and which carried the title *Beiträge zur Neurologie der Reptilien* (Neuchâtel, 1840). Cf. also Wilhelm Vogt, *op. cit.*, pp. 21, 31, where several articles published in Johannes Müller's *Archiv für Anatomie* from the Neuchâtel period are listed.
[27] The first volume, titled *Embryologie des Salmones*, was 'by C. Vogt,' and appeared in 1842. According to Vogt, his relatives in Boston forced Agassiz to make a written public retraction of his claim. (*AL*, p. 201). On the other hand, Vogt in the preface to the work acknowledged Agassiz's large share in it: "In confiding such an honorable task to me my celebrated friend has not at all remained a stranger to my researches. We have discussed together the major facts to the extent that observation has revealed them to me; often we have examined them again in common, and after I have drafted my work it is he who wishes to see it again." (*ES*, p. ii.)

In the autobiography Vogt recalls how upset Agassiz was to find out that he had, without Agassiz's knowledge, published his 130 page *Untersuchungen über die Entwicklungsgeschichte der Geburtshelferkröte* (Solothurn, 1842). Cf. *AL*, p. 201. Such stories about Agassiz were accepted at least among those who had little use for his religious conservatism; e.g., Haeckel, who, in response to a laudatory obituary article on Agassiz, repeated in 1876 the charge that Agassiz himself had not written most of his embryological works. Cited in Wilhelm Vogt, *op. cit.*, pp. 38–39.

One thing is certain. By the 1860's there definitely had been a parting of the ways between Vogt and Agassiz. Any question about it is settled by the critical notes pencilled in the margin of Agassiz's copy of the English translation of Vogt's *Vorlesungen über den Menschen*, deposited in the library of the Museum of Comparative Zoology at Harvard. Cf. pp. xv, 313, 314, 315, 317, 412, 437, 458, and 469.
[28] Wilhelm Vogt, *op. cit.*, pp. 32–33. The account of von Buch's total defeat and Vogt's glorious victory once more reflects the hero worship of the son Wilhelm. More on the encounter can be found in Alfred Wurzbach, *Zeitgenossen: Biographische Skizzen*, Heft II, *Carl Vogt* (Wien, 1871), pp. 18–22. For von Buch's alternative to

explaining the existence of erratic boulders by glaciers, see W. Nieuwenkamp, 'Leopold von Buch,' *Dictionary of Scientific Biography*, Vol. II, p. 556.
[29] Misteli's interpretation of Vogt as a romantic geologist is largely based on this work. See Misteli, *op. cit.*, pp. 22–30.
[30] *AL*, p. 198.
[31] Wilhelm Vogt, *op. cit.*, p. 36.
[32] *AL*, p. 201.
[33] Mistelli has cited one of the articles Vogt wrote as Paris correspondent for the *Allgemeine Zeitung* in the years immediately succeeding Neuchâtel in which Vogt refers to his close association "with one of the most outstanding paleontologists of our time." Misteli, *op. cit.*, p. 36. Cf. also the 'Vorrede' to Vogt's 1846 textbook on geology, *LG*, I, p. vii, and note that it was Agassiz who among others provided Vogt with a recommendation for the position in zoology he assumed at Giessen in 1847. A reprinting of this recommendation may be found in Wilhelm Vogt, *op. cit.*, p. 51.
[34] The friendship went smoothly as long as the two did not expect too much from each other. When Vogt and Bakunin decided to leave the hotel and set up common housekeeping, the experiment lasted two weeks. Bakunin, out of money and credit, spent funds that came in the mail for Vogt, payment for his work as correspondent for a German periodical. This, plus Bakunin's obsession with Hegel, soon drove Vogt out. The two did not split politically, however, until 1868. Cf. Wilhelm Vogt, *op. cit.*, pp. 41–43, 151–152.
[35] His articles are signed with ⌓. Cf. Misteli, *op. cit.*, p. 32.
[36] *LG*, I. pp. vii–viii. Second and third editions were published in Geneva in 1853 and 1866. Agassiz's argument that an ice age was responsible for many observable geological phenomena, e.g., the existence of erratic boulders, was not accepted by many of the older geologists of the day.
[37] Misteli, *op. cit.*, p. 32. I rely here on the work of Misteli, who has been able to examine these articles. Vogt's catastrophism is easily evident in the *LG*. Cf., for example, I, p. 3: "The wearing down of cliffs, the erosion of surfaces, the rocks which are torn away from their resting places, the sand which is deposited in deeper regions, all these became speaking tongues of catastrophes which were produced by forces now lost, but whose marks are preserved in solid portions of the earth's crust." Cf. also the section in volume two entitled 'Geschichte der Erde,' *passim*.
[38] The third edition of the *LG* (1866) contained the confession of his new attitude on catastrophism. Cf. pp. xiii–xiv. Three years earlier in the *VM*, Vogt had first publicly recorded his changed position. *VM*, II, pp. 82–83. For the fixity of species in the *LG* see II, pp. 296, 300–03. More on the species matter can be found below, pp. 175ff.

Gillispie has noted that Agassiz's work on glaciation provided a renewed impetus to catastrophism in a period when serious alternatives to it were being considered. C. Gillispie, *Genesis and Geology* (Cambridge, Mass., 1951), p. 151. Surely this helps to explain Vogt's early receptiveness to the use of catastrophes to account for the gaps in the fossil record.
[39] "And the same plan which can be recognized in the first appearance of animal life on the surface of our planet has spun itself out in harmonious sequence right down to the present. A series of important catastrophes marks a group of resting points in this pattern of development of the animal world, and every advance in the development of the original plan is given in a new characteristic creation which has nothing in common with the preceding one." Quoted from an article in the *Allgemeine Zeitung*, August 5, 1845 by Misteli, *op. cit.*, pp. 61–62. Cf. also n. 52. Misteli also cites references to the "infinite wisdom" brought to bear on the machine of the organism, and to the reverence one must have in the face of such a plan. These are to be found in the *Einleitung* of the 'Physiologische Briefe' series, also written in 1845. Cf. p. 61, also n. 50.

[40] *Werke*, VIII, p. 164.
[41] *LG*, II, p. 370.
[42] *Ibid.*
[43] Misteli, *op. cit.*, p. 61.
[44] *Ibid.*, p. 65.
[45] *PB*, pp. 112, 143 respectively.
[46] From January 14, 1845. Cited in Misteli, *op. cit.*, p. 56. Cf. also n. 31, and p. 63.
[47] *PB*, p. 206. Though Vogt was castigated for associating the soul with urine, it is only fair to point out that he had been making reference to the kidney throughout the book up to this point in his discussion of the nature of glandular secretion. For the statement of Cabanis, cf. his *Rapports du physique et du moral de l'homme et lettres sur les causes premieres*, 8 ed. (Paris, 1844), p. 137f.
[48] Cited in Misteli, *op. cit.*, p. 69, n. 15. His son confirms his disinclination to philosophy: "During his life he nourished a malice often comic against purely philosophical demonstrations, and he held philosophers at a distance even more than priests." W. Vogt, *op. cit.*, p. 14. Cf. also *W*, pp. 89, 310–11.
[49] Kamenka, *Philosophy of Feuerbach*, p. 156, n. 8.
[50] *PB*, p. 8. Such a declaration is common to scientism. Cf. Robert Darnton, *Mesmerism and the End of the Enlightenment in France* (New York, 1970), pp. 164–65.
[51] Quoted by W. Vogt, *op. cit.*, p. 48.
[52] *PB*, p. 326.
[53] *Ibid.*, pp. 325–26.
[54] Cf. above, p. 34. Note that Vogt found Herwegh very helpful as a partner in research. Cf. Misteli, *op. cit.*, p. 71, and n. 24.
[55] Misteli, *op. cit.*, pp. 67–68. Vogt had high regard for Milne-Edwards. Because he viewed him as the only one in the Academie des Sciences who had made serious progress in zoology, Vogt kept in communication with him during his trip. Cf. *OM*, I pp. 68, 81–82.
[56] For his description of the matter, see *OM*, I, pp. 168–81.
[57] Liebig had been in communication with Vogt's father in Bern concerning Karl's career. While he had experienced no difficulty with the Princes regarding the appointment, the council of ministers had been more difficult to convince. Vogt, after all, was known to be wearing a beard, a practice certainly against the custom of the day if not the rule. But the prestige of those recommending young Vogt for the post was simply too great to ignore. In addition to Liebig, they included Arago, Agassiz, and even his old opponent Leopold von Buch, who, besides praising highly his ability, mentioned in passing an equally favorable opinion of Vogt from Alexander von Humboldt. Cf. Wurzbach, *op. cit.*, pp. 40–43; W. Vogt, *op. cit.*, pp. 50–53.
[58] Cf. *OM*, I, p. 15. For the addressee of the letters, cf. Misteli, *op. cit.*, p. 66, n. 1, and *OM*, I, pp. 117, 171, 248; II, pp. 112, 125, 128.
[59] Vogt pointed to the similarity between the sea cucumber and the Negro, both of which he claimed were capable of suicide when taken into captivity. The theologians were wrong, therefore, to hold that man was the only creature that could commit suicide. (*OM*, II, pp. 29–30). Vogt had in view the protective habit displayed by sea cucumbers called auto-evisceration, in which the gut is ejected and offered to a predator. It is hardly suicide, since what is lost can be regenerated over a period of several months.
[60] *GK*, p. viii. Also in the *PB*: "A unique spirit flows through the natural science of our day." (p. 1)
[61] *OM*, I, p. 132.
[62] *Ibid.*, I, pp. 112–21.
[63] *Ibid.*, II, p. 43.

[64] *Ibid.*, I, pp. 55–56.
[65] *Ibid.*, II, p. 130.
[66] *HS*, p. 9.
[67] *Ibid.*, pp. 18, 21.
[68] *Ibid.*, p. 22.
[69] *Ibid.*, pp. 10–13, 16.
[70] This issue is treated in depth below in Chapter Seven, pp. 151ff.
[71] *Ibid.*, p. 41. Quoting from Leverrier, Vogt declared that fixed norms and peace were but abstractions found nowhere in nature. "Disturbances, revolutions are the rule." (p. 42)
[72] *OM*, I, p. 12. Misteli considers this preface to be the breakthrough to open materialism, which up to them, in his view, had been only a tendency. Misteli, *op. cit.*, p. 97.
[73] Misteli, *op. cit.*, p. 51.
[74] A. Büchner, *'Tolle' Jahr*, pp. 121–22.
[75] *Ibid.*, p. 181. For more of those pre-Frankfurt Giessen days, cf. Misteli, *op. cit.*, pp. 101–114.
[76] *Die Aufgabe der Opposition unserer Zeit* (Bern, 1849).
[77] *UT*, pp. 28–31, 153–54. More on this work is found below, pp. 195ff. In his bitterness he apparently had become susceptible to the ideas he had heard Bakunin *et al* expound in Paris.
[78] *Ibid.*, p. 199. The professors, "these priests of knowledge," wanted only the student's money. Once they have it, said Vogt, they let him drink beer.
[79] *NG*, p. vi.
[80] Ernst Krause, 'Karl Vogt,' *Allgemeine Deutsche Biographie,* **40**, 1896. Reprinted (Berlin, 1971), p. 185.
[81] *PM*, pp. 646–48.
[82] *Ibid.*, pp. 656–62. For Moleschott's version see discussion of his *Die Lehre der Nahrungsmittel* below, pp. 88ff.
[83] *ZB*, I, p. 5.
[84] *Ibid.*, p. 16.
[85] *Ibid.*, p. 7.
[86] *BT*, p. vi. My emphasis. Cf. also pp. 421–22.
[87] *Ibid.*, p. 355.
[88] *Ibid.*, p. 367. Wagner had been writing against Vogt's *Physiologische Briefe* in 'Physiologische Briefe' of his own in the *Allgemeine Zeitung*. Interestingly enough, Vogt also took the opportunity here to criticize in very sarcastic terms the creationist viewpoint of his former teacher Agassiz. Cf. pp. 368ff.
[89] *Ibid.*, p. 423.
[90] *Ibid.*, p. 418.
[91] W. Vogt, *op. cit.*, p. 86.
[92] *KW*, p. 23. Wagner was criticized by some of his colleagues, not for the quality of his work, but for writing in a common *Zeitung.* Cf. p. 35.
[93] *Ibid.*, pp. 27–30.
[94] R. Wagner, *Menschenschöpfung und Seelensubstanz* (Göttingen, 1854), p. 17.
[95] *Ibid.*, pp. 29–30.
[96] *KW*, pp. 41–43.
[97] Wagner, *op. cit.*, pp. v–vi.
[98] Quoted in R. Wagner, *Über Wissen und Glauben* (Göttingen, 1854), p. 5. Cf. also Vogt, *KW*, p. 45.
[99] The phrase was Lotze's, from his *Grundzüge der Psychologie*. Cf. Wagner's *Über Wissen und Glauben*, p. 10, n. The other notable personage Wagner cited in this pamphlet was Virchow. Vogt refers to a critical report of the lecture in *KW*, p. 39.

[100] *KW*, pp. 3–20.
[101] See below, pp. 200ff.
[102] His description of the trip is contained in *NF*. The excursion had been sponsored by a rich doctor from Frankfurt who was interested in geology. .ccompanying Vogt and the doctor were an old companion from Neuchâtel, and the son of Alexander Herzen. Cf. W. Vogt, *op. cit.*, pp. 142–50.
[103] On the zoological station see *OM*, I, pp. 134–35; 'Sea Butterflies,' *Popular Science Monthly* **35**, p. 313; R. Virchow, *Briefe an R. Virchow, Zum 100. Geburtstage*, ed. H. Meisner (Berlin, 1921), p. 63; W. Vogt, *op. cit.*, pp. 118–20. For some of Vogt's other activities in this period, see pp. 165–66.
[104] Cf. W. Vogt, *op. cit.*, p. 163 for a list of his articles on anthropology.
[105] Büchner unintentionally confirms that Vogt was thought by many to have been the originator of the idea of the simian origin of man. In Büchner's view he had been the first to lecture publicly on that subject. Cf. *MSN*, pp. 122–23.
[106] And hate him they did! Germany was reopened to him in 1867 and from 1867 to 1869 Vogt gave extensive public lectures there. More than once police were necessary to control the crowds gathered to oppose him, and once a rock crashed through a window, landing at his feet as he spoke. Cf. Wurzbach, *op. cit.*, pp. 83–85; W. Vogt, *op. cit.*, pp. 177–78; and Deubler, *op. cit.*, II, pp. 78–80.
[107] Quoted in W. Vogt, *op. cit.*, p. 167. Cf. also his opposition to romantic nationalism in *VT*, pp. 226–27.
[108] *PBK*, pp. 6ff. Most of this work is directed at this issue.
[109] *Ibid.*, pp. 10f, 15f, 41f.
[110] W. Vogt, *op. cit.*, p. 196.
[111] *VM*, I, p. 256. Cf. also p. 196. Throughout the chapter Vogt also brought negroes into the picture, listing the similarities between negroes and idiots, and negroes and apes. For him the white European male was the highest developed organism anywhere. Cf. *VM*, I, p. 7, 93–96.
[112] Otto Zöckler, *Die Geschichte der Beziehungen zwischen Theologie und Naturwissenschaft*, p. 628. Cf. also n. 39, pp. 807–08, and W. Vogt, *op. cit.*, pp. 172–74, 184.
[113] R. Mayer, *Die Mechanik der Wärme in Gesammelten Schriften*, ed. J. J. Weyrauch, 3d ed. (Stuttgart, 1893), p. 356. The lecture is reprinted here, pp. 347–57.
[114] *Ibid.*, p. 357. Weyrauch, editor of the *Mechanik*, supplies notes to the speech in which he identifies and elaborates on the location in Hirn's work to which Mayer referred. Cf. n. 13, p. 361. Weyrauch also adds a three page evaluation of Mayer's concept of religion, pp. 362–64.
[115] R. Mayer, *Kleinere Schriften und Briefe (1832–1877) von R. Mayer*, ed. Jacob Weyrauch (Stuttgart, 1893), p. 445.
[116] Weyrauch gives three such recollections, one of which specified Vogt and the materialists as the source of it. *Ibid.*, pp. 452–55.
[117] Cf. Weyrauch's account in *Ibid.*, pp. 447–49. For Mayer's reaction to the bad reviews his lecture received, cf. *Mechanik*, pp. 365–66.
[118] 'Lingering Barbarism,' *Popular Science Monthly*, pp. 638–39.
[119] *Ibid.*, pp. 639–40.
[120] W. Vogt, *op. cit.*, pp. 199–201.
[121] *FA*. In mind was also the antivivisectionist book of Friedrich Zöllner, the Leipzig astronomer who had singled out Vogt for particular derision. Zöllner's book resulted in a petition against vivisection in the *Reichstag*, and Wilhelm Vogt notes the great number of names from Prussian army officials on it! (W. Vogt, *op. cit.*, pp. 222–23. Cf. also n. 1, p. 223) Vogt never liked the English, and the *FA* was directed primarily at them: "John Bull in and of himself would not be so dangerous, but when his sister-in-law, Miss Threadneedle Snob, the old maid with the cats and dogs, when she

gets mixed up in things, they can be very unpleasant." (*FA*, p. 6) For Vogt's anti-English sentiments, see *OM*, I, p. 234, and II, pp. 4, 169, 172, 187–88.
[122] "Lingering Barbarism," pp. 640–43.
[123] Cf. W. Vogt, *op. cit.*, pp. 116–17.
[124] *Die Säugthiere in Wort und Bild* (München, 1883) with Specht; *Lehrbuch der praktischen, vergleichenden Anatomie* (Braunschweig, 1885) with Emile Jung.
[125] Cf. W. Vogt, *op. cit.*, p. 253, where the titles are listed.
[126] Krause, *op. cit.*, p. 188.
[127] W. Vogt, *op. cit.*, p. 182. Wurzbach noted that Vogt had married "a pretty highlander from Bern" who bore him three sons and a daughter, but said no more. Wurzbach, *op. cit.*, p. 91.
[128] Marx and Engels, *Werke*, XXXII, p. 15.

CHAPTER IV

[1] *FMF*, p. 6.
[2] *Ibid.*, p. 35.
[3] *Ibid.*, p. 57.
[4] *Ibid.*, p. 58.
[5] Ibid., p. 66. The influence of father on son was predominant. See, for example, *FMF*, p. 77. The German edition of the *LN* was dedicated to his father in 1850, but it was on the occasion of Moleschott's success at landing a teaching post that the reverent deference for his father came out. See the emotional dedication in the publication of his inaugural lecture, 'Licht und Leben,' 1856.
[6] Many of the strict values of the German middle class are evident in Moleschott's autobiographical account of his early experiences. The classic description of these values is of course Gustav Freytag's *Soll und Haben*, best selling novel in the Germany of the 1850's. One cannot help but compare Anton with the proud, energetic and aspiring young Jacob Moleschott of the *FMF*. Moleschott himself explained this combatitive streak in himself as the result of an exaggerated sensitivity that he had inherited from his father. (*FMF*, p. 103).
[7] *FMF*, p. 78.
[8] *Ibid.*, p. 84. Hence Moleschott turned to a Dr. Berg for directed reading in the *Phenomenology*, and later even lectured on Hegel at Heidelberg himself. (p. 118) He had only a short exposure to Carriere, for the latter became a *Privatdozent* at Giessen sometime during the same year in which Moleschott arrived at Heidelberg. Büchner, who entered Giessen in 1843, spoke enthusiastically of Carriere's teaching there. Cf. below, p. 102.
[9] For a popular image of the doctor in 1840, conforming more or less to such speculative and romantic notions, see the interesting short story 'Der Naturforscher,' by A. Winter in *Der Freihafen*, 3d Jahrgang, Heft 1 (1840), pp. 95–165. Moleschott agreed with his father's conviction in his Latin dissertation. Cf. Grützner, 'Jacob Moleschott,' *Allgemeine Deutsche Biographie*, **52**, 1906. Reprinted (Berlin, 1971), p. 436.
[10] *Ibid.*, p. 102. This open sympathy is all the more noteworthy when one remembers that this account was written at the end of the century by the old man, not by the young and enthusiastic student of science.
[11] "We all were thoroughly free thinkers, but no conspirators." *FMF*, p. 110.
[12] Moleschott criticized Liebig's inference that since the carbon dioxide content of the atmosphere was sufficient in amount to provide the carbon of plant life on earth, that therefore plants take no carbonaceous organic matter from the ground.
[13] *FMF*, pp. 115–16. The award did other things for him too. With the money

Moleschott won he bought his first microscope. More important, his future wife first learned his name as winner of the Teyler competition (pp. 116–17).

[14] Grützner, *op. cit.*, p. 436.

[15] M. A. Van Herwerden, 'Eine Freundschaft von drei Physiologen,' *Janus*, **20** (1915), pp. 175–176.

[16] Walter Moser, 'Der Physiologe Jakob Moleschott und seine Philosophie,' *Zürcher Medizinesgeschichtliche Abhandlungen.* Neue Reihe, No. 43 (1967), p. 13.

[17] G. J. Mulder, *Versuch einer allgemeinen physiologischen Chemie*, trans. J. Moleschott (Heidelberg, 1844), p. viii.

[18] *Ibid.*, pp. x–xi.

[19] Liebig held that proteins were the plastic elements of nutrition, that they made up the blood and flesh itself, particularly muscle tissue, and that they alone provided the molecular basis of muscular motion and vital activity. Cf. William Coleman, *Biology in the Nineteenth Century*, pp. 131–35. Mulder's position was not so exclusive.

[20] Moleschott confessed that Mulder's socialistically conceived work on nutrition and physiological chemistry had inspired him to treat *Stoffwechsel* as a central theme of the *KL*. Cf. *FMF*, p. 213.

[21] *FMF*, p. 148. His father, who visited him at Utrecht on occasion, had also recommended that he read *Das Wesen des Christentums*, which he had given to his son after graduation from Cleves. Cf. p. 149 and *Werke*, II, p. 227.

[22] *FMF*, p. 131.

[23] *FMF*, p. 147. Cf. his anger-filled letters to van Deen concerning the latter's rejection, reproduced in Herwerden, *op. cit.*, pp. 181–82. See also the letter to van Deen in which he discussed the significance of German political developments in 1848 for Holland. Herwerden, *op. cit.*, pp. 189–90.

[24] *HM*, pp. 10–11, 43; *FMF*, pp. 169–70.

[25] The story is found in Moleschott's letter to Feuerbach thanking him for reviewing his *LN*. Cf. *Werke*, II, p. 179.

[26] *HM*, pp. 70–71, 118; *FMF*, p. 236.

[27] Herwerden, *op. cit.*, p. 183. Cf. also *FMF*, pp. 181–91.

[28] Moleschott therefore succeeded where Vogt failed, for he betrayed how important social rank was when he confided that Sophie's father was among the better circles of Mainz. (*FMF*, p. 182) The irony is that Dr. Strecker was a friend of Karl Vogt! (p. 183)

[29] *HM*, pp. 5–6.

[30] *PN*, p.v. Only the first and third of Tiedemann's projected work appeared, both of them in the 1830's.

[31] Moleschott's *Physiologie der Nahrungsmittel* is an example of the "exceptional progress (which) had been made with regard to assaying the nature and proportions of foodstuffs required for the maintenance of life." Coleman, *op. cit.*, p. 11.

[32] *PN*, p. 2.

[33] Coleman, *op. cit.*, p. 129. For Vogt's views, cf. *PB*, pp. 31, 36. Coleman points to the work of Eduard Pflüger in 1872 as the "landmark in the history of respiratory physiology" that led to the downfall of this hegemony. Moleschott had been experimenting on the relation of various foods to the blood just prior to the *LN*. Cf. *FMF*, p. 196.

[34] *FMF*, p. 192.

[35] E. P. Evans, 'Sketch of Jacob Moleschott,' *Popular Science Monthly*, (1896), p. 402.

[36] *LN*, p. 4.

[37] *Ibid.*, p. 221. Cf. also p. 193.

[38] *Ibid.*, p. 215.

[39] *Ibid.*, pp. 241–42. Incidentally Moleschott denied in this connection that all

intellectuals were fat, though he admitted that monks, who he thought apparently never used their brains, may tend in that direction (p. 243).

[40] *Ibid.*, p. 242.

[41] *Ibid.*, p. 67.

[42] Even unlike Haeckel and the Monists, who feared alcohol and believed that only inferior beings became alcoholics. Cf. Daniel G. Gasman, *The Scientific Origins of National Socialism* (London, 1971), pp. 92–93.

[43] *LN*, p. 163. Cf. also pp. 165–66.

[44] *Ibid.*, sections 45, 47–65.

[45] *Ibid.*, pp. 230–36.

[46] *Ibid.*, pp. 50, 194–95. Although he held that matter alone sustains all mental activity, he allowed that there were less material impressions such as sound, which eluded our grasp, as well as the altogether intangible impressions of light and color. Wise men recognized this, he said. It was true piety to acknowledge this connection with the whole of things. According to Moleschott, Schleiermacher was correct to identify the sense of dependency as the true essence of all religion. (p. 256) This obviously was not the atheist Karl Vogt talking!

[47] *Ibid.*, p. 189.

[48] *Ibid.*, p. 69. Emphasis mine.

[49] *Ibid.*, pp. 102–03, 120, 170.

[50] *Ibid.*, p. 252.

[51] *Ibid.*, p. 129.

[52] *Ibid.*, p. 130. Cf. *The Chemistry of Food* (London, 1856), p. 350. The English translation, part of the series entitled *Orr's Circle of the Sciences*, omitted the *Vorwort* and *Einleitung*. Little was said of Moleschott himself, except that his book had met with great success in Germany (p. ix).

[53] The letter is reprinted in *FMF*, pp. 201–02.

[54] *Werke*, XIII, pp. 175–76.

[55] *Ibid.*, p. 176. This letter began a correspondence between the two that lasted until a few years before Feuerbach's death.

[56] *Werke*, X, pp. 3ff.

[57] *Ibid.*, p. 9. Feuerbach here played on the word "Revolution." Copernicus, in giving revolving motion to the earth, opened it up to other revolutions (p. 10).

[58] *Ibid.*, p. 12. Moleschott commented in the *FMF*: "In fact my book was intended as a materialistic and socialistic work." (p. 206) Cf. also p. 212. Those who had missed the social and political implications here in the *LN* would find them equally explicit in his article of 1852 in *Die Gegenwart*, reprinted in the *PSK*, pp. 78ff.

[59] *LN*, p. 15.

[60] Feuerbach's review of Moleschott has been a problem for Feuerbach scholars, since it contains such a superficial philosophical position. That it has to be a problem is not everyone's view. Bolin, for instance, allows that it was a satire against the reactionaries, but he thought that nineteenth century German materialism was so much more respectable and *wissenschaftlich* than the eighteenth century variety that he never really permitted himself to address the issue as such. (Bolin, *Ludwig Feuerbach*, pp. 28, 261–67). He does say that it was with this review that materialism explicitly joined atheism as the trademarks of Feuerbach's thought (*Werke*, XII, pp. 131, 179).

Nor do Kohut or Jodl interpret the review as an exception to Feuerbach's general position. (Kohut, *Feuerbach*, pp. 239, 246–49, 406–11; Jodl, *Feuerbach*, pp. 54–62, 120–30.)

Rawidowicz holds an intermediate position. On the one hand he rejects Treitschke's claim that the review was simply the logical outcome of the tendency of Feuerbach's ideas; hence he refuses to take what Feuerbach said in the review literally.

(Rawidowicz, *Feuerbachs Philosophie,* p. 203. Cf. also n. 1–5) Yet, on the other hand, he rejects the standard Marxist interpretation from Plekhanov that Feuerbach, like Marx and Engels, always denied scientific or "vulgar" materialism. (pp. 484–85) Lenin may have called Moleschott, Büchner, and Vogt pygmies besides Feuerbach, but in doing so he did not represent the real relationship between Feuerbach and Moleschott. (p. 495) He concludes that Feuerbach here was trying to build a bridge between philosophy and natural science, and to enlist the aid of natural science in reasserting after 1848 Feuerbach's hopes for the establishment of a "new philosophy." (p. 200)

The most extreme interpretation from the other direction is that of Sidney Hook, who cannot understand Feuerbach's "degenerate sensualism" here in the review. He considers the possibility that Feuerbach's sense of humor is responsible, but rejects it. Feuerbach is deadly serious in these absurd claims. Hook calls "Die Naturwissenschaft und die Revolution" one of the paradoxical features of Feuerbach's thought, and concludes that it represents Feuerbach grabbing for the nearest science at hand to justify his belief that his critique of religion and philosophy marks the turning point in the history of western thought. (*From Hegel to Marx*, pp. 237, 242, 267–71). Chamberlain (*Heaven Wasn't his Destination*, p. 158) follows Hook's analysis, and like Hook, emphasizes that Moleschott's *LN* was an unimportant scientific work in order to cast aspersions on Feuerbach's review.

Perhaps the most well balanced view is Eugene Kamenka's. Kamenka explains the review as the result of the extravagance of Feuerbach's style: "He was not so much a nonreductive materialist at one stage and a reductive materialist at another, as a confused interactionist all the way through, overemphasizing various aspects at various times." (Kamenka, *op. cit.*, p. 112.) Feuerbach had the habit of reducing everything to some X, which then served as the *sine qua non* of his argument. This he did with "man" in the *Wesen* of 1841, and then in 1850 in the *LN* to *Nahrungsmittel* (p. 113). Cf. also p. 171, n. 12.

As for Feuerbach himself, he had become more radical in his statements after 1848 (Cf. for example the *Vorlesungen* on religion in 1849, *Werke* VIII, pp. 179–180). Further, he did not retract what he had said in the review later. In fact, he said in 1862 that he was glad for what he had written. ('Das Geheimniss des Opfers, oder Der Mensch ist was er isst,' *Werke*, X, p. 41, n.) But in one place, cited by almost everyone, he did distinguish himself from the scientific materialists. Cf. above, p. 220, n. 82.

61 *Werke*, X, p. 22.

62 There was none in Moleschott's eyes. He thanked Feuerbach for the review, and added: "In short, I have received the best I could receive, the approval of the man who paved the way for us to bring about the incarnation (*Menschenwerdung*) of philosophy in natural science." *Werke*, XII, p. 179.

63 *KL*, pp. 3–8. Citations are from the second edition of 1855.

64 *Ibid.*, pp. 376–80. Corresponding references to the *Chemische Briefe* are provided in Moleschott's notes.

65 *FMF*, p. 219. The anti-Liebig position is also found in the 1852 article in *Die Gegenwart*, reprinted in *PSk*. Cf. pp. 2, 6.

66 *KL*, p. 42. This definition of life had been given earlier in the *PS*, p. xix.

67 Moleschott thought it one-sided to claim simply that plants prepare what animals consume; i.e., that plants, because they can live solely on inorganic nutriments while animals cannot, represent a lower form of life whose sole function is to serve animal needs. Cf. *KL*, pp. 70, 103, 105.

68 *KL*, pp. 103–06, 201–02.

69 *Ibid.*, pp. 85–86.

70 *Ibid.*, pp. 122–27. The Mulder-Liebig dispute was not due to this as much as it was due to the relationship between proteins and plants and animals. Cf. pp. 138f.

[71] *Ibid.*, Neunter Brief, "Ernährung und Athemung." "The pretence of a purpose always brings one-sidedness with it." (p. 117).
[72] Quoted in *KL*, p. 15.
[73] *KL*, 5th ed., pp. 3–4.
[74] *KL*, 2d ed., pp. 25, 404f. Cf. below, Chapter Seven, pp. 151f.
[75] *Ibid.*, pp. 18–19.
[76] *Ibid.*, p. 329.
[77] *Ibid.*, pp. 330–34. On the concept of force, cf. also the preface to the *PS* of one year later, pp. x–xii.
[78] *Ibid.*, pp. 349, 351, 353.
[79] *Ibid.*, p. 373. On vital force cf. pp. 358–72.
[80] *Ibid.*, p. 373.
[81] *Werke*, XII, p. 195.
[82] *KL*, p. 445. Büchner picked up this emphasis at several places in his work as well, again no doubt from Moleschott.
[83] *Ibid.*, p. 450. Cf. Albert Lévy, *La Philosophie de Feuerbach et son influence sur la litterature allemande* (Paris, 1904)' pp. 406, 409.
[84] *LN*, p. 256. The English translation distorted Moleschott to read: "It is true piety heartily to feel this connection with the mighty Creator of all." (*The Chemistry of Food*, p. 394)
[85] *KL*, p. 455.
[86] *Ibid.*, pp. 458–59.
[87] *Ibid.*, p. 480.
[88] Strauss moved to Heidelberg for a period, and while there the two met fairly regularly for conversation. While Strauss did not appreciate Moleschott's brand of materialism, Moleschott for his part later begrudged the fact that Strauss never once mentioned him in *Der Alte und der Neue Glaube*, in spite of the fact that there was much in the work which Moleschott had taught Strauss. Cf. *FMF*, pp. 238–39.
[89] *FMF*, p. 227. Moleschott himself contributed of course to *Die Natur*. In 1853, for example, he wrote a favorable review of Vogt's *Bilder aus dem Thierleben* and a sketch of Georg Forster. Cf. *Die Natur* 2 (1853), pp. 131–32, 140, 147–48, 196, 215–16; and 205–08.
[90] Cf. the open letter written in a Frankfurt periodical by students of Moleschott. *FMF*, pp. 255–56. Moleschott did not suffer alone. His troubles had been preceded by the dismissal from Heidelberg of Gervinus and Kuno Fischer. For Moleschott's discussion of the situation, see Herwerden, *op. cit.*, pp. 414ff, where a letter to van Deen is reprinted.
[91] Virchow thought well of Moleschott as a scientist, but the reason he recommended him over DuBois-Reymond was because of his ability to write popular material. DuBois-Reymond was a more of a specialist in Virchow's eyes; he belonged in an academy, not a university. Virchow's communication is reprinted in Moser, *op. cit.*, p. 22.
[92] Cf. Moleschott's account of Forster's conversion from religion to science in *GF*, pp. 104–10, and of his eventual conviction, similar to that of Moleschott in the *Kreislauf*, that science and religion were mutually exclusive endeavors (pp. 160, 164–65).
[93] "Humanity was his God, and humaneness his aim." *GF*, p. 2. Moleschott declared that Forster's views were identical to Feuerbach's. (p. 162)
[94] *Ibid.*, pp. 171–72.
[95] *Ibid.*, pp. 150–60.
[96] In addition to an annual lecture opening his course on physiology each fall, Moleschott also accepted speaking engagements from various groups requesting popular

lectures on the nature of science. Rather than deal with each one individually here, we shall draw on them as needed. For the complete list see the Bibliography. The *Physiologisches Skizzenbuch* of 1860 consisted of four essays published in *Die Gegenwart* and elsewhere, and is not therefore a book-length work on science.

[97] The manuscript of this work was given to the Medical Academy of Turin by Moleschott's children. Cf. Moser, *op. cit.*, p. 27. Talk of it crops up repeatedly in Moleschott's correspondence with Feuerbach and in the autobiography. He had told Feuerbach in 1850 that he envisioned the *LN* as "... a first small attempt at a general anthropology;" hence even at this point he had already begun the larger work. Cf. *Werke*, XIII, p. 176; *FMF*, pp. 250–51; *PSk*, p. vi; and *FW*, pp. 24–25.

[98] *Werke*, XIII, p. 211. Cf. also p. 151.

[99] *Ibid.*, pp. 212, 354.

[100] *LL*, pp. 42–43. And he succeeded very well, for of him it was said later by an Italian colleague: "Moleschott is a man of science and a student of the arts, yes, he is an artist as well as scholar." Quoted in Moser, *op. cit.*, p. 25.

[101] In addition to his native Dutch, he knew French, German, English and now Italian, and had once a reading knowledge of Danish. *FMF*, p. 300. Sophie Moleschott joined her husband and Emma Herwegh in studying Italian. Cf. *FMF*, pp. 298–302.

[102] Herwerden, *op. cit.*, p. 427. Herwerden reproduces a significant amount of the correspondence between Moleschott and Donders and Moleschott and van Deen from his years in Italy, which in many ways makes up for the lack of autobiographical material from this period.

[103] *FMF*, p. 275; *PSe*, p. 4.

[104] Moser, *op. cit.*, p. 27.

[105] *Ibid.*, p. 28.

CHAPTER V

[1] Cf. Anton Büchner, *Die Familie Büchner* (Darmstadt, 1963), pp. 10–16.

[2] A son Karl, born in 1818, died after a few months of life.

[3] *Ibid.*, p. 24.

[4] Ludwig Büchner, 'Georg Büchner,' pp. 1–50 in *Nachgelassene Schriften von Georg Büchner* (Frankfurt a.M., 1850), p. 7. Among the members of the society was August 'Red' Becker, who later would work with Ludwig at Giessen in 1848. The extent of Georg's commitment can be seen from a letter to Gutzkow in 1835, quoted by Ludwig: "I believe that in social things one must start with an absolute and fundamental position on justice, one must seek to form a new spiritual life in the people and let our decrepit modern society go to the devil." (p. 35)

[5] Ludwig reported that Oken had high hopes for Georg, and that he even sent his son to attend Georg's lectures. *Nachgelassene Schriften*, p. 38.

[6] *NH*, p. 190.

[7] Alexander Büchner, 'Vorwort,' *ID*, p. xxiii.

[8] *Ibid.*, pp. x–xi.

[9] *NH*, pp. 191–93.

[10] Anton Büchner, *op. cit.*, p. 65.

[11] Alexander Büchner, 'Vorwort,' p. xiii; and Alexander Büchner, *Das 'tolle' Jahr*, pp. 180–81.

[12] *HL*, p. 5.

[13] *Ibid.*, pp. 7, 35. Büchner added that his opponents, who did not appeal to material relationships as the basis of their explanation of the action of the spinal column, could not explain it at all, since it could not be concieved in any other way. (pp. 32–33).

[14] *Ibid.*, p. 13.
[15] *Ibid.*, pp. 7, 36. He later qualified his position due to the work of DuBois-Reymond, which, he believed, proved that each nerve was the source of an electrical current and not just a conductor or passive substratum. Cf. *KS*, p. 153; *KL*, p. 72.
[16] *HL*, p. 36, n. Büchner's nine theses are listed on pp. iii–iv.
[17] Georg's *Woyzeck* was completely omitted from the collection because it was too difficult for Ludwig to understand. Anton Büchner, *op. cit.*, p. 66.
[18] Alexander Büchner, 'Vorwort,' p. xiv.
[19] *Ibid.*, p. xv.
[20] *Ibid.*, p. xv; Alexander Büchner, *Das 'tolle' Jahr*, p. 199. Ludwig's vascillation must have been quite noticeable, for Alexander calls him at different times an inveterate optimist ('Vorwort,' p. xiv) and a born melancholic (*Das 'tolle' Jahr*, p. 209).
[21] *Das Od. Eine wissenschaftiche Skizze* (Darmstadt, 1854). Unavailable for perusal.
[22] That he did this so soon after receiving his degree brought him criticism for presumptuousness. Cf. the anonymous *Der Humor in Kraft und Stoff*, p. 73. We can surmise to some extent what Büchner's attitude was towards Reichenbach's ideas from his later articles on hypnotism. His main concern was to separate out hypnotism, which he believed had become scientific with the work of the Scot James Braid, from animal magnetism, and to deny the supernatural claims that were often associated with such phenomena. Cf. *NW*, II, pp. 73–75; *TT*, pp. 189–203; *FE*, pp. 96–111. Only in the *TT* did he spell out what he thought was responsible for hypnosis; viz., a functional disturbance of individual parts of the cerebrum. (*TT*, pp. 199–203).
[23] Alexander Büchner, 'Vorwort,' pp. xvi–xvii. The reader should not overlook that this recollection was written at the end of the century, therefore with full knowledge of the popularity that *Kraft und Stoff* had achieved. For Ludwig's own account of the circumstances surrounding the birth of *KS*, cf. his 'Selbst-Kritik' (more accurately 'Self-Praise') in *KS*, 12th ed. Of Büchner's brothers and sisters only one, Mathilde, never took up the pen. Luise Büchner, next oldest to Ludwig, wrote extensively on women's liberation. Cf. Anton Büchner, *op. cit.*, pp. 43–60.
[24] *KS*, pp. viii–ix.
[25] They were: English, French, Italian, Polish, Hungarian, Swedish, Spanish, Dutch, Greek, Russian, Danish, Armenian, Rumanian, Czech, Arabic, Bulgarian, and Lithuanian. Cf. *KS*, 19th ed., p. xvii.
[26] *KS*, pp. xii–xiii.
[27] *Ibid.*, pp. xiii–xiv.
[28] *Ibid.*, p. xi.
[29] In addition to Moleschott, Vogt too had preached this message prior to Büchner. Cf. *KW*, p 121.
[30] *KS*, p. 151. Cf. also *NG*, pp. 48–49.
[31] *KS* p. 202. Büchner wroted the chapter entitled "Der Gedanke" to clarify his own position vis-à-vis Vogt. He did not approve of the latter's pronouncement in the *PB* that thought was to the brain as urine was to the kidney. Cf. above, p. 64.
[32] *KS*, p. 137. In the chapter on the brain and soul Büchner spent most of his time listing the ways in which the material state of the brain affected mental activity; e.g., Newton's atrophied brain caused him in old age to become interested in studying the books of Daniel and Revelation in the Bible. (p. 123) Büchner distinguished between the soul and thought, the former being the broader category. In addition to the mental activity of thinking, it included also feelings, dispositions, etc.
[33] *KS*, pp. 3, 5, 151.
[34] *Ibid.*, 5th ed., p. 4.
[35] *KS*, p. 151. "We have defined force as a characteristic of matter, and we have seen

that both are inseparable; however, *conceptually* they are very far apart, even the negations of one another in a certain sense."

[36] *Ibid.*, p. 71, n. By the fifth edition he was bolder. Cf. *KS*, 5th ed., p. 67, n., and pp. 83–84.

[37] *KS*, pp. 60–63.

[38] *Ibid.*, p. 158.

[39] *Ibid.*, p. 173.

[40] *KS*, 5th ed., pp. 49–50. Elsewhere: "Either something is compatible with reason and experience – in which case it is true; or it is incompatible – in which case it is untrue and finds no place in philosophical systems. . . . Knowledge (*Wissenschaft*) does not have to do with what could be, but what is." (p. 203) An almost identical statement is found in the original edition, p. 205.

[41] *KS*, p. 261.

[42] *KS*, 5th ed., p. 247. Cf. also first ed., pp. 260–61 on hypothesis.

[43] *KS*, 5th ed., p. 247.

[44] *KS*, 5th ed., p. 249. This quotation had been used in the original conclusion, but not to communicate what he was now trying to say.

[45] Cf. the forewords to the third and fourth editions, where Büchner replies to his critics. The opposition of Gutzkow, friend and admirer of Ludwig's elder brother Georg, must have indeed discouraged him. Cf. *KS*, 5th ed., pp. xxix–xxx, lix–lx.

[46] *NH*, p. 186.

[47] Alexander Büchner, 'Vorwort,' p. xix.

[48] Cf. for example 'Arm und Reich,' and 'Bettler's Klage,' *NH*, pp. 34–36. Also pp. 135–36.

[49] *Ibid.*, p. 136.

[50] *Ibid.*, p. 194.

[51] *Ibid.*

[52] Hess, *Briefwechsel*, p. 355. Ule's review of the *KS* in *Die Natur* had not pleased Büchner. After he replied to Ule, the latter told Hess that some people were more motivated by emotion than by scientific facts. (p. 354) It would seem that everyone had possession of facts, but somehow their meaning depended on who was talking!

[53] *NG*, pp. 6, 15.

[54] *Ibid.*, pp. 39–40.

[55] Citing a work by Cornelius, *Über die Bildung der Materie* (Leipzig, 1856), Büchner argued that this had already been worked out. Cf. *NG*, pp. 45–48, 74–75.

[56] *NG*, pp. 53–57, 65.

[57] *Ibid.*, p. 70. Note that Büchner, like Feuerbach and Moleschott, was not opposed to religion as such. This remained his attitude throughout his life. On Büchner's religion see J. M. Robertson, *A History of Free Thought*, II, p. 607. Büchner conceded that he could speak of his "religion." He identified it as a belief that every event in the world happened, happens, and would happen according to the eternal and unchangeable laws of nature; i.e., according to the uninterrupted laws of cause and effect without the possibility of personal intervention. Man alone came under a law of higher potency. (*FE*, pp. 141–42) He spoke of his humanism as an *Ersatzreligion*: "The improvement of the individual and society in material, spiritual, and moral respects is the great goal towards which the new religion has to strive, the religion of the future, the religion of the love of mankind, which is destined to replace the old one." (*KL*, p. 140) He closed his discussion in this context with a quotation from Feuerbach: "Love of man is the only true love of God." (p. 141) On scientific materialism as an *Ersatzreligion*, see Hermelink, *Liberalismus und Konservatismus*, pp. 38–40, and Zöckler, *Geschichte*,

pp. 401–02. Zöckler quotes a poem of the minor materialist Eduard Löwnthal that captures the spirit of the new religion.

Die Zeit, sie naht, wo durch des Wissens Macht
Die Götter fallen, die Altäre wanken;
Wo die Natur allein das Heiligenbild
Als Offenbarung die Vernunft nur gilt –
Die Kirche stürzt im Sturme der Gedanken.

[58] *NG*, p. 73.
[59] *Ibid.*, pp. 87–88.
[60] *Ibid.*, p. 89.
[61] *Ibid.*, p. 90.
[62] *Ibid.*, pp. 165–67, 187–91.
[63] *Ibid.*, pp. 183–84, 190, 205. Yet Büchner himself confessed that he was not a real scientist. Anton Büchner, *op. cit.*, p. 78.
[64] *Ibid.*, p. 299.
[65] *Ibid.*, pp. 301–02. August did not find fault with this Hegelian idea, he merely identified it as speculation. There may well be substance to Siebeck's conclusion that Büchner in the *NG* took a stance very close to objective idealism. Cf. Hermann Siebeck, 'F. K. C. L. Büchner,' pp. 49–56 in *Hessische Biographien*, vol. I, ed. H. Haupt (Darmstadt, 1918), p. 52.
[66] *Werke*, XXXII, p. 579. This in a letter to Kugelmann in 1868. Marx's opinion of Büchner was not flattering. He did show some interest in Büchner's book on Darwin in 1868, mainly because he wanted to find out how the Germans were responding to Darwin. But Marx never failed to mock "the great Büchner" in his correspondence, and he delighted in pointing out Büchner's errors. (Cf. the letters to Engels of November 14, 18 and 23 of 1868 in *Werke*, XXXII, pp. 202–04, 206, and 213–14 respectively. Büchner's *NW*, I, p. 276 is relevant in this connection.) In 1870 Marx wrote to Paul and Laura Lafargue concerning a favorable French review of the tenth edition of the *KS* by one Henri Verlet: "In Germany one would wonder very much about Verlet's appreciation of Büchner. In our country he is viewed quite rightly as only a *vulgarisateur*." (*Werke*, XXXII, p. 671)
[67] *PB*, I, p. 147, n. Earlier in the *KS* Büchner had condemned Liebig for his speech against the dilettantes ot science, calling him an amateur in physiology. Cf. *KS*, 5th ed., pp. xliii–lv. Cf. also Büchner's letter to Virchow in 1856 urging him to speak out against Liebig. R. Virchow, *Briefe an R. Virchow*, p. 6.
[68] On realism and supernaturalism, cf. *PB*, I, p. 287. Volume one dealt with the heart, blood, heat and life, cells, air and lungs, and chloroform. Volume two, which appeared in 1875, covered the brain and nerves.
[69] *PB*, I, p. 281. Cf. also *NG*, p. 167.
[70] *PB*, II, pp. 59, 193–94.
[71] *NW*, II, pp. 108–09.
[72] Cf. 'Die Neugestaltung des Freien Deutschen Hochstifts zu Frankfurt a. M.,' *NW*, II, pp. 201–10. For his detailed chronicle of the ills of the German universities, see his denunciation of Eugen Dühring's dismissal from Berlin, *NW*, II, pp. 30–48.
[73] *Ibid.*, pp. 49–56, 355–76. Cf. also *ID*, pp. 192–200.
[74] *KS*, 19th ed., p. xviii. As a result of this tour Büchner had no more need to worry about his personal finances. Cf. F. Staudinger, 'L. Büchner,' pp. 459–61 in *Allgemeine Deutsche Biographie,* **55**, 1910. Reprinted (Berlin, 1971), p. 460.
[75] Büchner opposed freemasonry, however, because of its ritualistic and secretive practices. Cf. his review of M. Conrad's *Der Freimauer* in *FE*, pp. 269ff.
[76] *FE*, pp. 309–13. Cf. also 'Der Krieg und der Völkerfriede," pp. 196–200. This was not his first antiwar statement. In the *MSN* Büchner denied that the evolution of a new

and higher race of man was probable, but he did say that the struggle for existence would no longer be carried out as a physical battle. "The time is past when one people enslaves or exterminates another and puts itself in their place; not through *annihilating*, but only by *excelling* can one attain pre-eminence over another." *MSN*, p. 177. Cf. also p. 190. This of course was easy for a German to say in 1870. For Büchner's opinion of the trends of the 1870's, cf. *NW*, II, p. 84.

77 Anton Büchner, *op. cit*., p. 80. Cf. also pp. 27–42.

78 *LT*, pp. 1–2, 10–11. Büchner here seemed to be developing further the theme of the book he had completed just prior to this, the *Geistesleben der Thiere* of 1876. In that work he attempted to establish the link between man and animals by demonstrating the degree of mind present in animals. There his concern was to oppose the Cartesian view of animal behavior (blind mechanism without intelligence) by providing a naturalistic explanation for instinct. (*GT*, pp. 5–42) In the *LT* he appeared to be dealing with the argument of the French naturalist Quatrefages, who held that man's uniqueness was found in his moral capacity. Büchner's reply was to describe affection, compassion, marriage, jealousy, etc. in animal life, and to declare at the conclusion that only those who rejected the facts and purposely retained predetermined ideas could disavow the truth of the facts he had presented. (*LT*, p. 368) His glorification of the principle of love in the *LT* reminds one of Feuerbach, who did the same thing to Marx's disgust.

79 Quoted in Anton Büchner, *op. cit.*, p. 78.

80 *PB*, II, pp. 51–52.

81 From an obituary in the *Frankfurt Zeitung*, cited in Anton Büchner, *op. cit.*, p. 79.

82 *KS*, 19th ed., p. 495.

CHAPTER VI

1 Friedrich Nietzsche, 'David Strauss, the Confessor and the Writer,' in *Thoughts Out of Season. Complete Works*, ed. Oscar Levy, trans. Anthony Ludovici, vol. 4 (New York, 1924), p. 46; and W. Brazill, *The Young Hegelians*. pp. 127–130.

2 On Hess and his involvement with science, see Moses Hess, *Briefwechsel, 1825–1881*, ed. E. Silberner ('s-Gravenhage, 1959), pp 290–91, 305–310, 334; and E. Silberner, *Moses Hess: Geschichte seines Lebens* (Leiden, 1966), pp. 336–347, 637–638.

3 On Ule and Rossmässler, see above, p. 8.

4 Deubler's life and work is dealt with in Arnold Dodel-Port, ed. *Konrad Deubler: Tagebücher, Biographie und Briefwechsel*, 2 vols. (Leipzig, 1886).

5 For example by Karl Vorländer, *Geschichte der Philosophie*, 2 vols. (Leipzig, 1903), II, p. 427, and F. A. Lange, *History of Materialism*, II, pp. 284–94. It was Lange's historic work that first made Czolbe's thought widely known. Vörlander, *op. cit.*, p. 430.

6 Brazill, *Young Hegelians*, p. 5. Cf. also Eduard Johnson, *Heinrich Czolbe*, Heft 4, *Altpreussische Monatsheften*, **10** (1873), p. 339. Mundt is mentioned by Büchner as one of the five leaders of Young Germany. L. Büchner, *Nachgelassene Schriften von Georg Büchner*, p. 26.

7 Pinkas Friedmann, *Darstellung und Kritik der naturalistischen Weltanshauung H. Czolbes* (Bern, 1905), p. 2; Johnson, *op. cit.*, p. 340. A further incentive to retire came as the result of a condition he had contracted due to a fall from a horse during his years at Freiburg.

8 Wilhelm Dilthey, 'Zum Andenken an Friedrich Überweg,' *Preussiche Jahrbücher*, **28** (1871), p. 315. Dilthey reports that Czolbe and Überweg remained in almost daily contact during their years at Heidelberg. Cf. also Lange, *op. cit*., II, pp. 210, 308ff, and

Herbert Berger, *Begründung des Realismus bei J. H. von Kirchmann und Friedrich Überweg* (Bonn, 1958), pp. 42–47.
[9] I. H. Fichte, 'Übersicht der philosophischen Literatur,' *Zeitschrift für Philosophie und philosophische Kritik,* **23** (1853), pp. 136–43.
[10] H. Czolbe, 'Die Elemente der Psychologie vom Standpunkte des Materialismus,' *Zeitschrift für Philosophie und philosophische Kritik,* **26** (1855), p. 95. Fichte wrote a reply to Czolbe's article in the same volume. He commended Czolbe for his tolerance of other views, but opposed his criticism of scientists. (pp. 110–13)
[11] *NS*, p. vii.
[12] *Ibid.*, p. vi.
[13] *Ibid.*, p. 1.
[14] *Ibid.*, p. 6.
[15] *Ibid.*, p. 233. For Czolbe's acknowledgement of the equivalent logical status of speculative versus materialistic explanation, cf. also pp. vii, 2, 62, and the later *GU*, pp. 49, 53. Erdmann sees a resemblance to Kant in this. Johann Erdmann, *A History of Philosophy*, III, pp. 142–43.
[16] *NS*, p. 2.
[17] *Ibid.* Emphasis mine.
[18] *GU*, p. 51. Cf. also p. 57, where he declared that the major task of naturalism was to construct "an ideal statement of ethics." Cf. also Fabri, *Briefe*, pp. 91–92, n.
[19] *NS*, p. 5. Czolbe identified the poet Hölderin, whom he had visited during the period of Hölderlin's insanity, as the first impetus to naturalism in his life, and after him Strauss, Bruno Bauer and Feuerbach. Feuerbach's greatness lay in his elimination of the suprasensual from morality. Finally Lotze's attack on vital force had extended for Czolbe the principle of eliminating the suprasensual into the natural sciences. Cf. *NS*, pp. 203–04.
[20] *NS*, pp. 6–7.
[21] *Ibid.*, p. 4. Czolbe preferred the title "sensualist" to "materialist" because the latter's sole recognition of matter caused problems with entities like space and conceptual forms. Both of these things too had to be represented with absolute clarity. Cf. *ES*, pp. 28–29.
[22] *Ibid.*, p. 3. Frohschammer's reply to his Humean understanding of ideas was to argue that if thinking were merely a helping aid to perception, then animals understand better than men do when men reflect. Frohschammer, *Menschenseele und Physiologie*, pp. 202–03.
[23] *NS*, pp. 3–4. Nevertheless, his reference to Bishop Berkeley in the preface, and his willingness to restrict knowledge to that conveyed by the senses obliterated his disclaimer in the eyes of his critics.
[24] *NS*, p. 3. Cf. *GU*, p. v.
[25] *NS*, pp. 12ff. Incidentally Czolbe felt that his replacement of Müller's specific energy with his own specific elasticity was a replacement of a suprasensual concept with a clear one. (p. 15)
[26] *Ibid.*, p. 19. Cf. also pp. 19–21. Form, after all, was easy to visualize, or "*anzuschauen.*"
[27] *Ibid.*, p. 27.
[28] *Ibid.*, pp. 41–43. Cf. Friedmann, *op. cit.*, p. 13. Czolbe was never clear about the mechanical differentiation between sense perception and ideas. It would seem that sense perceptions were correlated with vibratory motion and ideas with the shape of the grooves or impressions carved out in the molecular structure by the sense stimuli, but he did not always make this clear.
[29] *NS*, p. 64. Cf. Friedmann, *op. cit.*, pp. 16–17 for an attempt to explain Czolbe's mechanical account of voluntary muscular activity.

[30] *Ibid.*, p. 72.
[31] *Ibid.*, pp. 86–87.
[32] "It is false to say . . . that men have in general a natural inclination to good or evil. They incline to both, or are capable of both." *NS*, pp. 88–89.
[33] *Ibid.*, p. 93.
[34] "Since economical concepts are immediately obvious, legal and moral concepts preserve mediately the same intuitiveness they are said to have according the principle of sensualism defended here." *NS*, p. 223. For the exposition that follows, cf. pp. 207–26.
[35] *NS*, pp. 220–21, 225. The later position came out in the *GU*, pp. 17, 277–78.
[36] *NS*, p. 232.
[37] *Ibid.*, pp. 229–31. Recall Moleschott's confidence in Christian values. (Above, pp. 95–96). Büchner, though he often spoke derisively of Christianity, defended Christ Himself. (*LT*, p. 6).
[38] *Ibid.*, pp. 98–99, 236. Cf. also p. 222. Czolbe later even recorded his respect for the Pope, with whom he once had a personal audience. *GU*, p. 276.
[39] *NS*, p. 5. Cf. pp. 207–226.
[40] *Ibid.*, p. 90.
[41] *Ibid.*, pp. 119–120. A decade later Czolbe declared that Subic's *Grundzüge einer Molecularphysik und einer mechanischen Theorie der Elektricität und des Magnetismus* of 1862 was the mathematical expression of his own explanation as given in the *NS* of physical and chemical phenomena, which was based solely on Newtonian mechanics. *GU*, p. 87.
[42] *Ibid.*, pp. 115–19, 138–39. Later he retreated back into the ether. *GU*, pp. 87–91.
[43] *Ibid.*, pp. 183–84. Sensualism and cosmogony, he said, were incompatible. (p. 184) For Lange's qualified agreement, cf. Lange, *op. cit.*, III, pp. 21, 50, 65, 74.
[44] *Ibid.*, pp. 146–47; 153–59; 166; 167–71; 161–62, 172 respectively.
[45] *Ibid.*, p. 193.
[46] *Ibid.*
[47] *Ibid.*, p. 167.
[48] *Ibid.*, p. 162.
[49] *Ibid.*, pp. 150–51.
[50] *Ibid.*, p. 182. Cf. the later *GE*, pp. 160–61.
[51] Cf. Johnson, *op. cit.*, p. 347. Ludwig Büchner, in a short biographical sketch of Friedrich Mohr, reported that among his literary remains was an answer to this same prize question in which Mohr defended his own stability theory. L. Büchner, *LL*, Note 33, p. xxxix. Little else is known about the competition.
[52] Hedwig Breilmann, *Lotzes Stellung zum Materialismus, unter besonderer Berücksichtigung seiner Controverse mit Czolbe* (Münster, 1925), p. 32.
[53] *NS*, pp. 203–04. In a sense Czolbe had written the *NS* as a materialistic alternative to Lotze's 1852 *Medicinische Psychologie.* Cf. *NS*, p. viii, and Merz, *History of Thought*, III, p. 563, n.
[54] *NS*, p. 250.
[55] Hermann Lotze, 'Recension von H. Czolbe, *Neue Darstellung des Sensualismus*, 1855,' pp. 238–50 in *Kleine Schriften*, 3 vols. (Leipzig, 1891), III, p. 239. Cf. also Hans Vaihinger, 'Die drei Phasen der Czolbeschen Naturalismus,' *Philosophisches Monatsheft,* **12** (1876), pp. 15–16.
[56] H. Lotze, 'Recension von H. Czolbe,' p. 240. As soon as one named something, said Lotze, one added the suprasensual to intuition (*Anschauung*). (p. 241) Cf. Merz, *op. cit.*, III, pp. 264, 268.
[57] Lotze, 'Recension von H. Czolbe,' pp. 242–43.
[58] Vaihinger, *op. cit.*, pp. 17–18. Cf. also pp. 8–9.

[59] *Ibid.*, p. 14.
[60] *NS*, p. 28, and *ES*, pp. 6–9. In the latter work he carried out the discussion in light of Schaller's pan-psychic position.
[61] *GU*, p. 214. For his recantation of the view that sensation could be derived from matter, cf. p. vi. With his idea of spatial sensation he claimed to be following Johannes Müller. Cf. *GE*, p. 7. On the pro and con of Czolbe's understanding of Müller, see Friedmann, *op. cit.*, p. 52, and Vaihinger, *op. cit.*, p. 21.
[62] *NS*, pp. 185–86. Czolbe confessed to there being idealistic elements in his system and excused it by saying, "Extremes touch one another and often overlap." *ES*, p. 2. Cf. also p. 53.
[63] *GU*, pp. 176–77.
[64] *Ibid.*, p. 4: "At the same time I will point out . . . that not only no opposition exists between the teleological and mechanical world views, but that . . . both principles necessarily condition one another, and the polemic of modern scientists, namely the materialists, against teleology rests only on misunderstanding." Vaihinger called his program here "mechanical spiritualism." Cf. Vaihinger, *op. cit.*, p. 3. Also Friedmann, *op. cit.*, p. 26.
[65] *NS*, p. 188.
[66] *NS*, p. 180. Culture was not an inevitable development, only a highly probable one. Czolbe may have been trying to oppose the Hegelian conception of history in this. Cf. also *GU*, pp. 149–50, 171–72.
[67] *GU*, p. 5. "The unity of the world consists . . . not in a primal substance, but in its ultimate purpose or ideal." (p. vi) Cf. also pp. 174–78, and Friedmann, *op. cit.*, p. 53.
[68] *Johnson, op. cit.*, p. 347. He also never relinquished his insistence on the eternity of life and the world.
[69] *GU*, pp. 71, 261.
[70] *Ibid.*, pp. 45–47. Kant, he explained, was a good example of how morality could be used as the basis of theology.
[71] *Ibid.*, p. 46.
[72] H. Czolbe, 'Die Mathematik als Ideal für all andere Erkenntniss,' *Zeitschrift für exacte Philosophie*, 7 (1866), pp. 242–43.
[73] *Ibid.*, p. 243. Mathematics ruled in natural science, psychology, and political economy. Natural science was the mechanics of atoms, psychology the mechanics of sensations and feelings, and politics the mechanics of individuals. In support Czolbe cited the work of Herbart and Fechner in psychology and Quetelet in political economy. (p. 238)
[74] *Ibid.*, pp. 219–220. One result of this position was Czolbe's fierce opposition to the projection theory of vision as defended mainly by Lotze and Helmholtz. At the root of this disagreement was Czolbe's belief that external sensations were spatially extended and mechanically transmitted as opposed to his opponents appeal to the soul's use of local signs in ordering sense impressions. Czolbe pointed out that he and Überweg had adopted space as irreducible, while Helmholtz tried to explain space in terms of nonspatial points. The controversy surfaces and resurfaces throughout the *GE*. Cf. *GE*, pp. 2–7, 13, 132–39, 153–55, 170–74. Also Vaihinger, *op. cit.*, pp. 28–30.
[75] *GE*, p. 219.
[76] *Ibid.*, pp. 221–22. Cf. also Vaihinger, *op. cit.*, pp. 24–25.
[77] Subtitled *Ein räumliches Abbild von der Entstehung der sinnlichen Wahrnehmung* (Plauen, 1875). For details regarding the motivation and publishing of the book, see pp. iii, 10–11, 14.
[78] *Ibid.*, pp. 7–8. Cf. Friedmann, *op. cit.*, pp. 54–55. Earlier Czolbe had criticized Spinoza for holding that a closed nexus excluded purposeful activity. ('Die Mathematik als Ideal,' p. 235.)

[79] For Czolbe's exposition of spatially extended force, which turned out to be more Aristotelian force than Newtonian, see *GE*, pp. 102–10. Czolbe of course rejected action at a distance. (p. 107,n.)
[80] *Ibid.*, p. 45.
[81] *Ibid.*, cf. also p. 4. Hints of this view can be found in the earlier *GU*, p. 83.
[82] *GE*, p. 53. Cf. also p. 109.
[83] *Ibid.*, p. 107.
[84] *Ibid.*, p. 4.
[85] *Ibid.*, pp. 64–65. Czolbe detailed a mechanical explanation of the brain's action in concentrating these sensations. The concentration was produced in a manner analogous to the production of magnetic force in an electrical coil. Cf. pp. 67–68, 117–18.
[86] *Ibid.*, p. 119. Cf. also pp. 39–40. For more on the various conditions for the combining of the two kinds of sensations, see pp. 57–58.
[87] There is something to be said for the idea that much of Marx's dogmatism resulted from a similar feeling of intuitive certainty. In his "Theses on Feuerbach" he complained that previous materialism confined its perception (*Anschauung*) to the external world. *He* would extend it to include human sensuous activity. Werke, III, p. 533.
[88] *GE*, pp. 50–51, 29–30.
[89] Vaihinger, *op. cit.*, p. 4, n. 9.
[90] *Ibid.*, p. 187.
[91] Feuerbach, *Werke*, II, p. 390.

CHAPTER VII

[1] *ID*, p. 10.
[2] *Ibid.*, pp. 28–29.
[3] Böhmer, *Naturwissenschaftliche Weltanschauung*, pp. 180–84, 191. Schleiden lamented the collapse of philosophy in the scientific-industrial age. Schleiden, *Über den Materialismus*, p. 57.
[4] Böhmer, *op. cit.*, pp. 143–44. Böhmer called for realism to replace idealism, but by realism he meant something in which the experiences of mind were primary. (pp. 214–16) For the perspective of the scientific materialists concerning the state of philosophy at mid-century see the following: Moleschott, *PSk*, p. viii; Büchner, *NW*, I, pp. 37, 179, 253. Cf. also Ernst Apelt, *Die Theorie der Induction* (Leipzig, 1854), pp. v–vi. Büchner in particular spoke of a crisis in philosophy precipitated by the rapid flowering of experimental science, and pointed to the founding of the *Zeitschrift für exacte Philosophie im Sinne des neueren philosophischen Realismus* in 1860. On the founding of this journal, cf. Merz, *History of Thought*, III, p. 69, n.
[5] I. Doedes, *Der Angriff eines Materialisten auf den Glauben an Gott* (Jena, 1875), p. 32. Doedes was a professor of theology at Utrecht.
[6] Fabri, *Briefe*, p. 72.
[7] *Ibid.*, p. 73.
[8] *UW*, p. 12. Moleschott distinguished between the philosophy of the past and that of the present. The former was spectulative, while modern philosophy had forsaken speculation in order to generalize the method of moving from facts to ideas. Modern philosophy operated under the conviction that the idea grew out of the facts, but that it did not create the facts. (*FW*, p. 18; *US*, p. 525).
[9] Hess, *Briefwechsel*, p. 635.
[10] To allay these fears Cornill had, for example, written a reply to a critic entitled 'Die Philosophie als Naturwissenschaft.' Cf. Cornill, *Materialismus*, pp. 9–21, especially p. 11.

[11] *NW*, I, p. 14.
[12] *SV*, p. 400.
[13] *PB*, I, p. 253. Cf. also *NW*, I, p. 262, and note.
[14] *FE*, p. 353; *PB*, II, p. 234, n.
[15] Cf. Büchner's reviews of anti-Kantian books in *NW*, II, pp. 317ff; *SJ*, chapter 3, pp. 59–101; *ID*, pp. 443–68. Cf. also Czolbe, *GU*, p. v.
[16] *MV*, pp. 91ff; *ID*, pp. 55–56. Cf. also Czolbe, *NS*, p. 60; *GU*, pp. 96–97. The most Büchner would admit was that some forms of thought, transmitted by evolution, might have become *a priori* for the individual, but never, he said, for the race. (*MV*, p. 92; *ID*, p. 161) Even in the *KS* he had included a chapter opposing the notion of innate ideas.
[17] *ID*, p. 167. The argument that man can know nature because he is a product of nature is repeated *ad nauseum* throughout Büchner's many works. It is also found in Moleschott, *GM*, p. 305.
[18] *SJ*, p. 70; *ID*, p. 454.
[19] *ID*, pp. 464–65.
[20] Quoted in *ID*, pp. 465–66.
[21] *GU*, p. iii. For more on Czolbe's understanding of the overcoming of Kantianism, see pp. 76, 101–07, 215–53. Johnson, *Czolbe*, p. 349, quotes an extremely interesting statement from Czolbe that demonstrates his distrust, on grounds similar to those of Hegel, of those who feel obligated to test the capacity of the apparatus of knowledge before explaining the world.
[22] Czolbe spoke of the theological bent dominating Kant's logic (*GE*, p. 51), and Büchner equated Kant's *Ding-an-sich* with Hegel's *Geist*, Schopenhauer's *Wille*, Hartmann's *Unbewusstsein*, Spencer's Unknowable and DuBois-Reymond's "Ignorabimus." (*ID*, p. 270) Cf. also Czolbe, *GU*, p. 255, and "Die Mathematik als Ideal für alle andere Erkenntniss," pp. 247–48. This way of seeing Kant might have been motivated by a passage in Feuerbach's *Philosophie der Zukunft*: "Kantian idealism, in which things are arranged according to the understanding, not the understanding according to things, is therefore nothing other than the realization of the theological idea of the divine understanding, which is not determined by things, but instead determines them." (*Werke*, II, p. 271)
[23] *RWW*, p. 65.
[24] *NW*, II, p. 252.
[25] Czolbe, *GU*, p. 271; *NS*, pp. 4, 38. Moleschott, *PS*, p. xii. Moleschott says here that even Hegel and Schelling had overcome Kant.
[26] *ID*, p. 170.
[27] *Ibid.*, p. 268. In his book on man's place in nature, Büchner included a poem which captures his attitude best of all.

O Ding an sich,	Ob jung, ob alt,
Wie lieb ich Dich,	Ob warm, ob kalt,
Du Aller Dinge Ding!	Ob gerade oder krumm,
Nur blinder Wahn	Ob Du voll Zwist,
Sieht schief Dich an	Ob sanft Du bist,
Und achtet Dich gering.	Ob pfiffig oder dumm?
Zwar weiss ich nicht,	Doch einerlei!
Ob Dein Gesicht	Dir bleib ich treu
Ist hässlich oder schön?	Und unveränderlich,
Und ob Du wohl,	Und thue dar,
Fest oder hohl,	Dass nichts ist wahr,
Magst liegen oder stehn?	Als nur "das Ding an sich!"

MSN, p. lxxxvi, n. 82.
[28] Paul E. Cranefield, 'The Organic Physics of 1847 and the Biophysics of Today,' *Journal of the History of Medicine*, **12** (1957), p. 417.

[29] Paul Volkmann, *Die materialistische Epoche des 19. Jahrhunderts und die phänomenologisch-monistische Bewegung der Gegenwart* (Leipzig, 1909), p. 8.
[30] Lange, *History of Materialism*, II, p. 161. In addition the reductionists were more sceptical of radical political positions, remaining far more conservative than most of the scientific materialists.
[31] Merz, *op. cit.*, III, p. 560. Cf. also Treitschke, *History of Germany*, VII, p. 198.
[32] Merz, *op. cit.*, II, p. 584.
[33] *Ibid.*, IV, p. 343, n.
[34] *Ibid.* Lange too referred to the "men of science, who, so far as materialism failed to satisfy them, have inclined for the most part to a way of thinking which in very essential points agrees with that of Kant." Lange, *op. cit.*, II, pp. 154–55.
[35] E. DuBois-Reymond, 'Über die Grenzen des Naturerkennens,' pp. 8–57 in *Zwei Vorträge von E. DuBois-Reymond* (Leipzig, 1882), pp. 24–25.
[36] *Ibid.*, pp. 27, 33–34.
[37] *Ibid.*, pp. 34–35. Cf. also Hermann Lotze, *Medicinische Psychologie* (Leipzig, 1852), p. 181: "An analysis of the light wave provides nowhere in it a reason to perceive why it must be sensed as luminous."
[38] E. DuBois-Reymond, *Untersuchungen über Thierische Elektricität* (Berlin, 1848), pp. xxxv–xxxvi. "... Analytical mechanics reaches up to the problem of personal freedom, whose settlement must remain a matter of the gift of abstraction of every individual." (*Die analytische Mechanik reicht bis zum Problem der persönlichen Freiheit, dessen Erledigung Sache der Abstractionsgabe jedes Einzelnen bleiben muss.*) Temkin's interpretation of "bis zum" to mean "including" alters the sense of DuBois-Reymond's statement completely and, in my judgement, misrepresents his position. (Temkin, "Materialism in French and German Physiology of the Early Nineteenth Century," p. 326) In his second lecture on the topic, "Die Sieben Welträthsel," DuBois-Reymond sharpened the focus on the logical priority of other problems over the freedom of the will. (Cf. *Zwei Vorträge*, p. 104.) The seven world riddles were classified as transcendent or not transcendent in this lecture. The freedom of the will was listed as a transcendent riddle, after the essence of matter, the origin of motion, and consciousness, while the origin of life, the purposeful constitution of nature, and, given consciousness, the origin of reasoning were identified as not transcendent.
[39] DuBois-Reymond, 'Über die Grenzen,' p. 38. He says here "*Sinnesempfindungen*," but earlier he had specified this as equivalent to consciousness as far as the problem before him was concerned. (p. 27).
[40] *Ibid.*, p. 45.
[41] 'Die Sieben Welträthsel,' in *Zwei Vorträge*, p. 62. He was flattered, he said, that the speech had been seen as Kantian in flavor.
[42] *PB*, II, p. 190.
[43] *Ibid.*, p. 193.
[44] *Ibid.*, pp. 193–94. Büchner reported that DuBois-Reymond had said that one could deny that man would ever understand mental processes by way of material conditions without denying that these processes were produced by material conditions. (p. 193. Cf. also *KL*, p. 83) But DuBois-Reymond's intent was not at all to defend that matter was the source of consciousness; rather, to show that the mechanical explanation of consciousness left the question of the source of consciousness untouched. ('Die Grenzen des Naturerkennens,' pp. 41–42.)
[45] *KL*, p. 82. Cf. also *SJ*, p. 120.
[46] *KL*, pp. 82, 91.
[47] *ID*, pp. 236–37. Cf. also *KL*, pp. 32–33.
[48] *KL*, p. 83; *ID*, 237.
[49] Lange, *op. cit.*, III, p. 325.

[50] Cf. Böhmer, *op. cit.*, p. 185.
[51] Cf. Büchner, *PB*, I, p. 252; *LL*, p. 238.
[52] Examples may be found in the following: Büchner, *KS*, chapter 14, *TT*, pp. 337ff; Czolbe, *NS*, p. 235, 'Die Mathematik als Ideal,' pp. 249f; Vogt, *OM*, II, pp. 196–97, *BT*, pp. 426–27; Moleschott, *KL*, chapter 2.
[53] Vogt, *KW*, 4th ed., p. lxii; Büchner, *NW*, I, p. 96.
[54] *NS*, p. 235. Emphasis mine.
[55] *Ibid.*
[56] He continued to oppose hypotheses in spite of the maturation that occurred between his early references to community agreement and the later ones. In his first work he noted that community agreement played a selective role, but added that it was nevertheless in no way infallible. (*NS*, p. 64) Later he said that that theory was true which the majority agreed upon, noting that it was not enough simply to be consistent with the laws of thought, but that all ideas must be verified by experience. ('Die Mathematik als Ideal,' p. 252) Still, like Ule, he denied that false hypotheses had any part in natural science. ('Die Mathematik als Ideal,' p. 255) For Ule's judgement concerning hypotheses, see the letter to Moses Hess in Hess, *op. cit.*, p. 349.
[57] *NS*, p. 59. Cf. also p. 7.
[58] 'Die Elemente der Psychologie vom Standpunkt des Materialismus,' pp. 95–96.
[59] *ES*, p. 30.
[60] 'Die Mathematik als Ideal,' p. 241. Czolbe's position on atomic theory was a good example of the variance and uncertainty of his ideas on scientific theory in general. On the one hand he declared that atoms were a necessary conclusion from the facts of optical, thermal, magnetic and chemical phenomena. (*NS*, p. 105; *GU*, pp. 92–93; *GE*, pp. 94–95) On the other hand he pointed out that atoms were not indivisible since they were divisible in thought. (*NS*, p. 105; *GE*, p. 100) Further, he acknowledged that atoms would never be seen (*NS*, p. 18), yet he held that although they arose in our knowledge *per analogiam* with mechanical relationships, they were not merely a symbol, but objective truth. (*GU*, p. 94) Czolbe's basic conviction that the real must be thinkable, but that the thinkable need not necessarily be real defined, he thought, a consistent position. (*GU*, p. 78) The least his position means is that one must mitigate Vaihinger's conclusion that atoms were for Czolbe, as they were for Vaihinger, "starting points of calculation." Vaihinger, *Die Philosophie des Als Ob* (Berlin, 1911), p. 103.
[61] *GK*, p. viii.
[62] *Ibid.*, p. ix.
[63] *Ibid.* Note here the resemblance of Vogt's position to that of Karl Popper.
[64] *Ibid.* Emphasis mine. Recall that Vogt did change his mind about species change and geological catastrophism.
[65] R. Biber, *Karl Vogt's naturwissenschaftliche Vorträge über die Urgeschichte des Menschen* (Berlin, 1870), pp. 6, 15.
[66] For his discussion, see *UW*, pp. 14–37.
[67] *Ibid.*, p. 8. Cf. also *PN*, p. 570.
[68] *PP*, p. 37. Tyndall was more straightforward. "I call a theory a principle or conception of the mind which accounts for observed facts, and which helps us to look for and predict facts not yet observed. Every new discovery which fits into a theory strengthens it. The theory is not a thing complete from the first, but a thing which grows, as it were asymptotically, towards certainty." John Tyndall, 'Virchow and Evolution,' *Popular Science Monthly* (1878–1879), p. 290.
[69] *KS*, p. xii.
[70] *NW*, I, p. 352.
[71] *Ibid.*, p. 353.

[72] *Ibid.* Büchner quoted from Whewell: "Without laws facts have no connection and no relationship, without facts law has no reality." (p. 360)
[73] *Ibid.* Cf. also *NW*, II, pp. 288, 317.
[74] *Ibid.*, pp. 353–54.
[75] *Ibid.*, p. 354. In addition to induction and deduction analogy, abstraction, theory, criticism, and history were all mentioned as aids in arriving at the truth.
[76] *Ibid.*, p. 355.
[77] Philip Frank, *Einstein: His Life and Times*, trans. George Rosen (New York, 1947), p. 13.
[78] Einstein, replying to the question whether there was a "correct" explanation of nature, put it as follows: "To this I answer with complete assurance that in my opinion there is the correct path and, moreover, that it is in our power to find it." Frank, *op. cit.*, p. 282. Cf. pp. 282–288 for more on what Frank calls Einstein's "religious" conception of the world.
[79] *TT*, p. 283.
[80] *Ibid.* For Büchner on hypotheses, cf. *KS*, 5th ed., p. 247.
[81] *PB*, I, p. 251. In fact the mental eye can convince us of things that the corporeal eye cannot see. (*TT*, p. 338)
[82] *ID*, p. 407.
[83] *Ibid.*, p. 12. Büchner's position on atomism further exemplified his attempt to be pragmatic in the face of difficulties. An atom was an expression of a concept necessary for certain purposes. It had not been, nor would it ever be observed, and we should not be offended at the idea that atoms are infinitely divisible. (*KS*, pp. 22–23; *LT*, p. 3; *TT*, p. 284) In spite of this Büchner blithely referred to molecules in motion, not allowing such an (implicit) assumption of indivisible atoms to bother him in the least, as it did thinkers like Merz and Lange. (Merz, *op. cit.*, III, pp. 568–69; Lange, *op. cit.*, II, pp. 376–77).
[84] Karl Fortlage noted this dependence on *Naturphilosophie*. Rebuking the materialists and others who denounced Schelling and his school, he acknowledged that *Naturphilosophie* might have affected the development of natural science adversely, but not for the reason usually given. Schelling, he said, instituted an era of imprecision in language. This was the real danger of *Naturphilosophie*, "and not in the assumption of an identity between reason and nature, which is more verified every day." Fortlage, "Materialismus," Nr. 12, p. 206.
[85] *GU*, p. 69.
[86] *Ibid.*, pp. 64–65.
[87] *Ibid.*
[88] *Ibid.*, pp. 67–68.
[89] *ID*, p. 56.
[90] *Ibid.*, p. 160.
[91] *KS*, p. 24.
[92] *ID*, p. 56.
[93] *Ibid.*, p. 407.
[94] *Ibid.*, pp. 410–440.
[95] *PB*, pp. 42–43. Vogt also insisted on a strictly mechanical explanation for glandular secretion. The glands acted as filters to remove the already present "secretion" from the blood. This would suggest, in accordance with Vogt's pronouncement on the relationship between thought and the brain, that the brain somehow filtered out thought from the blood! (Cf. pp. 84, 101–08).
[96] *Ibid.*, pp. 7, 26–27, 158.
[97] *Ibid.*, p. 8.
[98] *US*, p. 522; *FW*, p. 11.

[99] *NH*, p. 48. Cf. also *PP*, p. 38. On the uniqueness of physiology and the internal relationship of the sciences, cf. *EL*, pp. 11–16, and *FW*, pp. 4–15.
[100] *Werke*, VI, pp. 228–237.
[101] *Ibid.*, II, p. 345. "The understanding, like death, separates [real things] into their elements; but *they are what they are* only as long as their elements are received into the union of the senses (*in den Bund der Sinne*)." (Emphasis mine.)
[102] *KRD*, p. 32. Cf. also pp. 31–33, 37. Elsewhere he added to the outstanding achievements of the century spectral analysis and cell theory. (*EL*, p. 39)
[103] *SLM*, pp. 11–18.
[104] *UW*, pp. 37–38.
[105] *GU*, pp. 199–200. Cf. also p. 92.
[106] *KL*, p. 73; *PB*, I, pp. 130–37; *PB*, II, p. 45; *TT*, p. 302.
[107] *NW*, I, pp. 54–68. The chapter in *KS* with the same title did not appear in the first edition of 1855.
[108] In fact he consciously rejected Helmholtz's formulation, "Erhaltung der Kraft." (*NW*, I, p. 68.)
[109] *Ibid.*, p. 54. Another way of putting it would be to say that all energy was kinetic, a doctrine Stallo said was "irrecusable by any consistent advocate of the mechanical theory." J. B. Stallo, *The Concepts and Theories of Modern Physics*, ed. P. W. Bridgemann (Cambridge, Mass., 1960), p. 95.
[110] Perhaps the reader can make sense of it. "The work force of the steam engine, transformed into heat by friction, plus the dissipated heat is equal to the heat of combustion of the coal." (*NW*, I, pp. 64–65).
[111] "What one designates with the name 'conservation of force' is in reality only the conservation of motion." (*FE*, p. 28. Cf. also *ID*, pp. 115–16).
[112] *LL*, p. 193.
[113] *Ibid.*, Note 43, pp. il–l. Büchner suggested that gravity would someday be explained as a wave phenomenon. (p. 1)
[114] *LL*, pp. 87–88, 191; *TT*, p. 306. On Prout, cf. *TT*, p. 70.
[115] Mohr had then passed the article on to Baumgartner's *Zeitschrift für Physik* in Vienna, where it was published. Mohr himself did not learn until 1868 that Baumgartner did publish the paper, and he thereupon made a bid to establish his priority in the matter. In his *Allegemeine Theorie der Bewegung und Kraft als Grundlage der Physik und Chemie* of 1869 he reprinted the 1837 article. The book was subtitled: *Ein Nachtrag zur mechanischen Theorie der chemischen Affinität* (Braunschweig, 1869). The reprint is found on pp. 84–106.
[116] *FE*, p. 361.
[117] *Ibid.*, p. 363.
[118] Mohr, *op. cit.*, p. 103: "[Force] can come to the fore under the appropriate circumstances as motion, chemical affinity, cohesion, electricity, light, heat and magnetism, and from each of these kinds of phenomena all the others can be produced." Earlier Mohr had said of force that it could be handled just like ponderable matter. One could divide, subtract from or add to it without the original quantity being lost or changed. (p. 89) Büchner's history of conservation of energy is found in *LL*, pp. 107–21, and his defense of Mohr in *LL*, Note 33, pp. xxxiii–xli. For additional material on Mohr, see Merz, *op. cit.*, II, pp. 106ff, and T. Kuhn, 'Energy Conservation as an Example of Simultaneous Discovery,' pp. 321–56 in M. Clagett, ed. *Critical Problems in the History of Science* (Madison, 1969), pp. 321–22, 325, 337, 353–54. Kuhn points up the influence of *Naturphilosophie*, which Büchner despised, on the formulation of Mohr's views.
[119] *LL*, p. 125 and Note 35, p. xliv. On Joule's lesser role, see pp. 128, 132.
[120] He translated, for example, Tait's use of "energy" as *Spannkraft* in *NW*, I, p. 453.

Cf. also *LL*, p. 136.
[121] *LL*, pp. 193–94. Elsewhere he repeated this and said that all forms of force were "only modifications of the vibrations of atoms, molecules, or the smallest particles." (*TT*, pp. 306–07).
[122] *TT*, p. 307.
[123] Cf. Yehuda Elkana, 'Helmholtz's "Kraft": An Illustration of Concepts in Flux,' *Historical Studies in the Physical Sciences*, **2** (1970), pp. 263–298. Cf. especially p. 270 for a list of premises common to Helmholtz and Büchner.
[124] *LL*, pp. 73–74.
[125] *Ibid.*, pp. 33–36. Cf. also Note 10, pp. xii–xviii. One receives the impression that Kirchhoff's significance for Büchner was in establishing the existence of the solar core. (p. xvi)
[126] *Ibid.*, pp. 78–79.
[127] *Ibid.*, pp. 100–01.
[128] *Ibid.*, pp. 205–08; *TT*, pp. 317–19.
[129] *LL*, pp. 209–10; *TT*, pp. 318–19.
[130] *NW*, II, pp. 393–94, 399.
[131] *LL*, pp. 219, 317.
[132] *Ibid.*, p. 222.
[133] *Ibid.*, p. 221. The only thing he admitted was that we would be forgotten. (p. 224)
[134] *Ibid.*, p. 228.

CHAPTER VIII

[1] Temkin, 'Materialism in French and German Physiology of the Early Nineteenth Century," p. 324; Büchner, *PB*, I, p. 193.
[2] Since the work of Ernst von Baer in the 1820's, epigenetic explanations of the development of the embryo had gained ground on the formerly predominant preformation theory. Epigeneticists brought with them the problem of explaining the direction the development took, which the preformationist avoided by openly accounting for form as the result of a miracle. Vital force was the answer for many epigeneticists, at least for von Baer. Cf. Coleman, *Biology in the Nineteenth Century*, p. 42.
[3] Temkin, 'Materialism in French and German Physiology,' pp. 324–25.
[4] *Ibid.*, p. 325.
[5] *PB*, p. 3.
[6] 'Physiologie des Menschen,' *Die Gegenwart*, **4** (1850), p. 673; *PB*, p. 139. For his discussion and evaluation of the various explanations given, see *PB*, pp. 128ff. Moleschott, a critic of Liebig, faulted him for restricting his explanation only to oxidation. Cf. *KL*, chapter 14, especially pp. 231, 238, 254. For Liebig's role in the development of the theory of animal heat, see June Goodfield, *The Growth of Scientific Physiology* (London, 1960), pp. 72, 100–12, 119–24, and F. L. Holmes, 'Introduction,' in Justus Liebig, *Animal Chemistry* (New York, Johnson Reprint Corp., 1964). Büchner became involved in the animal heat-vital force issue only secondarily. In the first volume of his *Physiologische Bilder* he quoted Helmholtz in favor of vital force, not realizing that the passage he cited, when taken in context, was actually a disassociation from vitalism on the part of Holmholtz. This can be seen by comparing Büchner's claim in the *PB*, I, pp. 129–30 with Everett Mendelsohn's reference to Helmholtz's remarks in 'Revolution and Reduction,' in *The Interaction Between Science and Philosophy*, ed. Yehuda Elkana (New York, 1973), pp. 423–424.
[7] *PB*, p. 143.
[8] *PS*, p. xxi.

[9] *Ibid.* Elsewhere Molschott described the three historical eras of physiology, the first two of which had depended on personification for explanation, "because the impact of physics had not yet penetrated the whirlpool of life." (Cf. *EL*, p. 14)

Before Galileo personification was common in explanations of nature. With the iatromechanical tradition of the second period scientists brought physics into physiology, but they retained a dualistic attitude vis-à-vis certain vitalistic processes. Only now in the modern era, said Moleschott, had a unified position been achieved. (*EL*, pp. 6–57)

[10] *KS*, p. 215.

[11] *Ibid.*, p. 216.

[12] *Ibid.*, pp. 218–19.

[13] *Ibid.*, p. 228. This came from Vogt's *PB*, p. 143.

[14] Schaller, *Leib und Seele*, p. 146, n.

[15] *Ibid.*, p. 152.

[16] *Ibid*, p. 158.

[17] Goodfield, *op. cit.*, p. 148; Coleman, *op. cit.*, p. 148. Cf. also Goodfield, *op. cit.*, pp. 137–39, and Coleman, *op. cit.*, pp. 148–50. Claude Bernard's solution was intermediate between the reductionists and Liebig, for he did claim that functions of a special character could be associated with organic phenomena, but he saw no need to look for a unique mechanism behind them. A machine rising in the air, he said, may seem to violate natural laws, but on closer examination it is found to be a result of them. (Goodfield, *op. cit.*, p. 158) Hence Bernard called himself a "physical vitalist." (p. 162)

[18] Timothy Lippmann, 'Vitalism and Reductionism in Liebig's Physiological Thought,' *Isis*, **58** (1967), p. 185.

[19] Goodfield, *op. cit.*, p. 162.

[20] *Ibid.*, p. 163.

[21] *PB*, p. 460. He also denied that there was an "idea" behind species. (p. 478) For more on species, see below, pp. 175ff.

[22] *PB*, p. 451; *ZB*, pp. 14–18. His critic Frohschammer called him to account for this. Frohschammer, *op. cit.*, pp. 90–91.

[23] *PS*, p. xxii.

[24] *NG*, p. 152. As for the origin of this principle, August admitted ignorance and unconcern. (pp. 152f)

[25] *PB*, I, p. 264.

[26] *ES*, pp. 56–57.

[27] *GU*, p. 122.

[28] *EL*, p. 57. Cf. also p. 42.

[29] John Farley, 'The Spontaneous Generation Controversy (1859–1880): British and German Reactions to the Problem of Abiogenesis,' *Journal of the History of Biology*, **5** (1972), p. 285.

[30] Czolbe's unusual position is of course an exception. He rejected abiogenesis as unclear, and although he did not concern himself with heterogenesis, it is clear from the nature of his argument that this too would be equally unacceptable. Cf. above, pp. 130ff.

[31] *LG*, I, p. 300; *PB*, p. 300.

[32] *PB*, pp. 301–302.

[33] *Ibid.*, pp. 302–313.

[34] Ibid., p. 313. Those who refused to accept physical proofs in physiology, he said, might go on believing in the spontaneous generation of infusorians, but in doing so they only prove their blind foolishness. (p. 303)

[35] *NG*, pp. 135, n., 136, n.

[36] *BT*, p. 108.
[37] *Ibid.*, pp. 109–12. In the notes to the *NG* a year earlier, Vogt had commented that these reagents "leave the air completely unchanged chemically." (p. 135, n)
[38] *Ibid.*, p. 112.
[39] *Ibid.*, p. 113. But cf. p. 120, where he wrote that the reproduction of infusorians by division was undoubted.
[40] *Ibid.*, p. 114. He even called for more experiments on the origin of mites through electrical sparks, which, he said, had received little attention in Germany. (pp. 114–15).
[41] *VM*, II, p. 253.
[42] For more on Vogt's stance on descension, see below, pp. 175ff.
[43] Above, p. 60.
[44] *LG*, II, p. 303.
[45] *BT*, pp. 104–05. Cf. also *LG*, II, p. 245.
[46] *Ibid.*, p. 106.
[47] *Ibid.*, p. 107.
[48] *KS*, p. 67.
[49] *Ibid.*, p. 70.
[50] *Ibid.*, pp. 73–74.
[51] He thought "heterogenesis" was a term the French used to refer to Ur-Zeugung, or the original production of life. Cf. *NW*, I, p. 431.
[52] *KS*, p. 76.
[53] *NG*, p. 228.
[54] *Ibid.*, p. 230.
[55] *Ibid.*, pp. 234–36.
[56] *PB*, I, pp. 228, 232.
[57] *SV*, pp. 101–03.
[58] *Ibid.*, p. 105. Hence Büchner did not believe that man would ever artificially produce higher organic forms, although he might well create organic matter. (pp. 123–24)
[59] *NW*, I, pp. 442–43.
[60] *NG*, p. 230; *SV*, p. 106.
[61] *SJ*, p. 31.
[62] Owsei Temkin, 'The Idea of Descent in Post-Romantic German Biology: 1848–1858,' pp. 345–51.
[63] *NG*, p. 124, n.
[64] *Ibid.*, pp. 98, n.; 170, n.
[65] Cited in Temkin, 'Idea of Descent,' p. 348.
[66] The statement Temkin refers to is found in *BT*, p. 366. For other references to his stand against transmutation, cf. pp. 362, 365, 397, and *ZB*, p. 24.
[67] *NG*, pp. 147–48, n; 156, n; 58, n; 170, n; 182, n; 209, n., respectively.
[68] *Ibid.*, pp. 124, n; 229, n.
[69] *NG*, 124, n; 224, n. Cf. also *BT*, pp. 376–77.
[70] *NG*, pp. 229, n; 267, n. Cf. also *BT*, pp. 376–77.
[71] *NG*, 2nd ed., p. 138, n.
[72] *LG*, II (1846), p. 296. Essentially the same definition appears in *ZB*, (1851), p. 20; *LG*, II, 2d ed. (1854), p. 382; *VM*, I, (1863), pp. 274–75.
[73] *ZB*, pp. 20–23. Nevertheless Vogt inserted a qualification to the effect "in so far as we know at present."
[74] *LG*, II, 2d ed., p. 383.
[75] Cf. above, p. 38.
[76] *VM*, I, pp. 275–79; II, pp. 207, 272–74.

[77] *Ibid.*, II, pp. 273–74.
[78] *Ibid.*, I, p. 279.
[79] *Ibid.*, I, p. 284.
[80] *Ibid.*, II, pp. 221, 248. Cf. also pp. 208–21.
[81] *Ibid.*, II, pp. 207, 248–49.
[82] *Ibid.*, II, pp. 261–62, 272–74.
[83] Cf. *Lectures on Man: His Place in the Creation and in the History of the Earth*, trans. James Hunt (London, 1863), p. xv. Hunt felt that this was one point on which Vogt dissented from Darwin. The fact that monogenism had become the orthodox position was almost by itself enough to make Vogt a polygenist. Cf. *VM*, II, pp. 250–51.
[84] Cf. *De la Variation des animaux et des plantes a l'état domestique*, trans. J. Moulime. Preface by C. Vogt (Paris, 1868), pp. viii–ix.
[85] 'Quelques Heresies Darwinistes,' *Revue Scientifique*, **38** (1886), p. 481. Vogt's anti-Haeckelian thesis in the article is that zoological classification does not reflect entities that actually exist in nature. Rather, the relationship among phylum, class, order, family, etc., merely expresses groups of similar characters found in beings derived from different stocks. One is reminded again of Vogt's understanding of species as an abstraction. Cf. also Wilhelm Vogt, *op. cit.*, pp. 135–37, 245, and Karl Vogt, 'Pope and Anti-Pope,' *Popular Science Monthly*, **14** (1878–1879), p. 320 for more on Vogt's disagreement with Haeckel. Haeckel in the latter article was depicted as the anti-pope to Pope Virchow, whose position on Darwinism ("an unproven hypothesis") also differed from Vogt's. (Cf. pp. 320–24) Vogt was here assessing the Haeckel-Virchow dispute. Cf. Gasman, *Scientific Origins*, pp. 12, 23, and Ackerknecht, *Rudolph Virchow*, pp. 201–02.
[86] *VM*, I, pp. 144–45.
[87] Cf. *KL*, 5th ed., pp. 81ff, 124–30. Haeckel is mentioned far more than Darwin.
[88] *KRD*, pp. 15–16.
[89] *Ibid.*, p. 18.
[90] *Ibid.*, pp. 23ff.
[91] *Ibid.*, p. 21.
[92] *Ibid.*, pp. 21–22.
[93] *Ibid.*, p. 23. Moleschott delighted in quoting the Frenchman Quatrefages, who was no Darwinist, against Darwin's assumption of a creative act in a scientific work. (pp. 22–23. Cf. also p. 45, n. 17).
[94] *KL*, 5th ed., p. 81.
[95] *KRD*, p. 39.
[96] *KL*, 5th ed., p. 613.
[97] *GU*, p. 143.
[98] *Ibid.*, p. 147. Cf. also Friedmann, *Weltanschauung Czolbes*, p. 49.
[99] *KS*, p. xiii.
[100] It *is* true that Büchner also rejected pantheism, if by pantheism is meant an impersonal creative intelligence. He himself believed firmly that the laws of nature and those of reason were identical (*KS*, p. 52), but he *contrasted* the laws of nature to "eternal reason." If the world was governed according to the laws of nature, then it was determined, but if creative reason was behind the world's operation, then these laws could be altered or even negated. Hence Büchner represented the situation as an either/or, either the laws of nature governed or eternal reason governed, and if the latter, then something extra-natural became involved.
[101] *KS*, p. 71.
[102] *Ibid.*, pp. 74–89.
[103] *Ibid.*, p. 83.

[104] *Ibid.*, p. 85.
[105] *Ibid.*, p. 87.
[106] *Ibid.*
[107] *Ibid.*, pp. 186ff. He also denied personal immortality (Chapter 16) and the freedom of the will (Chapter 19), although with regard to the latter issue his position remained qualified both here and for the rest of his life. He allowed that man was free, but only within very narrow limits, so that much of the significance of the so-called freedom of the will was, in his view, nonsense. (Cf. especially p. 257.)
[108] *Ibid.*, p. 97.
[109] *Ibid.*, pp. 80–83.
[110] *Ibid.*, p. 96.
[111] *NG*, p. 250. Büchner noted that the recapitulation observable in embryonic development supported his idea. (p. 251) It has been pointed out to me that this scheme bears a considerable resemblance to the ideas of von Baer.
[112] *NW*, I, pp. 220–27.
[113] *Ibid.*, p. 271. Cf. also the second edition of *NG*, which had been updated by footnotes, p. 246, n; pp. 247–48, n.
[114] *Ibid.*, pp. 279–80.
[115] *Ibid.*, pp. 277–78. This position he continued to hold throughout his life.
[116] *Ibid.*, p. 278. But then Darwin's theory dealt "more with development than origin." Many years later, on a visit to Darwin at Down, Büchner brought up the matter of Darwin's reference to a Creator in the *Origin*. Cf. 'Ein Besuch bei Darwin,' pp. 386–97 in *FE*; especially pp. 390–92. Cf. also *SV*, pp. 96–101, 128.
[117] The discussions are similar in both places. Here we shall follow that of the *FNG*. For the corresponding location in the *SV*, cf. pp. 223–69.
[118] *FNG*, p. 31.
[119] *Ibid.*, pp. 19–20, 29. Cf. Vogt's *BT*, p. 338.
[120] *Ibid.*, p. 20. Further it helps to make sense of a remark Büchner once made in a different appeal to his tree model. Each race, he said, must begin its education from the beginning. In other words the origin of every race was the common stem of all races, not the race preceding it in time. (*SJ*, p. 4). There were new branches that went *beyond* the older ones, but they were new branches, "which to be sure sprouted their first eye at a much deeper place than that achieved by the older branch at its highest point." (*FNG*, p. 31.)
[121] *FNG*, p. 19.
[122] *NG*, p. 244. Cf. also pp. 261–64.
[123] *ID*, p. 350.
[124] *NG*, pp. 255–57; *FNG*, p. 21. Cf. also *NW*, I, pp. 308–09.
[125] *LT*, p. 14.
[126] *LL*, pp. 20–21.
[127] *Ibid.*, pp. 21, 24–25.
[128] *Ibid.*, p. 22. Other deficiencies were given. Lamarck had not the proper notion of species, nor did he realize sufficiently the role played by extended periods of time. (pp. 22, 24)
[129] *Ibid.*, pp. 18–40. Goethe was included in this list (pp. 28–30), no doubt because of his sympathy to the linking of man with animal. Büchner therefore joined with Haeckel and D. F. Strauss against DuBois-Reymond, who did not think Goethe a forerunner of Darwin. Cf. F. Magnus, *Goethe as a Scientist*, trans. H. Norden (New York, 1949), p. xi. Cf. also pp. 81–84.
[130] *LL*, 79–81.
[131] *Ibid.*, pp. 24–26. Species in Darwin, said Büchner, were nothing real, only an abstraction of the mind. (p. 62)

[132] *Ibid.*, p. 65.
[133] *MV*, p. 2.
[134] *Ibid.*, pp. 26–27.
[135] *Ibid.*, pp. 40–41. Cf. also *LL*, p. 285.
[136] *ID*, pp. 234–35. This in an essay, "Können während des Lebens erworbene Eigenschaften vererbt werden?"
[137] *Ibid.*, p. 239.
[138] *Ibid.*, p. 238.
[139] *Ibid.*, p. 349.
[140] *EL*, p. 25.
[141] *US*, p. 525.
[142] *Nachgelassene Schriften*, p. 293.
[143] *NW*, II, p. 126.
[144] *GB*, p. 18.
[145] *Ibid.*, p. 31. Cf. also the discussion on purpose in the *NG*, pp. 270–301.

CHAPTER IX

[1] *Werke*, II, pp. 219, 345–46, 350–53.
[2] *Ibid.*, XII, pp. 133–34.
[3] Brazill, *Young Hegelians*, p. 265.
[4] Kamenka, *Philosophy of Feuerbach*, p. 160, n. 1. Cf. also p. 36.
[5] *Ibid.*, pp. 41, 59.
[6] Feuerbach, *Werke*, II, p. 218.
[7] Ludwig Feuerbach, *Sämtliche Werke*, 10 vols. (Leipzig, 1846), I, p. 10ff.
[8] *Werke*, VIII, p. 174.
[9] Rau, *Feuerbachs Philosophie*, pp. 11, 60.
[10] *Ibid.*, p. 11.
[11] Ferdinand Lassalle, *Nachgelassene Briefe und Schriften*. ed. Gustav Mayer, 5 vols., 1921–1925. Reprinted (Berlin, 1967), V, p. 246.
[12] Lange, *History of Materialism*, III, p. 333.
[13] Ivan Jessin, *Die materialistische Philosophie Ludwig Feuerbachs*, trans. from the Russian by U. Kuhirt (Berlin, 1956), pp. 5, 40.
[14] *KL*, p. 456.
[15] *Ibid.*, pp. 456ff.
[16] *Ibid.*, p. 458. Emphasis mine. Cf. also p. 479.
[17] *Ibid.*, pp. 458–59, 477.
[18] *Ibid.*, p. 459.
[19] *FMF*, p. 213.
[20] *US*, p. 527.
[21] *UW*, p. 21. Cf. also *UW* pp. 10–11.
[22] *OM*, I, pp. 19–21.
[23] *Ibid.*, p. 21.
[24] Mistelli, *Carl Vogt*, p. 89.
[25] *Ibid.*, p. 90.
[26] *Ibid.*, p. 224. Cf. also Hermelink, *Liberalismus und Konservatismus*, pp. 70–71.
[27] Quoted from the stenographic report by Wilhelm Vogt, *Carl Vogt*, p. 66, n. 1.
[28] *Ibid.*
[29] *Ibid.*
[30] Wurzbach. *op. cit.*, pp. 50–51.
[31] For evidence of Vogt's use of catastrophism at Frankfurt, cf. Misteli, *op. cit.*,

pp. 186–203. Misteli claims that the conservation of energy was well known and used by different members of the *Pauluskirche*. On closer examination of his claim one may conclude that deliberate allusions to force in a scientific context were made, but that they had anything to do with conservation of energy is simply not verified in any way. Cf. Misteli, *op. cit.*, pp. 156–86, and especially pp. 170ff.

[32] Quoted in Georg Mollat, *Reden und Redner des ersten deutschen Parlaments* (Osterwieck, 1895), p. 768.

[33] *Ibid.*, p. viii.

[34] Quoted from the stenographic report by Mollat, *op. cit.*, p. 768.

[35] *Ibid.*, p. 769. For Vogt's position on various issues at Frankfurt, see the following: Mollat, *op. cit.*, pp. 540, 676, 767–69; A. W. Ward, "The Revolution and the Reaction in Germany and Austria, 1848–1849," pp. 142–201 in A. Ward, G. Prothero, and S. Leathes, *eds. Cambridge Modern History*, Vol. XI (New York, 1934), p. 166; W. Vogt, *op. cit.*, pp. 61–62, 65, 70, 72, 78; Misteli, *op. cit.*, pp. 129, n. 1, 203–04, 214, 221, 224, 228; and Karl Vogt, *Der 18. September in Frankfurt am Main, passim.*

[36] 'Physiologie des Menschen,' pp. 651. 706–707.

[37] *UT*, pp. v–xvi. "They spoke of freedom and desired to subjugate other peoples." (p. xii)

[38] *Ibid.*, p. 13.

[39] *Ibid.*, pp. 21–23. Vogt even gave the absurd example of one Herr Risser, who changed his views after arriving in Frankfurt because, Vogt declared, of a change in diet. (p. 24)

[40] *Ibid.*, p. 5. As did Moleschott, Vogt also drew implications of the effect of diet on nationality. (p. 101, n.)

[41] *Ibid.*, p. 140.

[42] *Ibid.*, pp. 88–89.

[43] *Ibid.*, pp. 183–84. Cf. also pp. 188–90, 211–12. Later in the book Vogt demonstrated that this form of social organization could result in a thoroughly conservative government. (p. 239)

[44] *Ibid.*, p. 28.

[45] Cited in Mistelli, *op. cit.*, p. 214. Cf. also Mollat, *op. cit.*, p. 769.

[46] *UT*, pp. 28–29.

[47] *Ibid.*, p. 31.

[48] *Ibid.*, p. 30. The unfortunate aspects of free competition were a lesser evil in Vogt's mind than the suppression of the individual, and could be avoided with cleverness. (pp. 153–54). Later Vogt did refer sentiments similar to these to the SPD. Cf. *W*, p. 98.

[49] *Ibid.*, p. 31.

[50] "Pope and Anti-pope," p. 324.

[51] *Ibid.* This in response to the attempts of Virchow and Haeckel to demonstrate the political implications of Darwinism.

[52] *UT*, p. 212.

[53] Cited from Vogt's diary by Misteli, *op. cit.*, p. 82. Cf. n. 78.

[54] *AS*, pp. 21–22. 41, 48–51, 68.

[55] Misteli, *op. cit.*, p. 122.

[56] Quoted in Misteli, *op. cit.*, p. 124.

[57] *Ibid.*, p. 126.

[58] *UT*, p. 191.

[59] Marx's opposition to Vogt had begun already in 1848. When the *Märzverein*, to which Vogt belonged, gave the *Neue Rheinische Zeitung* its highest rating, Marx protested immediately (according to his later account). (Marx and Engels, *Werke*, XXX, p. 508.) On December 29, 1848 the *Neue Rheinische Zeitung* explicitly came out against Vogt in an article entitled 'Der Marzverein.' (Cf. Marx, *Herr Vogt*, *Werke*,

XIV, p. 462. Cf. also p. 772, n. 377.) Again in March, 1849 Marx referred to Vogt as a "rowdy from the university and a misplaced baron of the Reich," (*Herr Vogt*, p. 463) and mockingly as "the loyal sentry of the revolution." (p. 415) Evidence of the disgust Marx and Engels felt towards Vogt is further clear from the correspondence from 1852. (*Werke*, XXVIII, pp. 28, 64, 82, 493)

For Vogt's part, he joined in denouncing Marx's man Wilhelm Wolff, the editor of the *Neue Rheinische Zeitung* who came to Frankfurt as a replacement for a Silesian delegate. Wolff had dared to call Archduke Johann and all his ministers traitors, and received for his efforts almost unanimous condemnation. (*Herr Vogt*, pp. 464–67) (For the amusing personal encounter between Vogt and Wolff following this incident, see Marx's letter to Lassalle, *Werke*, XXIX, p. 624). Later, in assembling material for his case against the *Allgemeine Zeitung*, Vogt depicted Marx and his group as "Die Schwefelbande," a highly intolerant band of fugitives dedicated to the dictatorship of the proletariat at any cost. Lying, blackmailing, etc., was not beneath them in their attempt to gain power. (*MP*, pp. 136–39) He even went so far as to suggest that Marx was in league with the Reaction, for not only did his radicalism fracture the genuine efforts of the Left, but too many people invariably ended up in prison after workers' congresses like that in Cologne. (*MP*, pp. 166–77)

[60] *MP*, p. 178.

[61] Wilhelm Vogt, *op. cit.*, p. 78.

[62] For Vogt's mockery of the *kleindeutsch* position at Frankfurt, cf. Misteli, *op. cit.*, p. 129, n. 1. For Marx's position on German unification, see Marx's letter to Engels concerning Lassalle's 1859 *Der italienische Krieg und die Aufgabe Preussens*, and Vogt's *SGL* in *Werke*, XXIX, p. 432. Cf. also p. 708, n. 380, and Marx to Engels in *Werke*, XXX, p. 463. Cf. George Lichtheim, *Marxism: An Historical and Critical Study*, 2d ed. (New York, 1965), pp. 65–75.

[63] "I regret that our relations are not as satisfactory as formerly." Cf. E. Masi, 'Cavour and the Kingdom of Italy,' pp. 366–92 in *The Cambridge Modern History*, ed. by A. Ward, G. Prothero and S. Leathes, vol. XI (New York, 1934), p. 381.

[64] The program is reprinted as document 8 in Vogt's *MP*, section 3, pp. 33–36.

[65] Among the prominent Germans was included Ludwig Feuerbach. Vogt's letter to Feuerbach is found in *Ludwig Feuerbach in seinem Briefwechsel und Nachlass sowie in seiner philosophischen Charakterentwicklung dargestellt von K. Grün*, I (Heidelberg, 1874), pp. 118–19. Cf. also Feuerbach, *Werke*, XIII, p. 251, n.

[66] *SGL*, pp. 128–33.

[67] *Ibid.*, pp. vii, 76.

[68] *MP*, pp. v, 223–24.

[69] See Marx's letter to Engels, April 22, 1859 in *Werke*, XXIX, p. 426. Marx saw Vogt's program, as described to Freiligrath, as the epitome of ignorance. (p. 426)

[70] *Herr Vogt*, p. 473.

[71] *Ibid.*, p. 475. The article from *Das Volk* is reprinted as document 4 in Vogt's *MP*, section 3, pp. 17–20. Regarding the journal *Das Volk,* cf. pp. 3–4, 58. That Marx completely believed Blind is evident from his May 16 letter to Engels, *Werke,* XXIX, p. 430.

[72] The speech was delivered to a Swiss workers society on May 23, and printed in Vogt's *Schweizer Handels-Kourier*. It is summarized in the *MP*, section 3, pp. 31–33. Cf. also *Herr Vogt*, p. 431.

[73] Reprinted in *MP*, section 3, as document 2, pp. 3–5.

[74] The formal complaint is printed in section 3 of *MP*, pp. 1–3.

[75] Cf. *MP*, p. 100. Other things mentioned included Vogt's support of revolution as a scientist while at Frankfurt (pp. 63–64), his extreme anti-authoritarianism in the *KW* (p. 56), and even the charge that the France Vogt was working to support was an immoral nation, caught like all Roman races "in decline." (p. 48)

[76] *MP*, pp. 194–96, 227, 233. In reply the editors said they were only using Marx, not agreeing with him. As far as they were concerned, Vogt and Marx were one of a kind, (p. 133) a remark that angered Vogt. (p. 136)

[77] One of the editors, Hermann Orges, had been hired earlier specifically to fight the new philosophical tendency supported by Büchner's *Kraft und Stoff.* Cf. *Der Humor in Kraft und Stoff*, pp. 67–68. Vogt's opinion of Orges is found in *MP*, pp. 199–207. For the admission that there were no documents, see *MP*, p. 106.

[78] The ruling is found in *MP*, pp. 114–30.

[79] On the history of the term "Schwefelbande," see Marx and Engels, *Werke*, XIV, p. 757, n. 280. Vogt erred in associating Marx with a fugitive band of 1849–1850 with this name. (*MP*, p. 136, 174–75) That group was a beer-drinking, nonpolitical clique centered in Geneva. Cf. *Herr Vogt*, pp. 390–93.

[80] Marx worked on the *Herr Vogt* between January and December of 1860. Cf. *Werke*, XIV, p. 756, n. 275.

[81] Marx to Engels, February 3, 1860. In *Werke*, XXX, p. 22. Engels letter is found on p. 15.

[82] *Werke*, XXX, p. 459.

[83] See, for example, the *Vorwort* to volume II of the *Werke*, pp. xvi–xvii. On Ruge and Herwegh see *Werke*, XXX, p. 509.

[84] Cf. the letters of Marx to Engels, June 14 and September 25, 1860 in *Werke*, XXX, pp. 62, 96, and Rudolf Franz, 'Einleitung,' pp. iii–viii in *Herr Vogt* by Karl Marx (Leipzig, 1927), p. iii.

[85] Engels, *Werke*, III, p. 554. Yet later the same year he wrote to Marx that the completed book was "the best polemical writing you have penned yet." (vol. XXX, p. 129)

[86] Franz, *op. cit.*, p. iii. Cf. also the exchange between Marx and Engels, January 28 and 29, 1869 in *Werke*, XXXII, pp. 250, 252, and p. 758, n, 288.

[87] R. Reichard, *Crippled from Birth: German Social Democracy, 1844–1870* (Ames, Iowa, 1969), p. 134. Cf. also p. 303, n. 95–96.

[88] Cf. chapter 8, 'Da-Da Vogt und seine Studien,' *Herr Vogt*, pp. 490–540.

[89] *Ibid.*, pp. 406, 616.

[90] Cf. Franz, *op. cit.*, p. v, and Engels letter to Laura Lafargue in 1886, *Werke*, XXXVI, p. 572.

[91] Cf. André Lefèvre, ed., *Papiers et correspondance de la famille impériale*, 2 vols. (Paris, 1870), II, section xxvii, p. 166. The civil list contained private funds dispensed by the Emperor for personal reasons, and they could be used for political purposes. Cf. Lefèvre's 'Notes sur les dépenses de la liste civile de Napoleon III,' in Lefèvre, *op. cit.*, pp. 106–07.

[92] *Herr Vogt*, p. 549.

[93] The "Abermals Herr Vogt" may be found in the Franz edition of the *Herr Vogt*, pp. 193–99.

[94] Cf. the letters between Marx and Liebknecht in *Werke*, XXXIII, pp. 203, 214, 220. Cf. also p. 744, n. 234, and p. 748, n. 256. A year later one of Vogt's defenders told Engels that the "Abermals" had completely silenced Vogt. (*Werke*, XXXIII, p. 378.)

There is evidence that Vogt was capable of lying under such conditions. In the copy of the English translation of Vogt's *Vorlesungen über den Menschen* in the Museum of Comparative Zoology at Harvard, there is pencilled in the margin on page 432 the following reply to the translator's comment that it was not like Vogt to accept something without reliable evidence: "– unless he can make something by the lie, for he is one of those who approves of lying when profitable. I have heard him say so in the presence of Professors Guyot and Quenstadt." The author of these remarks wished thereby to make a formal testimony to this effect, for he appended his signature to the remark: 'L. Agassiz.'

[95] *LA*, p. 408.
[96] Lassalle's letter is reprinted in part in Büchner's *BFL*, pp. 5–10.
[97] This earlier evaluation of Lassalle's ideas by Büchner has been lost. Cf. *BFL*, p. 5.
[98] *LA*, pp. 16–18.
[99] *Ibid.*, p. 8
[100] *Ibid.*, pp. 19–24.
[101] For an account of this visit and Büchner's portrait of Lassalle's personality, see *BFL*, pp. 22–30.
[102] *Ibid.*, pp. 32–35.
[103] *Ibid.*, p. 36.
[104] *BFL*, pp. 23, 36. Büchner was not unaware of Marx. He had made use of Marx's criticism of Lassalle (*BFL*, p. 20), but still in 1894 he saw Lassalle as "the messiah or prophet of the new worker-religion." (p. 14) In his *DS* of the same year, he tempered his omission of Marx somewhat, calling him the intellectual father of social democracy, and referring to both Lassalle and Marx as holy men of the movement, in spite of the fact that Lassalle's views were no longer followed. (*DS*, p. 53).
[105] *LA*, p. 26. Büchner denied that he was being bourgeois in this. He did not think there was a real bourgeoisie in Germany. (p. 26).
[106] *Ibid.*, p. 29.
[107] *DS*, p. 54.
[108] Alexander Büchner, *Das 'tolle' Jahr*, p. 374.
[109] *MSN*, p. 211.
[110] *Ibid.*, Anmerkung 93, p. xciv. Cf. also *FE*, pp. 214–33, where the Manchester theory is examined in light of its American application. Büchner's major point is that it is impossible to exist without some elements of social control. In *SJ*, pp. 253ff, Büchner deals with anarchism, although by that term he did not mean what Vogt did. Stirner and Nietzsche, both worshippers of the individual, were both representative anarchists in his eyes, and both were deplorable because they sanctioned violence and did not view man as a social creature.
[111] *Ibid.*, pp. 181–82. Cf. also *DS*, pp. 6–8.
[112] *MSN*, p. 182.
[113] *Ibid.*, pp. 183–84.
[114] *MSN*, pp. 200–06. Cf. also Anm. 93–95, pp. xciv–xcviii; *DS*, pp. 22–46; *ID*, pp. 333–43.
[115] Büchner shared the view that a period of decadence had hit Germany in the late nineteenth century. As he put it, it was an era bent "very much on standing pat or regressing." Ironically, he listed Nietzsche and Langbehn as *signs* of the decadence, not as plaintiffs against it. For Büchner the revival of mysticism and religion was the reason for the loss of German nationalism. (Cf. *SJ*, pp. 12–14, 350.)
[116] *MSN*, pp. 217–20; DS, pp. 13–14.
[117] Ludwig's sister Luise was active in the German movement for women's rights in the nineteenth century, authoring several books on the subject. As for Ludwig, he explained woman's inferior performance mostly as the result of her suppressed social position, which in turn was due to the male fear of female abilities. The smaller female brain, he said, was larger in proportion to body size than a man's. Yet Büchner too opposed complete political emancipation of women because women were still intellectually immature, as shown among other things by their predilection for religion. Besides, nature had equipped woman to be a housewife and mother. Cf. *MSN*, pp. 224–32; *PB*, II, p. 90, n., 119–28; *TT*, pp. 61–81; *SJ*, pp. 309, 319–33.
[118] *DS*, p. 41.
[119] *FNG*, pp. 24–26.
[120] *Ibid.*, p. 56.

[121] *Ibid.*, p. 47.
[122] *FE*, p. 260. Cf. also *GZ*, p. 335, and *KL*, p. 129.
[123] *DS*, pp. 48–49.
[124] *GZ*, pp. 278–79.
[125] *GB*, p. 7, n. In fact Büchner reprimanded modern Christians for ignoring what Christ had said concerning social responsibility. (pp. 48–49)
[126] *DS*, p. 54.
[127] Ibid., pp. 61–62. Cf. pp. 63ff for his critique of the Erfurt Program, which ran along similar lines. Also cf. two articles on social democracy in *FE*, pp. 234–41, and 242–51.
[128] Cf. Engels letter to Liebknecht in *Werke*, XXXIII, p. 456.
[129] *MSN*, p. 172.
[130] *Ibid.*, p. 198.
[131] *Ibid.*, p. 191.
[132] *NW*, II, p. ix.
[133] *GB*, p. 45.
[134] *Ibid.*, p. 50.
[135] *MV*, p. 73.
[136] *Ibid.*, p. 70.
[137] *KL*, pp. 145–46.
[138] *FE*, p. 281. Emphasis mine. For Büchner's belief that these conclusions came from natural science, cf. also p. 293, and his words in *DS*, p. 17: "Here again the proper cue is given by natural science, which these days might be cited as affecting not only the mental, but also the social emancipation of humanity."
[139] *DS*, p. 18. Cf. also pp. 20–21.
[140] *FE*, p. 329.

CONCLUDING REMARKS

[1] Merz, *History of Thought*, IV, pp. 230–31.
[2] Otto Finger and Friedrich Herneck, *eds. Von Liebig zu Laue: Ethos und Weltbild grosser deutscher Naturforscher und Aerzte*, 2d ed. (Berlin, 1963), pp. 7–8.
[3] *Ibid.*, pp. 171–73.
[4] *Ibid.*, pp. 174–75.
[5] Dieter Wittich, *ed., Vogt, Moleschott, Büchner: Schriften zum kleinbürgerlichen Materialismus in Deutschland* (Berlin, 1971), pp. vi–vii.
[6] *Ibid* p. lxiv.
[7] Cf. especially *ibid.*, xxiii–xxiv, xxvi, xxviii, xl, lxv–lxvi.
[8] Julian V. L. Casserley, *The Retreat from Christianity in the Modern World* (London, 1952), p. 36.

BIBLIOGRAPHY

Ackerknecht, Erwin, 'Beiträge zur Geschichte der Medicinalreform von 1848,' *Archiv der Geschichte der Medizin*, **XXV** (1932), pp. 61–109, 113–183.

Ackerknecht, Erwin, *Rudolf Virchow: Doctor, Statesman, Anthropologist*. Madison, 1953.

Bachmann, Carl F., *Anti-Hegel. Antwort an Herrn Professor Rosenkranz in Königsberg*. Jena. 1835.

Bachmann, Carl F., *Über Hegels System und die Notwendigkeit einer nochmaligen Umgestaltung der Philosophie*. Leipzig, 1833.

Berger, Herbert, *Begründung des Realsimus bei J. H. von Kirchmann und Friedrich Ueberweg*. Bonn, 1958.

Berlin, Isaiah, *The Life and Opinions of Moses Hess*. Cambridge, 1959.

Biber, R., *Karl Vogts naturwissenschaftliche Vorträge über die Urgeschichte des Menschen*. Berlin, 1870.

Billing, Sidney, *Scientific Materialism and Ultimate Conceptions*. London, 1879.

Binkley, Robert C., *Realism and Nationalism, 1852–1871*. New York, 1963.

Böhmer, Heinrich, *Geschichte der Entwicklung der naturwissenschaftlichen Weltanschauung in Deutschland*. Gotha, 1872.

Böhner, August N., *Naturforschung und Culturleben in ihren neuesten Ergebnissen zur Beleuchtung der grossen Fragen der Gegenwart über Christentum und Materialismus, Geist und Stoff*. Hannover, 1859.

Bolin, Wilhelm, Biographische Einleitung to *Ausgewählte Briefe von und an Ludwig Feuerbach*. Pp. 1–211 in Ludwig Feuerbach, *Sämtliche Werke*. 2d ed. Edited by Wilhelm Bolin and Friedrich Jodl, vol. XII, Stuttgart, 1964.

Bolin, Wilhelm, *Ludwig Feuerbach, sein Wirken und seine Zeitgenossen*. Stuttgart, 1891.

Brazill, William J., *The Young Hegelians*. New Haven, 1970.

Breilmann, Hedwig, *Lotzes Stellung zum Materialismus, unter besonderer Berücksichtigung seiner Controverse mit Czolbe*. Münster, 1925.

Büchner, Alexander, *Das 'tolle' Jahr: Vor, Während, und Nach 1848*. Giessen, 1900.

Büchner, Anton, *Die Familie Büchner. Georg Büchners Vorfahren, Eltern und Geschwister*. Darmstadt, 1963.

Büchner, Georg, *Nachgelassene Schriften von Georg Büchner*. Edited by Ludwig Büchner, Frankfurt am Main, 1850.

Büchner, Ludwig, *Am Sterbelager des Jahrhunderts. Blicke eines freien Denkers aus der Zeit in die Zeit*. 2d ed. Giessen, 1900. (1st edition, 1898)

Büchner, Ludwig, *Aus dem Geistesleben der Thiere, oder Staaten und Thaten der kleinen*. Berlin, 1879. (Originally published, 1876) Translated as *Mind in Animals* by Annie Besant. London, 1903.

Büchner, Ludwig, *Aus Natur und Wissenschaft. Studien, Kritiken und Abhandlungen in allgemeinen verständlicher Darstellung*. 2 vols. 3d ed. Leipzig, 1874, 1884. (1st edition, vol. 1, 1862)

Büchner, Ludwig, *Beiträge zur Hall' schen Lehre von einem excito-motorischen Nerven System*. Giessen, 1848.

Büchner, Ludwig, *Darwinismus und Sozialismus, oder der Kampf um das Dasein und die moderne Gesellschaft*. Leipzig, 1894.

Büchner, Ludwig, *Der Fortschritt in Natur und Geschichte im Lichte der Darwin'sche Theorie*. Stuttgart, 1884.

Büchner, Ludwig, *Fremdes und Eigenes aus dem geistigen Leben der Gegenwart*. Leipzig, 1890.

Büchner, Ludwig, 'Georg Büchner', pp. 1–50 in *Nachgelassene Schriften von Georg Büchner*. Edited by Ludwig Büchner. Frankfurt am Main, 1850.

Büchner, Ludwig, *Das goldene Zeitalter, oder, Das Leben vor der Geschichte*. Berlin, 1891.

Büchner, Ludwig, *Der Gottesbegriff und dessen Bedeutung in der Gegenwart. Ein allgemein-verständlicher Vortrag*. Leipzig, 1874.

Büchner, Ludwig, *Herr Lassalle und die Arbeiter. Bericht und Vortrag über das Lassalle'sche Arbeiterprogram*. Frankfurt am Main, 1863.

Büchner, Ludwig, *Im Dienst der Wahrheit. Ausgewählte Aufsätze aus Natur und Wissenschaft*. Giessen, 1900. Translated in part as *Last Words on Materialism* by Joseph McCabe. London, 1901.

Büchner, Ludwig, *Kraft und Stoff. Empirisch-naturphilosophische Studien. In allgemein verständlicher Darstellung*. Frankfurt am Main, 1855. (5th German edition, 1858; 19th German edition, 1898) Translated as *Force and Matter* by Frederick Collingwood. London, 1864.

Büchner, Ludwig, *Das künftige Leben und die moderne Wissenschaft. Zehn Briefe an eine Freundin*. Leipzig, 1889.

Büchner, Ludwig, *Licht und Leben. Drei allgemeinverständliche naturwissenschaftliche Vorträge als Beiträge zur Theorie der natürlichen Weltordnung*. Leipzig, 1882.

Büchner, Ludwig, *Liebe und Liebesleben in der Thierwelt*. Berlin, 1879.

Büchner, Ludwig, *Die Macht der Vererbung und ihr Einfluss auf der moralischen und geistigen Fortschritt der Menschheit*. Leipzig, 1882.

Büchner, Ludwig, *Meine Begegnung mit Ferdinand Lassalle. Ein Beitrag zur Geschichte der sozialdemokratischen Bewegung in Deutschland. Nebst fünf Briefen Lassalles*. Berlin, 1894.

Büchner, Ludwig, *Der Mensch und seine Stellung in der Natur in Vergangenheit, Gegenwart und Zukunft, oder, Woher kommen wir? Wer sind wir? Wohin gehen wir? Allgemein verständlicher Text mit zahlreichen wissenschaftlichen Erläuterungen und Anmerkungen*. 2d ed. Leipzig, 1872. (1st edition, 1870) Translated as *Man in the Past, Present and Future*, by W. S. Dallas. London, 1872.

Büchner, Ludwig, *Natur und Geist. Gespräche zweier Freunde über den Materialismus und über die real-philosophischen Fragen der Gegenwart. In allgemein verständlicher Form*. Frankfurt am Main, 1857.

Büchner, Ludwig, *Der neue Hamlet. Poesie und Prosa aus Papieren eines verstorbenen Pessimisten*. Giessen, 1885.

Büchner, Ludwig, *Physiologische Bilder*. 2 vols. Leipzig, 1861, 1875.

Büchner, Ludwig, *Sechs Vorlesungen über die Darwin'sche Theorie von der Verwandlung der Arten und die erste Entstehung der Organismenwelt, sowie über die Anwendung der Umwandlungstheorie auf den Menschen, das Verhältniss dieser Theorie zur Lehre vom Fortschritt und den Zusammenhang derselben mit der materialistischen Philosophie der Vergangenheit und Gegenwart*. 3d ed. Leipzig, 1872. (1st edition, 1868)

Büchner, Ludwig, *Thatsachen und Theorien aus dem naturwissenschaftlichen Leben der Gegenwart*. Berlin, 1887.

Büchner, Ludwig, *Über religiöse und wissenschaftliche Weltanschauung. Ein historisch-kritischer Versuch*. Leipzig, 1887.

Büchner, Ludwig, *Zwei gekrönte Freidenker. Ein Bild aus der Vergangenheit als Spiegel für die Gegenwart. Dem deutschen Volke gewidmet.* Leipzig, 1890.

Chamberlain, William, *Heaven Wasn't His Destination.* London, 1941.

[Chambers, Robert] *Natürliche Geschichte der Schöpfung des Weltalls.* Translated by Karl Vogt. Braunschweig, 1851.

Coleman, William, *Biology in the Nineteenth Century: Problems of Form, Function, and Transformation.* New York, 1971.

Cornhill, Adolph, *Materialismus und Idealismus in ihren gegenwärtigen Entwicklungskrisen beleuchtet.* Heidelberg, 1858.

Cranefield, Paul, 'The Organic Physics of 1847 and the Biophysics of Today,' *Journal of the History of Medicine*, **XII** (1957), pp. 407–423.

Czolbe, Heinrich, 'Die Elemente der Psychologie vom Standpunkte des Materialismus," *Zeitschrift für Philosophie und philosophische Kritik*, **XXVI** (1855), pp. 91–109.

Czolbe, Heinrich, *Entstehung des Selbstbewusstseins. Eine Antwort an Lotze.* Leipzig, 1856.

Czolbe, Heinrich, *Die Grenzen und der Ursprung der menschlichen Erkenntnis im Gegensatz zu Kant und Hegel. Naturalistisch-teleologische Durchführung des mechanischen Princips.* Jena, 1865.

Czolbe, Heinrich, *Grundzüge einer extensionalen Erkenntnistheorie. Ein räumliches Abbild von der Entstehung der sinnlichen Wahrnehmung.* Plauen, 1875.

Czolbe, Heinrich, 'Die Mathematik als Ideal für alle andere Erkenntniss und das Verhältniss der empirischen Wissenschaften zur Philosophie," *Zeitschrift für exacte Philosophie*, **VII** (1866), pp. 217–278.

Czolbe, Heinrich, *Neue Darstellung des Sensualismus. Ein Entwurf.* Leipzig, 1855.

Darwin, Charles, *De la variation des animaux et de plantes à l'état domestique.* Translated by J. J. Moulinie. Preface by Karl Vogt. 2 vols. Paris, 1868.

Deubler, Konrad, *Tagebücher, Biographie und Briefwechsel (1848–1884) des oberösterreichschen Bauernphilosophen.* Edited by Arnold Dodel-Port. 2 vols. Leipzig, 1886.

Diesch, Carl, *Bibliographie der Germanistischen Zeitschriften.* Leipzig, 1927.

Dietrich, Reinhard (ed), *Bibliographie der deutschen Zeitschriftenliteratur. Ergänzungsbände*, vols. 78a–86a. 5 vols. Leipzig, 1937–1942.

Doedes, J. Isaak, *Der Angriff eines Materialisten, Dr. L. Büchner, auf den Glauben an Gott.* Jena, 1875.

Dörpinghaus, Hermann, *Darwins Theorie und der deutsche Vulgärmaterialismus im Urteil deutscher katholischer Zeitschriften zwischen 1854 und 1914.* Freiburg, 1969.

DuBois-Reymond, Emil, *Untersuchungen über thierische Elektricität.* Berlin, 1848.

DuBois-Reymond, Emil, *Zwei Vorträge von E. DuBois-Reymond.* Leipzig, 1882.

Elkana, Yehuda, 'Helmholtz's "Kraft": An Illustration of Concepts in Flux,' *Historical Studies in the Physical Sciences*, **II** (1970), pp. 263–298.

Evans, E. P., 'Sketch of Jacob Moleschott,' *Popular Science Monthly*, **XLIX** (1896), pp. 399–406.

Fabri, Friedrich, *Briefe gegen den Materialismus.* 2d ed. Stuttgart, 1864. (1st edition, 1855)

Farley, John, 'The Spontaneous Generation Controversy (1859–1880): British and German Reactions to the Problem of Abiogenesis,' *Journal of the History of Biology*, **V** (1972), pp. 285–319.

Feuerbach, Ludwig, *The Fiery Brook: Selected Writings of Ludwig Feuerbach.* Translated with an introduction by Zawar Hanfi. Garden City, New York, 1972.

Feuerbach, Ludwig, *Sämtliche Werke.* 2d ed. Edited by Wilhelm Bolin and Friedrich Jodl. 10 vols. 1903–1911. Reprint (13 vols. in 12) Stuttgart, 1959–1964. (1st edition, Leipzig, 1846–1866)

Fichte, I. H., 'Übersicht der philosophischen Literatur,' *Zeitschrift für Philosophie und philosophische Kritik*, **XXIII** (1853), pp. 134–169.

Finger, Otto and Friedrich Herneck (eds.), *Von Liebig zu Laue: Ethos und Weltbild grosser deutscher Naturforscher und Aerzte*. 2d ed. Berlin, 1963.

Fischer, Karl, *Die Unwahrheit des Sensualismus mit besonderer Rücksicht auf die Schriften von Feuerbach, Vogt und Moleschott bewiesen*. Erlangen, 1853.

Fortlage, Karl, 'Materialismus und Spiritualismus,' *Blätter für literarische Unterhaltung*. No. 30 (1856), pp. 541–548; no. 49 (1856), pp. 889–899; no. 19 (1857), pp. 337–347; no. 12 (1858), pp. 205–215.

Franz, Rudolf, Einleitung to *Herr Vogt* by Karl Marx. Leipzig, 1927.

Frauenstädt, Julius, *Der Materialismus. Eine Erwiderung auf Dr. L. Büchner's 'Kraft und Stoff.'* Leipzig, 1856.

Frauenstädt, Julius, *Die Naturwissenschaft in ihrem Einfluss auf Poesie, Religion, Moral und Philosophie*. Leipzig, 1855.

Friedmann, Pinkas, *Darstellung und Kritik der naturalistischen Weltanschauung Heinrich Czolbes*. Bern,1905.

'Friedrich Carl Christian Ludwig Büchner.' Pp. vii–xxi in *Kraft und Stoff*. 19th ed. Leipzig, 1898.

Frohschammer, Jakob, *Menschenseele und Physiologie. Eine Streitschrift gegen Professor Karl Vogt*. Munich, 1855.

Gabriel, Leo, 'Konrad Deubler.' *Neue Deutsche Biographie*, vol. 3. Berlin, 1957, pp. 620–621.

Gasman, Daniel G., *The Scientific Origins of National Socialism*. London, 1971.

Goodfield, June, *The Growth of Scientific Physiology*. London, 1960.

Graham, Loren, *Science and Philosophy in the Soviet Union*, New York, 1972.

Grün, Karl, *Ludwig Feuerbachs philosophische Charakterentwicklung, sein Briefwechsel und Nachlass*. 2 vols. Leipzig, 1874.

Grützner, P., 'Jacob Moleschott,' *Allgemeine Deutsche Biographie*, vol. 52, 1906. Reprint. Berlin, 1971, pp. 435–438.

Harless, Adolf [Dr. Mantis], *Goethe im Fegfeuer. Eine materialistisch-poetische Gehirnsekretion*. Stuttgart, 1856.

Hayes, Carlton, *A Generation of Materialism, 1871–1900*. New York, 1941.

Heinze, M., 'Jakob Frohschammer.' *Allgemeine Deutsche Biographie*, vol. 49, 1904. Reprint. Berlin, 1971, pp. 172–177.

Herwegh, Georg, *Georg Herweghs Briefwechsel mit seiner Braut*. Edited by Marcel Herwegh. Stuttgart, 1906.

Herwegh, Georg, *Werke*. Edited by H. Tardel. 3 parts. Berlin, 1909.

Herwerden, M. A. van, 'Eine Freundschaft von drei Physiologen,' *Janus*, **XX** (1915), pp. 174–201, 409–436.

Hess, Moses, *Briefwechsel, 1825–1881*. Edited by Edmund Silberner. 's-Gravenhage, 1959.

Holborn, Hajo, *A History of Germany, 1840–1945*. New York, 1969.

Hook, Sidney, *From Hegel to Marx*. Ann Arbor, 1966. (Original publication, 1936).

Hook, Sidney, 'What is Materialism?' *Journal of Philosophy*, **XXXI** (1934), pp. 235–242.

Der Humor in Kraft und Stoff, oder die exacten Ungereimtheiten der modernen Realphilosophie. Darmstadt, 1856.

Janet, P. A. R., *The Materialism of the Present Day. A Critique on Dr. Büchner's System*. Translated by G. Masson. London, 1865.

Jessin, Ivan, *Die materialistische Philosophie Ludwig Feuerbachs*. Translated by U. Kuhirt. Berlin, 1956.

Jodl, Friedrich, *Ludwig Feuerbach*. 2d ed. Stuttgart, 1921.

Johnson, Eduard, 'Heinrich Czolbe,' *Altpreussische Monatsheften*, **X** (1873), pp. 338–352.

Jung, Johannes, *Karl Vogts Weltanschauung. Ein Beitrag zur Geschichte des Materialismus im neunzehnten Jahrhundert*. Studien zur Philosophie und Religion. Edited by Stölzle. Heft 17. Paderborn, 1915.

Kamenka, Eugene, *The Philosophy of Ludwig Feuerbach*. London, 1970.

Klencke, Hermann, *Die Naturwissenschaft der letzten fünfzig Jahre und ihr Einfluss auf das Menschenleben. In Briefen an Gebildete aller Stände.* Leipzig, 1854.

Kohut, Adolph, *Ludwig Feuerbach, sein Leben und seine Werke*. Leipzig, 1909.

Krause, Ernst, 'Karl Vogt,' *Allgemeine Deutsche Biographie*, vol. 40, 1896. Reprint. Berlin, 1971, pp. 181–189.

Krause, Herbert, *'Die Gegenwart: Eine enzyklopädische Darstellung der neuesten Zeitgeschichte für alle Stände, 1848–1856.' Eine Untersuchung über den deutschen Liberalismus*. Saalfeld, 1936.

Kuhn, Thomas, 'Energy Conservation as an Example of Simultaneous Discovery.' Pp. 221–256 in *Critical Problems in the History of Science*. Edited by Marshall Clagett. Madison, 1969.

Lange, Friedrich A., *The History of Materialism and Criticism of its Present Importance*. Translated by E. C. Thomas. 3d ed. 3 vols. in 1. London, 1925. (1st German edition, 1865)

Langer, William, *Political and Social Upheaval, 1832–1852.* New York, 1969.

Lassalle, Ferdinand, *Nachgelassene Briefe und Schriften*. Edited by Gustav Mayer. 5 vols. Stuttgart, 1921–1925. Reprint. Berlin, 1967.

Lefèvre, André, (ed.), *Papiers et correspondance de la famille impériale*. 2 vols. Paris, 1870.

Lévy, Albert, *La Philosophie de Feuerbach et son influence sur la litterature allemande*. Paris, 1904.

Lichtheim, Georg, *Marxism: An Historical and Critical Study*. 2d ed. New York, 1965.

Lippmann, Timothy, 'Vitalism and Reductionism in Liebig's Physiological Thought,' *Isis*, LVIII (1967), pp. 167–185.

Livingstone, Rodney, Introduction to *The Cologne Communist Trial* by Karl Marx. Translated by R. Livingstone. London, 1971.

Lotze, Hermann, *Kleine Schriften*. 3 vols. Leipzig, 1891.

Lotze, Hermann, *Medicinische Psychologie*. Leipzig, 1852.

Löwith, Karl, Einleitung to *Sämtliche Werke* by Ludwig Feuerbach. 2d ed., vol. 1. Stuttgart, 1959, pp. vii–xxxv.

Marx, Karl and Friedrich Engels, *Werke*. Edited by the Institut für Marxismus-Leninismus beim ZK der SED. 39 vols. Berlin, 1957–1968.

Mascall, W., *The Newest Materialism*. London, 1873.

Masi, E., 'Cavour and the Kingdom of Italy.' Pp. 366–392 in *The Growth of Nationalities*. The Cambridge Modern History, vol. 11. Edited by A. Ward, G. Prothero and S. Leathes. New York, 1934.

Mayer, J. Robert, *Kleinere Schriften und Briefe (1932–1877) von R. Mayer.* Edited by Jacob J. Weyrauch. Stuttgart, 1893.

Mayer, J. Robert, *Die Mechanik der Wärme in Gesammelten Schriften*. Edited by Jacob J. Weyrauch. 3d. ed. Stuttgart, 1893.

McLellan, David, *Karl Marx: His Life and Thought*. New York, 1973.

Mendelsohn, Everett, 'The Emergence of Science as a Profession in Nineteenth Century Europe.' Pp. 3–48 in *The Management of Scientists*. Edited by Karl Hill. Boston, 1964.

Mendelsohn, Everett, 'Revolution and Reduction.' In *The Interaction Between Science and Philosophy*. Edited by Yehuda Elkana. New York, 1973.

Merz, John T., *A History of European Thought in the Nineteenth Century.* 4 vols. 1896–1912. Reprint. New York: Dover, 1965.

Meyer, Jürgen Bona, *Philosophische Zeitfragen. Populäre Aufsätze*. 2d ed. Bonn, 1874. (1st edition, 1870).

Meyer, Jürgen Bona, *Zum Streit über Leib und Seele. Sechs Vorlesungen am Hamburger akademischen Gymnasium gehalten*. Hamburg, 1856.

Michelis, Friedrich, *Der kirchliche Standpunkt in der Naturforschung. Ein Wort zur Verständigung über den Verhältniss der Naturforschung zu dem Glauben und der Hoffnung des Christen. Sendschreiben an M. J. Schleiden*. Münster, 1855.

Misteli, Hermann, *Carl Vogt. Seine Entwicklung vom angehenden naturwissenschaftlichen Materialisten zum idealen Politiker der Paulskirche, 1817–1849*. Zürich, 1938.

Mohr, Friedrich, *Allgemeine Theorie der Bewegung und Kraft, als Grundlage der Physik und Chemie. Ein Nachtrag zur mechanischen Theorie der chemischen Affinität*. Braunschweig, 1869.

Moleschott, Jacob, *Die Einheit des Lebens*. Giessen, 1864.

Moleschott, Jacob, *Für meine Freunde. Lebens-Erinnerungen von Jacob Moleschott*. Giessen, 1894.

Moleschott, Jacob, 'Georg Forster,' *Die Natur*, **II** (1853), pp. 205–208.

Moleschott, Jacob, *Georg Forster, der Naturforscher des Volks. Zur Feier des 26 November, 1854*. Frankfurt am Main, 1854.

Moleschott, Jacob, *Die Grenzen des Menschen*. Giessen, 1863.

Moleschott, Jacob, *Hermann Hettner's Morgenroth*. Giessen, 1883.

Moleschott, Jacob, *Karl Robert Darwin. Denkrede gehalten im Collegio Romano im Namen der Studierenden der Hochschule zu Rom*. Giessen, 1883.

Moleschott, Jacob, *Der Kreislauf des Lebens. Physiologische Antworten auf Liebig's 'Chemische Briefe.'* 2d ed. Mainz, 1855. (1st edition, 1852).

Moleschott, Jacob, *Die Lehre der Nahrungsmittel. Für das Volk*. Erlangen, 1850. English translation, *The Chemistry of Food and Diet*. Translated by E. Bronner. Pp. 305–394 in vol. 8 of *Orr's Circle of the Sciences*. Edited by J. S. Bushmann. 9 vols. London, 1854–1856.

Moleschott, Jacob, *Licht und Leben. Rede beim Antritt des öffentlichen Lehramts zur Erforschung der Natur des Menschen, an der Züricher Hochschule*. 2d ed. Frankfurt am Main, 1856.

Moleschott, Jacob, *Natur und Heilkunde. Vortrag bei der Wiedereröffnung der Vorlesungen über Physiologie an der Turiner Hochschule am 28. November, 1864 gehalten*. Giessen, 1865.

Moleschott, Jacob, *Pathologie und Physiologie. Vortrag bei der Wiedereröffnung der Vorlesungen über Physiologie an der Turiner Hochschule am 28. November, 1864 gehalten*. Giessen, 1866.

Moleschott, Jacob, *Die Physiologie der Nahrungsmittel*. 2d ed. Giessen, 1859. (1st edition, 1850)

Moleschott, Jacob, *Die Physiologie des Stoffwechsels in Pflanzen und Thieren*. Erlangen, 1851.

Moleschott, Jacob, *Eine physiologische Sendung*. Giessen, 1864.

Moleschott, Jacob, *Physiologisches Skizzenbuch*. Giessen, 1861.

Moleschott, Jacob, 'The Unity of Science,' *Popular Science Monthly*, **XXXIII** (1888), pp. 520–527.

Moleschott, Jacob, *Ursache und Wirkung in der Lehre vom Leben*. Giessen, 1867.

Moleschott, Jacob, *Von der Selbststeuerung im Leben des Menschen. Rede zur Wiedereröffnung der Turiner Hochschule am 16. November, 1870 gehalten*. Giessen, 1871.

Moleschott, Jacob, *Zur Feier der Wissenschaft. Rede gehalten bei Wiedereröffnung der Universität zu Rom am 3. November, 1887*. Giessen, 1888.

Mollat, Georg, *Reden und Redner des ersten deutschen Parlaments*. Osterwieck, 1895.
Moser, Walter, *Der Physiologe Jakob Moleschott und seine Philosophie. Zürcher Medizinesgeschichtliche Abhandlungen*. Neue Reihe. Nr. 43. Zürich, 1967.
Mosse, George L., *The Culture of Western Europe*. New York, 1961.
Mulder, G. J., *Versuch einer allgemeinen physiologischen Chemie*. Translated by Jacob Moleschott. Heidelberg, 1844.
Mullen, Pierce C., *The Preconditions and Reception of Darwinian Biology in Germany, 1800–1870*. Ann Arbor: University Microfilms, Inc., 1971.
Nietzsche, Friedrich, 'David Strauss, the Confessor and the Writer.' *Complete Works*. Edited by Oscar Levy. Translated by Anthony Ludovici, vol. 4. New York, 1924.
Prantl, 'Karl C. Planck,' *Allgemeine Deutsche Biographie*, vol. 26, 1888. Reprint. Berlin, 1970, pp. 228–231.
Prantl, 'Ludwig Feuerbach,' *Allgemeine Deutsche Biographie*. vol. 6, 1888. Reprint. Berlin, 1968, pp. 747–753.
Rau, Albrecht, *Ludwig Feuerbachs Philosophie, die Naturforschung und die philosophische Kritik der Gegenwart*. Leipzig, 1882.
Rawidowicz, S., *Ludwig Feuerbachs Philosophie: Ursprung und Schicksal*. Berlin, 1931.
Reichard, Richard W., *Crippled from Birth: German Social Democracy, 1844–1870*. Ames, Iowa, 1969.
Reuschle, Carl, *Philosophie und Naturwissenschaft*. Bonn, 1874.
Richter, 'Karl Friedrich Bachmann,' *Allgemeine Deutsche Biographie*, vol. 1, 1875. Reprint. Berlin, 1967, pp. 753–754.
Robertson, J. M., *A History of Free Thought in the Nineteenth Century*. 2 vols. New York, 1930.
Rosenberger, Ferdinand, *Die Geschichte der Physik in Grundzügen*. 3 vols. in 2. Braunschweig, 1882–1890.
Schaller, Julius, *Leib und Seele. Zur Aufklärung über Köhlerglaube und Wissenschaft*. 2d ed. Weimar, 1858. (1st edition, 1855)
Schilling, Gustav, *Beiträge zur Geschichte und Kritik des Materialismus*. Leipzig, 1867.
Schlawe, Fritz, *Die Briefsammlungen des neunzehnten Jahrhunderts. Repertorien zur deutschen Literaturgeschichte*. Edited by Paul Raabe, vol. 4. Stuttgart, 1969.
Schleiden, Matthias J., *Über den Materialismus der neueren deutschen Naturwissenschaft, sein Wesen und seine Geschichte*. Leipzig, 1863.
Schnabel, Franz, *Die katholischen Kirchen in Deutschland. Deutsche Geschichte im neunzehnten Jahrhundert*, vol. 7. Basel: Herder-Taschenbuch, 1965. (Original publication, 1937)
Schnabel, Franz, *Die protestantischen Kirchen in Deutschland. Deutsche Geschichte im neunzehnten Jahrhundert*, vol. 8. Basel: Herder-Taschenbuch, 1965. (Original publication, 1937)
Schulz-Bodmer, Wilhelm, *Der Froschmauskrieg zwischen den Pedanten des Glaubens und Unglaubens*. Leipzig, 1856.
Schweitzer, Albert, *The Quest of the Historical Jesus: A Critical Study of its Progress from Reimarus to Wrede*. Translated by W. Montgomery. New York, 1961.
Siebeck, Herman, 'F. K. C. L. Büchner.' Pp. 49–56 in *Hessischen Biographien*. vol. 1. Edited by H. Haupt. Darmstadt, 1918.
Silberner, Edmund, 'Moses Hess.' *Neue Deutsche Biographie*, vol. 9. Berlin, 1972, pp. 11–12.
Silberner, Edmund, *Moses Hess: Geschichte seines Lebens*. Leiden, 1966.
Staudinger, F., 'L. Büchner,' *Allgemeine Deutsche Biographie*, vol. 55, 1910. Reprint. Berlin, 1971, pp. 459–461.
Sternfeld, Willi, 'German Students and their Professors.' Pp. 128–161 in *In Tyrannos*. Edited by Hans Rehfisch. London, 1944.

Temkin, Owsei, 'The Idea of Descent in Post-Romantic German Biology.' Pp. 323–355 in *Forerunners of Darwin: 1745–1859.* Edited by Bentley Glass. Owsei Temkin, and W. L. Strauss, Jr. Baltimore, 1968. (Original publication, 1959)
Temkin, Owsei, 'Materialism in French and German Physiology of the Early Nineteenth Century,' *Bulletin of the History of Medicine,* **XX** (1946), pp. 322–327.
Tittmann, F. W., *Geist und Materialismus. Zur Verwahrung gegen die Antrittsrede des Herrn Prof. Moleschott: Licht und Leben.* Dresden, 1856.
Treitschke, Heinrich, *History of Germany in the Nineteenth Century.* Translated by E. and C. Paul. 7 vols. London, 1919.
Tyndall, John, 'Virchow and Evolution,' *Popular Science Monthly,* **XIV** (1878–1879), pp. 266–290.
Ule, Otto, 'Die Aufgabe der Naturwissenschaft,' *Die Natur,* **I** (1852), pp. 1–4.
Vaihinger, Hans, 'Die drei Phasen der Czolbeschen Naturalismus,' *Philosophisches Monatsheft,* **XII** (1876), pp. 1–31.
Vidler, Alexander R., *The Church in an Age of Revolution: 1789 to the Present Day.* Baltimore, 1965.
Virchow, Rudolph, *Briefe an Rudolph Virchow. Zum 100. Geburtstage.* Edited by H. Meisner. Berlin, 1921.
Vogt, Karl, *Der achtzehnte September in Frankfurt am Main. Im Auftrage der Clubbs der Linken.* 2d ed. Frankfurt am Main, 1848.
Vogt, Karl, *Altes und Neues aus Thier–und Menschenleben.* 2 vols. Frankfurt am Main, 1859.
Vogt, Karl, *Aus meinem Leben. Erinnerungen und Rückblicke.* Stuttgart, 1896.
Vogt, Karl, *Beiträge zur Neurologie der Reptilien.* Neuchâtel, 1840.
Vogt, Karl, *Bilder aus dem Thierleben.* Frankfurt am Main, 1852.
Vogt, Karl, *Eduard Desor. Lebensbild eines Naturforschers.* Breslau, 1883.
Vogt, Karl, *Embryologie des Salmones.* vol. 1. Neuchâtel, 1842.
Vogt, Karl, *Ein frommer Angriff auf die heutige Wissenschaft.* Breslau, 1882.
Vogt, Karl, *Im Gebirge und auf den Gletschern.* Solothurn, 1843.
Vogt, Karl, *Köhlerglaube und Wissenschaft. Ein Streitschrift gegen Hofrath R. Wagner in Göttingen.* Giessen, 1855. (4th edition also 1855)
Vogt, Karl, *Lehrbuch der Geologie, und Petrafactenkunde. Theilweise nach E. de Beaumont's Vorlesungen an der École des Mines.* 2 vols. Braunschweig, 1846. (2d edition, Geneva, 1854; 3d edition, Geneva, 1866)
Vogt, Karl, 'Lingering Barbarism,' *Popular Science Monthly,* **XVIII** (1881), pp. 638–643.
Vogt, Karl, *Mein Prozess gegen die Allgemeine Zeitung. Stenographischer Bericht. Dokumente und Erläuterungen.* Frankfurt am Main, 1859.
Vogt, Karl, *Nord-Fahrt entlang der Norwegischen Küste, nach dem Nordkap, den Inseln Jan Mayen und Island, auf dem Schooner Joachim Hinrich.* Frankfurt am Main, 1863.
Vogt, Karl, *Ocean und Mittelmeer. Reisebriefe.* 2 vols. Frankfurt am Main, 1848.
Vogt, Karl, 'Die Physiologie des Menschen,' *Die Gegenwart,* **IV** (1850), pp. 646–707.
Vogt, Karl, *Physiologische Briefe für Gebildete aller Stände.* 3 parts. Stuttgart, 1845–1847.
Vogt, Karl, *Politische Briefe an Friedrich Kolb. Separatabdruck aus dem Schweizer Handels-Courier.* Biel, 1870.
Vogt, Karl, 'Pope and Anti-Pope,' *Popular Science Monthly,* **XIV** (1878–1879), pp. 320–325.
Vogt, Karl, 'Quelques heresies darwinistes,' *Revue Scientifique,* **XXXVIII** (1886), pp. 481–488.
Vogt, Karl, 'Sea Butterflies,' *Popular Science Monthly,* **XXXV** (1889), pp. 313–322.
Vogt, Karl, *Studien zur gegenwärtigen Lage Europas.* Geneva, 1859.

Vogt, Karl, *Über den heutigen Stand der beschreibenden Naturwissenschaften*. Giessen, 1847.
Vogt, Karl, *Untersuchungen über die Entwicklungsgeschichte der Geburtshelferkröte*. Solothurn, 1842.
Vogt, Karl, *Untersuchungen über die Thierstaaten*. Frankfurt am main, 1851.
Vogt, Karl, *Vorlesungen über den Menschen, seine Stellung in der Schöpfung und in der Geschichte der Erde*. 2 vols. Giessen, 1863. Translated as *Lectures on Man: His Place in Creation and in the History of the Earth* by James Hunt. London, 1864.
Vogt, Karl, 'Weltanschauungen,' *März: Halbmonatschrift für deutsche Kultur*, I (1910), pp. 89–98, 183–189, 305–311.
Vogt, Karl, *Zoologische Briefe. Naturgeschichte der lebenden und untergegangenen Thiere*. 2 vols. Frankfurt am Main, 1851.
Vogt, Karl, 'Zur Anatomie der Amphibien.' Inaugural Dissertation. Bern, 1839.
Vogt, Wilhelm, *La Vie d'un homme: Carl Vogt*. Paris, 1896.
Volkmann, Paul, *Die materialistische Epoche des neunzehnten Jahrhunderts und die phänomenologisch-monistische Bewegung der Gegenwart*, Leipzig, 1909.
Wagner, Andreas, *Naturwissenschaft und Bibel, im Gegensatz zu dem Köhlerglauben des Herrn Carl Vogt, als des wiedererstandenen und aus dem Französischen ins Deutsche ubersetzten Bory*. Stuttgart, 1855.
Wagner, Rudolf, *Menschenschöpfung und Seelensubstanz*. Göttingen, 1854.
Wagner, Rudolf, *Über Wissen und Glauben*. Göttingen, 1854.
Ward, A. W., 'The Revolution and the Reaction in Germany and Austria, I (1848–1849).' Pp. 142–201 in *The Growth of Nationalities*. The Cambridge Modern History, vol. 11. Edited by A. W. Ward, G. W. Prothero and S. Leathes. New York, 1934.
Wittich, Dieter, (*ed.*), *Vogt, Moleschott, Büchner: Schriften zum kleinbürgerlichen Materialismus in Deutschland*. 2 vols. Berlin, 1971.
Wurzbach, Alfred, *Zeitgenossen: Biographische Skizzen*. Heft 2. *Carl Vogt*. Vienna, 1871.
Zeller, Eduard, *David Friedrich Strauss in his Life and Writings*. Authorized translation. London, 1874.
Ziegler, Theobald, 'Zur Biographie von D. F. Strauss,' *Deutsche Revue*, **XXX** Jahrgang. Vol. II (1905).
Zöckler, Otto, *Die Geschichte der Beziehungen zwischen Theologie und Naturwissenschaft*. 2 vols. Gütersloh, 1879.

OTHER WORKS CITED

Apelt, Ernst F., *Die Theorie der Induction*. Leipzig, 1854.
Ben-David, Joseph, *The Scientist's Role in Society*. Englewood Cliffs, New Jersey, 1971.
Bruford, W. H., *Germany in the Eighteenth Century: The Social Background of the Literary Revival*. Cambridge, 1935.
Casserley, Julian V. L., *The Retreat from Christianity in the Modern World*. London, 1952.
Clerke, Agnus, 'Alexander von Humboldt,' *Encyclopedia Britannica*, 11th ed., vol. XIII. Cambridge, 1910.
Darnton, Robert, *Mesmerism and the End of the Enlightenment in France*. New York, 1970.
Dilthey, Wilhelm, 'Zum Andenken an Friedrich Ueberweg,' *Preussiche Jahrbücher*, **XXVIII** (1871), pp. 309–322.

Dostoyevsky, Fyodor, *The Possessed*. Translated by Constance Garnett. New York, 1936.
Erdmann, Johann, *A History of Philosophy*. Translated by W. S. Hough. 3 vols. London, 1880.
Frank, Philip, *Einstein: His Life and Times*. Translated by George Rosen. New York, 1947.
Gay, Peter, *The Dilemma of Democratic Socialism*. New York, 1952.
Gillispie, Charles, *Genesis and Geology. A Study in the Relations of Scientific Thought, Natural Theology, and Social Opinion in Great Britian, 1790–1850*. Cambridge, Mass., 1951.
Hegel, G. F. W., *Phenomenology of Mind*. Translated by J. B. Baillie. New York: Harper Torchbooks, 1967.
Helmholtz, Hermann, *Popular Lectures on Scientific Subjects*. Translated by E. Atkinson. New York, 1873.
Hermelink, Heinrich, *Liberalismus und Konservatismus, 1835–1870. Der Christentum in der Menschheitsgeschichte von der Französischen Revolution bis zur Gegenwart*, vol. 2. Stuttgart, 1953.
Hirsch, Emanuel, *Geschichte der neuern evangelischen Theologie im Zusammenhang mit den allgemeinen Bewegungen des europäischen Denkens*. 5 vols. Gütersloh, 1954.
Huges, H. Stuart, *Consciousness and Society: The Reorientation of European Social Thought, 1890–1930*. New York, 1958.
Kohut, Adolph, *Justus von Liebig. Sein Leben und Wirken*. 2d ed. Giessen, 1905.
Kuntz, Paul G., Introduction to *Lotze's System of Philosophy* by George Santayana. Bloomington, Ind., 1971.
Magnus, Rudolf, *Goethe as a Scientist*. Translated by Heinz Norden. New York, 1949.
Paoloni, Carlo, *Justus von Liebig. Eine Bibliographie sämtlicher Veröffentlichungen*. Heidelberg, 1968.
Schelling, Friedrich, 'Introduction to a first Sketch of a System of Natural Philosophy.' Translated by Tom Davidson. *Journal of Speculative Philosophy*, I (1867), pp. 193–220.
Stallo, J. B., *The Concepts and Theories of Modern Physics*. Edited by P. W. Bridgemann. Cambridge, Mass., 1960. (Original edition, 1882)
Vaihinger, Hans, *Die Philosophie des 'Als Ob.'* Berlin, 1911.
Vorländer, Karl, *Geschichte der Philosophie*. 2 vols. Leipzig, 1903.
Winter, Amalie, 'Der Naturforscher.' (Novelle) .*Der Freihafen*. 3rd Jahrgang. Heft 1 (1840), pp. 95–165.

INDEX

STUDIES IN THE HISTORY
OF MODERN SCIENCE

forthcoming titles:

Stephen G. Brush, C. W. F. Everitt, and Elizabeth Garber:
Maxwell on Kinetic Theory and Statistical Mechanics
G. Canguilhem:
The Normal and the Pathological
Edward Manier:
The Young Darwin and His Cultural Circle